Lecture Notes in Mathematics

Edited by A. Dold and B. Eckmann

1271

A. M. Cohen W. H. Hesselink
W. L. J. van der Kallen J. R. Strooker (Eds.)

Algebraic Groups Utrecht 1986

Proceedings of a Symposium in Honour of T. A. Springer

Springer-Verlag
Berlin Heidelberg New York London Paris Tokyo

Editors

Arjeh M. Cohen
Stichting Mathematisch Centrum, Centrum for Wiskunde en Informatica
Kruislaan 413, 1098 SJ Amsterdam, The Netherlands

Wim H. Hesselink
Subfaculteit Wiskunde en Informatica, Rijksuniversiteit Groningen
Postbus 800, 9700 AV Groningen, The Netherlands

Wilberd L. J. van der Kallen
Jan R. Strooker
Rijksuniversiteit Utrecht, Mathematisch Instituut
Budapestlaan 6, 3508 TA Utrecht, The Netherlands

Mathematics Subject Classification (1980, revised 1985): 11FXX, 14LXX, 15A69, 17BXX, 18GXX, 20GXX, 22EXX, 35A27, 43A90, 57T10

ISBN 3-540-18234-9 Springer-Verlag Berlin Heidelberg New York
ISBN 0-387-18234-9 Springer-Verlag New York Berlin Heidelberg

Printing and binding: Druckhaus Beltz, Hemsbach/Bergstr.
2146/3140-543210

PREFACE

A symposium on Algebraic Groups took place at the University of Utrecht from 1 - 4 April, 1986. It was organized to celebrate two birthdays: the 350th anniversary of the university and the 60th birthday of its distinguished member, Professor T.A. Springer.

The university celebrated with a series of international scientific symposia and congresses, of which 'Algebraic Groups' was the first one. Our symposium was funded by the 'Stichting 350 jaar Rijksuniversiteit Utrecht', the 'Koninklijke Nederlandse Academie van Wetenschappen', and the Dutch research organisation ZWO. To the first of these bodies and to the convention bureau QLT employed by them, we are also indebted for help with the organization, as we are for secretarial help to the Mathematics Department of Utrecht University.

To honour Professor Springer, we felt it would be appropriate to invite a number of leading experts in the field of algebraic groups to lecture on their current research; in this way a wide spectrum of topics would be covered and the central rôle of algebraic groups in mathematics emphasized. It is a tribute to the active part which Springer has played in the development of the subject, that all of the speakers have had close scholarly and personal contacts with him at one time or another.

Of the fifteen invited speakers thirteen were able to come, while another mathematician graciously accepted to be a last-minute stand-in. Fourteen manuscripts were contributed to these Proceedings, which often, but not always, cover the subject of the talk delivered (cf. the list of talks and the table of contents below). They have been put in alphabetical order with respect to author's name(s) rather than in an order determined by subject. We briefly touch upon them here.

As the reader will notice, there are contributions on various topics centered around algebraic groups. Now that algebraic groups have been with us for about three decades, much is known about their structure (nevertheless, Tits contributes new information on unipotent subgroups of reductive groups in positive characteristic). Thus, attention has gone to subsequent questions, such as a structure theory for finite-dimensional algebraic groups. Popov's contribution shows that there is still little grip on a 'standard' example, such as Aut $\mathbb{A}_n$ (he constructs infinitely many nontriangular subgroups of $\mathbb{A}_n$ $(n\geqslant 3)$ isomorphic to the 1-dimensional additive group $\mathbb{G}_a$). Nevertheless, many new insights have been obtained for special classes of infinite-dimensional groups such as Kac-Moody groups. The contribution by Kac & Peterson illustrates this. It also reflects the interest regained in invariant theory, a classical aspect of algebraic groups. In their paper, Le Bruyn & Procesi pay attention to this subject by studying the GL(n)-orbit space of the affine space of m-tuples of complex n by n matrices (on which the GL(n)-action is componentwise by conjugation). Richardson uses modern techniques of invariant theory to derive an elementary necessary condition for normality of a closed subvariety of the Lie algebra of a semisimple group which is stable under the adjoint action of the group. Piatetski-Shapiro employs a Poincaré series for split reductive groups to produce Langlands L-functions. Geometric invariant theory is concerned with a description of the quotient space of a variety by a group acting on it. As a set on which this group acts, the variety can then be recovered from the quotient space and certain group data (the stabilizers). In the case where the group is an algebraic torus, Goresky & MacPherson indicate what kind of data suffice to reconstruct the variety as a topological space on which the group acts.

Invariant theory usually starts with a known group action on a variety. Representation theory on the other hand tries to describe all linear representations of a given group. The (finite-dimensional) representation theory of algebraic groups where the characteristic of the representation space coincides with that of the group is reasonably well understood, especially in characteristic 0. Part of the present interest in this area is directed to questions concerning special functions related to series of group representations. In his paper, Macdonald deals with a class of polynomial symmetric functions including the 'classical' Schur functions and zonal spherical functions related to various real forms of GL(n). In the cross characteristic representation theory, enormous progress has been made. In particular, for the finite subgroups of algebraic groups which are the fixed points of a (Frobenius type) automorphism, many class functions have been constructed which lead to characters. Lusztig, in his contribution, describes a Lie algebra variant of a special kind of class function he needed in his theory of character sheaves (in fact, functions vanishing outside the nilpotent variety, whose Fourier transforms have the same property). At the origin of the class functions in this theory are certain bundles on the flag variety. Borho exploits the theory of D-modules on the flag variety to study the orbits in the nilpotent variety and the classification of primitive ideals in the enveloping algebra of a semisimple Lie algebra. Relations with D-modules are also present in Brylinski's contribution, describing examples of cyclic homology of certain noncommutative algebras. The nilpotent variety, which turns up in so many contributions related to representation theory, also appears in Jantzen's survey of the determination of the cohomology of a restricted Lie algebra in positive characteristic.

There are also contributions on Lie groups, the ancestors of algebraic groups. Borel proves a theorem in unitary representations theory of Lie groups. It concerns the vanishing of relative Lie algebra cohomology, and such theorems are of importance for the cohomology of cocompact discrete subgroups. The latter groups plays a central role in the contribution of Mostow & Yau. They use Morse theory to compute homological invariants of the quotient of the unit ball - viewed as the positive cone in natural PU(1,2) space - by a discrete subgroup arising from the monodromy of multivariate hypergeometric functions.

As most of the lectures were excellent, they permitted the audience to gain a good impression of several areas of mathematics around the central theme of algebraic groups. The rather leisurely schedule of the symposium permitted many personal exchanges, and we hope the occasion stimulated and furthered the cause of research in the area.

We wish to thank the invited speakers for their talks and their manuscripts, and also all the other mathematicians who attended the symposium and helped to make it into a success. Contributors and editors alike take pleasure in offering this volume of Proceedings to Professor Springer as a token of esteem and friendship.

The Editors:

A.M. Cohen
W.H. Hesselink
W.L.J. van der Kallen
J.R. Strooker

List of Talks

Tuesday, April 1,

W. Borho, *Nilpotent orbits, primitive ideals and characteristic classes*

R.W. Richardson, *Invariant vector fields on a semisimple Lie algebra*

C. Procesi, *Matrices and invariant theory*

Wednesday, April 2,

J.L. Tits, *On rational unipotent elements of simple algebraic groups*

G.D. Mostow, *Some surfaces covered by the ball and a problem in finite groups*

G. Harder, *Cohomology and special values of L-functions*

I. Piatetski-Shapiro, *L-functions and automorphic forms on classical groups with Whittaker model*

Thursday, April 3,

G. Lusztig, *Fourier transforms on a semisimple Lie algebra over* $\mathbb{F}_q$

I.G. Macdonald, *Commuting differential operators and zonal functions*

J.C. Jantzen, *Restricted Lie algebra cohomology*

Friday, April 4,

J.-L. Brylinski, *Some examples of Hochschild and cyclic homology*

R.D. MacPherson, *The variety of complete quadrics*

V. Kac, *Unitary representations of Diff* S^1, *exceptional Lie algebras and statistical mechanics*

A. Borel, *A vanishing theorem in Lie algebra cohomology*

TABLE OF CONTENTS

Authors' Addresses

A. BOREL, Institute for Advanced Study, Princeton, NJ 08540, USA

W. BORHO, Gesamthochschule Wuppertal, Fachbereich 7 Mathematik, Gaußstr. 1,
4600 Wuppertal, Federal Republic of Germany

J.-L. BRYLINSKI, Mathematics Department, Brown University, Providence, R.I. 02912, USA

M. GORESKY, Department of Mathematics, Northeastern University, Boston, MA 02115,
USA

J.C. JANTZEN, Mathematisches Seminar, Universität Hamburg, Bundesstr. 55
2000 Hamburg 13, Federal Republic of Germany

V.G. KAC, M.I.T., Mathematics Department, Cambridge, MA 02139, USA

L. LE BRUYN, Universitaire Instellingen Antwerpen, Departement Wiskunde
Universiteitsplein 1, 2610 Wilrijk, Belgium

G. LUSZTIG, M.I.T., Mathematics Department, Cambridge, MA 02139, USA

I.G. MACDONALD, School of Mathematical Sciences, Queen Mary College, Mile End Road,
London E1 4NS, England

R.D. MACPHERSON, M.I.T., Mathematics Department, Cambridge, MA 02139, USA

G.D. MOSTOW, Department of Mathematics, Yale University, New Haven, CT 06520, USA

D.H. PETERSON, Department of Mathematics, Universitv of British Columbia,
Vancouver V6T 1W5, British Columbia, Canada

I. PIATETSKI-SHAPIRO, School of Mathematical Sciences, Tel-Aviv University,
Ramat-Aviv, 69978 Tel-Aviv, Israel
and Department of Mathematics, Yale University,
New Haven, CT 06520, USA

V.L. POPOV, MehMat, MGU, per. B. Vuzovskii 3/12, Moscow 109028, USSR

C. PROCESI, Dipartimento di Matematica, Istituto "Guido Castelnuovo"
Università degli Studi di Roma "La Sapienza", Piazzale Aldo Moro 5,
00185 Roma, Italy

R.W. RICHARDSON, Department of Mathematics, The Australian National University,
GPO Box 4, Canberra A.C.T. 2601, Australia

J.L. TITS, Collège de France, 11, Place Marcelin Berthelot, 75231 Paris Cedex 05, France

S.T.S. YAU, Department of Mathematics, University of Illinois at Chicago,
Chicago, IL 60680, USA

§1. *Preliminaries. Notation. Assumptions.*

1.1. Let L be a reductive group, $K(L)$ a maximal compact subgroup of L and $\mathfrak{h}$ a Cartan subalgebra of $\mathfrak{l}$. For $\mu \in \mathfrak{h}_c^*$, let F_μ denote an irreducible finite dimensional representation of L with infinitesimal character χ_μ (i.e. highest weight $\mu-\rho_L$ if μ is dominant). Let $C(L,\mu)$ be the greatest q for which there exists $H \in \hat{L}$ such that $H^q(\mathfrak{l},K(L);H\otimes F_\mu) \neq 0$. If there is no such q, then set $C(L,\mu) = -\infty$. This is the case if and only if $F_\mu \neq \bar{F}_\mu^*$ [V; VZ].

1.2. In the sequel G is a connected real simple Lie group, with finite center, K a maximal compact subgroup, θ the associated Cartan involution, E an irreducible representation of G with infinitesimal character χ_λ, P a proper maximal parabolic subgroup and $P = N.A.M.$ a Langlands decomposition of P, with A,M stable under θ. We fix a Cartan subalgebra $\mathfrak{h} = \mathfrak{a} \oplus \mathfrak{h}_M$ invariant under θ, where $\mathfrak{h}_M$ is a *fundamental* Cartan subalgebra of $\mathfrak{m}$. We choose an ordering on the set $\Phi(\mathfrak{g}_c)$ of roots of $\mathfrak{g}_c$ with respect to $\mathfrak{h}_c$ such that

(1)
$$\Phi^+|_{\mathfrak{a}} \subset \Phi(P,A) \cup \{0\}$$

where $\Phi(P,A)$ denotes the weights of $\mathfrak{a}$ in $\mathfrak{n}$. We let W^P be the usual set of representatives of $W(\mathfrak{m}_c\oplus\mathfrak{a}_c)\backslash W(\mathfrak{g}_c)$. An element ν of $\mathfrak{a}^*$ is >0 if it is a positive real multiple of some element in $\Phi(P,A)$.

1.3. We assume given a decomposition of M into an almost direct product $M = L.G_P$ of θ-invariant closed subgroups and write accordingly $\mathfrak{h}_M = \mathfrak{h}_L \oplus \mathfrak{h}_{G_P}$. The groups $K(L) = K \cap L$ and $K(G_P) = K \cap G_P$ are maximal compact in L and G_P respectively. The following assumption is basic

(A)
$$\mathrm{rk}\, G = \mathrm{rk}\, K \quad\text{and}\quad \mathrm{rk}\, G_P = \mathrm{rk}\, K(G_P).$$

It follows first that $\mathfrak{h}_{G_P}$ is a Cartan subalgebra of $K(G_P)$. We may write $\mathfrak{h}_L = \mathfrak{a}_0 \oplus \mathfrak{t}_0$, with $\mathfrak{a}_0$ split and $\mathfrak{t}_0$ a Cartan subalgebra of $K(L)$. Then $\mathfrak{t}_0 \oplus \mathfrak{h}_{G_P} = \mathfrak{t}$ is a Cartan subalgebra of a maximal compact subgroup $K(M)$ of M. The centralizer $\mathfrak{z}_{\mathfrak{g}}(\mathfrak{t})$ of $\mathfrak{t}$ in $\mathfrak{g}$ can be written as $\mathfrak{t} \oplus \mathfrak{z}_1$. We have $\mathfrak{z}_1 \cap \mathfrak{m} = 0$ since $\mathfrak{t}$ contains regular elements of $\mathfrak{m}$. From (A) it follows that $\mathfrak{z}_1$ has a split Cartan subalgebra $(\mathfrak{a}_0\oplus\mathfrak{a})$ and a compact one. Therefore $\mathfrak{z}_1$ is semi-simple and split over $\mathbb{R}$.

At the very end of the proof, I shall need to use the following assumption, which is easily checked in the cases of interest (see §6) and which may well be true whenever (A) is and G_P is the greatest factor of M satisfying the second equality of (A).

(B) *The intersection of* $\mathfrak{l}$ *with* $\mathfrak{z}_{\mathfrak{g}}(\mathfrak{z}_1)$ *is the Lie algebra of a compact subgroup of* L.

Note that $\mathfrak{z}_1$ contains $\mathfrak{a}_0$. Therefore, if $\mathfrak{a}_0$ is maximal $\mathbb{R}$-split in $\mathfrak{l}$, (e.g. if $\mathfrak{l}$ is a complex semi-simple Lie algebra, viewed as a real Lie algebra), then (B) is fulfilled. It so happens that this stronger condition is often fulfilled.

§2. *The vanishing theorem.*

2.0. *Notation.* For $s \in W^P$ we write $\nu(s\lambda)$ for $s\lambda|\mathfrak{h}_L$. For H reductive, X_H is the quotient of H by a maximal compact subgroup.

2.1. THEOREM. *We keep the previous notation and assume* (A),(B) *of* 1.3 *to hold. Let* $s \in W^P$. *Then*

(1) $$C(L,\nu(s\lambda)) + \ell(s) < (\dim X_G - \dim X_{G_P})/2, \quad \text{if } s\lambda|\mathfrak{a} > 0.$$

2.2. By [C: 2.6] the condition $s\lambda|\mathfrak{a} > 0$ is equivalent to $2\ell(s) < \dim N$. By [K], $F_{s\lambda}$, viewed as a M.A. module, occurs in $H^{\ell(s)}(\mathfrak{n};E_\lambda)$, with multiplicity one, and nowhere else in $H^{\bullet}(\mathfrak{n};E_\lambda)$. We can therefore also write (1) as:

(2) $$H^q(\mathfrak{l},K(L);H^i(\mathfrak{n};E_\lambda)\otimes H) = 0 \quad \text{for } i < (\dim N)/2 \text{ and } q+i \geq m,$$

where

(3) $$m = (\dim X_G - \dim X_{G_P})/2.$$

2.3. We now sketch two (related) applications of 2.1.

(a) Combined with the results of [BC1], 2.1 implies that if Γ is an arithmetic subgroup of L, for some given $\mathbf{Q}$-structure, then

(1) $$H^q(\mathfrak{l},K(L);F_{\nu(s\lambda)} \otimes L^{2,\infty}(\Gamma\backslash L)) = 0 \quad \text{for } q \geq m-\ell(s) \text{ and } \ell(s) < (\dim N)/2.$$

(b) We consider the setup of Zucker's conjecture [B; BC2; Z]. Let $\mathbf{G}$ be a $\mathbf{Q}$-simple connected linear algebraic group, and Γ an arithmetic subgroup. Assume that the symmetric space X of maximal compact subgroups of $G = \mathbf{G}(\mathbb{R})$ is a bounded symmetric domain and let V^* be the minimal compactification of $V = \Gamma\backslash X$ constructed in [BB]. Assume that P is the normalizer of a rational boundary component X_P of maximal dimension. We have an almost direct product decomposition $M = L.G_P$ where L is the greatest subgroup of M acting trivially on X_P, (possibly up to a compact factor), and X_P is the symmetric space of maximal compact subgroups of G_P. Modulo various reductions, 2.1 allows one to prove that the local L^2-cohomology of the intersection with V of a suitable neighborhood U of a point of the image V_P of X_P in V^* vanishes from the complex codimension of X_P on. This is the main point needed to show that the L^2-cohomology sheaf on V^* is homology isomorphic to the middle intersection cohomology sheaf along V_P. In this case, it is a rather direct application of 2.1. For the next stratum (level 2), the situation is much more complicated and 2.1 has to be used in conjunction with results of Casselman on automorphic forms (cf. [BC 2]).

§3. *Results of Vogan-Zuckerman. Reformulation of 2.1.*

3.1. In the sequel, I write $\mathfrak{h}_o$ for $\mathfrak{h}_L$. I recall that under the first assumption of 1.3(A), $-w_G$ gives the effect of complex conjugation on the Dynkin diagram (see [BC1]). It leaves the set $\Delta(\mathfrak{l}_c)$ of simple roots of $\mathfrak{l}_c$ with respect to $\mathfrak{h}_{o,c}$ stable. Also w_G leaves $\mathfrak{a}$ stable. Moreover the transformation τ_I (cf. [BC1]), which assigns to a representation its complex conjugate contragredient one, is given by $w_L w_G$. This is the same as $w_M w_G$, because $w_{G_P} \cdot w_G$ is the identity of $\Delta(\mathfrak{g}_P, \mathfrak{h}_{G_P})$ in view of (A). We write also α^* for $w_m w_G \alpha$ $(\alpha \in \Delta(\mathfrak{l}_c))$. We have then

(1) $$C(L,\nu(s\lambda)) \neq -\infty \Leftrightarrow (s\lambda,\alpha) = (s\lambda,\alpha^*) \text{ for all } \alpha \in \Delta(\mathfrak{l}_c) \Leftrightarrow s\lambda|\mathfrak{a}_o = 0.$$

3.2. We have $\mathfrak{h}_o = \mathfrak{t}_o \oplus \mathfrak{a}_o$ and $\mathfrak{t}_o$ contains regular elements of $\mathfrak{l}$. We may assume that $\Phi^+(\mathfrak{l}_c)$ is given by $\alpha(it) > 0$ for some regular $t \in \mathfrak{t}_o$. We let also θ denote the Cartan involution of L with respect to $K(L)$. It is indeed the restriction of θ. Its restriction to $\mathfrak{h}_o$ is also the linear transformation defined by τ_I. The algebra $\mathfrak{b}_c = \mathfrak{h}_{o,c} \oplus_{\alpha>0} \mathfrak{l}_{c,\alpha}$ is θ-stable. A *standard θ-stable parabolic subalgebra* $\mathfrak{q}$ of $\mathfrak{l}_c$ is one which contains $\mathfrak{b}_c$ and has a Levi subalgebra $\mathfrak{m}_q$ which is the centralizer of some element in $\mathfrak{t}$. It is in particular θ-invariant, and $\mathfrak{m}_{q,o} = \mathfrak{m}_q \cap \mathfrak{l}$ is a real form of $\mathfrak{m}_q$. We let $M_{q,o}$ be the corresponding analytic subgroup of L and X_q the symmetric space $M_{q,o}/K(M_{q,o})$.

Let $\nu \in \mathfrak{h}_o^*$ be dominant regular. We say that $\mathfrak{q}$ and ν are *compatible* if $\nu - \rho_L$ is the differential of a unitary character of $M_{q,o}$. This is the case if and only if ν is zero on $\mathfrak{a}_o$ and on the derived algebra of $\mathfrak{m}_q$. Given a compatible pair $(\mathfrak{q},\nu)$, [VZ] defines an irreducible representation $A_q(\nu-\rho_L)$ of L, which is unitary by [V], such that

(1) $$H^\cdot(\mathfrak{l},K(L);A_q(\nu-\rho_L)\otimes F_\nu) = H^\cdot(\mathfrak{m}_{q,o},K(M_{q,o});\mathbf{C}), \text{ suitably translated.}$$

"Suitably translated" means so that the left-hand side satisfies Poincaré duality. It follows that the top cohomology occurs in dimension $(\dim X_L+\dim X_{\mathfrak{q}})/2$. Moreover, by [VZ], any irreducible unitary representation H such that $H^{\cdot}(\mathfrak{l},K(L);H\otimes F_\nu) \neq 0$ is so obtained. Therefore

(2) $$2.C(L,\nu) = \mathrm{Max}_{\mathfrak{q}}(\dim X_L+\dim X_{\mathfrak{q}}) \quad \text{if} \quad \nu|\mathfrak{a}_o = 0,$$

where $\mathfrak{q}$ runs through the standard θ-stable parabolic subalgebras which are compatible with ν. The biggest possible $M_{\mathfrak{q},o}$ is the one which is generated by $\mathfrak{h}_o$ and a semi-simple group $M'_{\mathfrak{q},o}$ whose roots are all the $\beta \in \Phi(\mathfrak{l}_c)$ such that

(3) $$(\nu,\beta) = (\rho_L,\beta).$$

We can write $\beta = \Sigma\, c_\alpha \alpha$ (with $\alpha \in \Delta(\mathfrak{l}_c)$) and the c_α integers all of the same sign. We have then

(4) $$\sum_\alpha c_\alpha(\nu,\alpha) = \sum c_\alpha(\rho_L,\alpha).$$

But $(\rho_L,\alpha) = (\alpha,\alpha)/2$ and, since ν is dominant regular, $(\nu,\alpha) \geq (\alpha,\alpha)/2$. Equality holds, therefore, if and only if

(5) $$c_\alpha \neq 0 \Rightarrow (\nu,\alpha) = (\rho_L,\alpha).$$

The roots of $M'_{\mathfrak{q},o}$ are therefore all the linear combinations of the elements of

(6) $$\Delta_\nu = \{\alpha \in \Delta(\mathfrak{l}_c) \,|\, (\nu,\alpha) = (\rho_L,\alpha)\}.$$

Let us write M_ν, M'_ν, X_ν for $M_{\mathfrak{q},o}, M'_{\mathfrak{q},o}, X_{\mathfrak{q}}$ for this choice of $\mathfrak{q}$. We have

(7) $$\text{If} \quad \nu|\mathfrak{a}_o = 0, \quad \text{then} \quad 2.C(L,\nu) = \dim X_L + \dim X_\nu.$$

Moreover, it is clear that

(8) $$\dim X_\nu \leq \dim X'_\nu + \dim \mathfrak{a}_o,$$

where X'_ν is the symmetric space of maximal compact subgroups of M'_ν.

In all this, it was assumed that $\tau_1\nu = \nu$. Then $\Phi(M'_\nu)$ and Δ_ν are automatically τ_1-stable. However the assumption $\tau_1\nu = \nu$ does not play a role in the implication (5). Generalizing the above slightly, for any dominant regular ν we let Δ_ν be the greatest τ_1-stable subset of $\Delta(\mathfrak{l}_c)$ whose elements are orthogonal to $\nu - \rho_L$ and M'_ν, M_ν the corresponding groups. Then $\Phi(M'_\nu)$ is the greatest τ_1-stable subset of $\Phi(1_c)$ all of whose elements are orthogonal to $\nu - \rho_L$ and Δ_ν is a basis of $\Phi(M'_\nu)$.

3.3. We now come back to 2.1. We have

$$\dim X_G = \dim X_L + \dim X_{G_P} + \dim N + 1.$$

Therefore by 3.2(7), 2.1(1) can be written

(1) $$\ell(s) + (\dim X_L + \dim X_{\nu(s\lambda)})/2 < (\dim X_L + \dim N+1)/2$$

or

(2) $$\ell(s) + \dim X_{\nu(s\lambda)}/2 < (\dim N+1)/2.$$

The map $s \mapsto s' = w_M w_G s$ is an involution of W^P. We have

(3) $$\ell(s') + \ell(s) = \dim N$$

(4) $$s'\lambda|\mathfrak{a} + s\lambda|\mathfrak{a} = 0$$

(5) $F_{\nu(s'\lambda)}$ is complex conjugate contragredient to $F_{\nu(s\lambda)}$.

[This is the s' occurring in the proof of 2.6 in [C].] The relation (2) is equivalent to

$$\ell(s) + \dim X_{\nu(s\lambda)}/2 \leq (\dim N)/2 = (\ell(s)+\ell(s'))/2,$$

hence to

(6) $$\ell(s') - \ell(s) \geq \dim X_{\nu(s\lambda)} \quad \text{if} \quad \ell(s) < (\dim N)/2 \quad \text{and} \quad s\lambda|\mathfrak{a}_o = 0.$$

The left-hand side depends only on s, and we want to prove 2.1 for any dominant regular λ. Therefore we are reduced to showing

(7) *Fix* $s \in W^P$ *such that* $\ell(s) < (\dim N)/2$. *Then* $\ell(s') - \ell(s) \geq \max \dim X_{\nu(s\lambda)}$, *where* λ *runs through the regular dominant weights of* $\mathfrak{g}$ *such that* $s\lambda|\mathfrak{a}_o = 0$.

§4. *Further reductions.*

4.1. Let μ,ν be regular dominant such that $\sigma = \mu - \nu$ is dominant. Then

$$(s\mu,\alpha) \geq (s\nu,\alpha) \quad \text{for} \quad \alpha \in \Delta(1_c).$$

In fact $(s\sigma,\alpha) = (\sigma,s^{-1}\alpha) \geq 0$ since σ is dominant and $s^{-1}\alpha > 0$ by definition of W^P. We also know that $\nu(s\mu)$ is regular dominant (for 1_c), hence

(1) $$(s\mu,\alpha) \geq (s\nu,\alpha) \geq (\rho_L,\alpha) \qquad (\alpha\in\Delta(1_c)).$$

Since $\mu - \rho$ is dominant, this yields in particular

(2) $$(s\mu,\alpha) = (\rho_L,\alpha) \Rightarrow (s\rho,\alpha) = (\rho_L,\alpha) \qquad (\alpha\in\Delta(1_c)),$$

therefore

(3) $$M_{\nu(s\mu)} \subset M_{\nu(s\rho)}, \qquad \dim X_{\nu(s\mu)} \leq \dim X_{\nu(s\rho)}.$$

Note that we have not assumed τ_I-stability, and have used the convention made at the end of 3.2. In the sequel we replace the index $\nu(s\rho)$ in $M_{\nu(s\rho)}, M'_{\nu(s\rho)}, X_{\nu(s\rho)}, X'_{\nu(s\rho)}, M'_{\nu(s,\rho)}$ by s. In view of (3) we see that, in order to prove 3.3(7), and hence 2.1, it suffices to establish

4.2. PROPOSITION. *If* $\ell(s) < (\dim N)/2$, *then*

(1) $$\ell(s') - \ell(s) \geq \dim X_s.$$

Remark. It is not clear to me that $\nu(s\rho)$ is τ_I-stable. If it is, then (I) is equivalent to 3.3(7), because $\dim X_s$ is then the maximum of the right-hand side of 3.3(7). Otherwise, it is conceivably stronger. In fact, we shall prove a still slightly stronger inequality, namely

(2) $$\ell(s') - \ell(s) \geq \dim X'_s + \dim \mathfrak{a}_o + 1.$$

4.3. We let $\Phi_{\mathfrak{n}}$ be the set of weights of $\mathfrak{h}_c$ in $\mathfrak{n}_c$. Therefore $\Phi^+ = \Phi^+_{\mathfrak{m}} \coprod \Phi_{\mathfrak{n}}$. Let

(1) $$A_s = \left\{\alpha \in \Phi^+ | s^{-1}\alpha > 0\right\}, \quad B_s = \left\{\alpha \in \Phi^+ | s^{-1}\alpha < 0\right\}.$$

Then

(2) $$\Phi^+ = A_s \coprod B_s, \quad A_s \supset \Phi^+_{\mathfrak{m}}, \quad \ell(s) = \text{Card } B_s.$$

(3) $s\rho = \rho - \langle B_s \rangle$, where $\langle B_s \rangle$ is the sum of the elements in B_s.

The discussion in [C: 2.6] and standard facts about reduced decompositions show that $B_s \subset B_{s'}$. Let $C_s = B_{s'} - B_s$. Then

$$\ell(s') - \ell(s) = \text{Card } C_s.$$

To prove 2.1, it suffices, in view of 4.2(1), (2), to show:

4.4. PROPOSITION. *We have the inequality*

(1) $$\text{Card } C_s \geq \dim X'_s + \dim \mathfrak{a}_o + 1.$$

§5. *Proof of Proposition* 4.4.

5.0. *Notation.* Let $\alpha,\beta \in \Phi$. We write $\alpha \perp\!\!\!\perp \beta$ if α is strongly orthogonal to β, i.e. if neither $\alpha + \beta$ nor $\alpha - \beta$ is a root. This implies in particular that $(\alpha,\beta) = 0$. We have $\alpha \perp\!\!\!\perp \beta$ if and only if $[\mathfrak{g}_{\pm\alpha},\mathfrak{g}_{\pm\beta}] = 0$. Recall that if $\alpha \perp \beta$ but $\alpha \not\perp\!\!\!\perp \beta$, then $\alpha + \beta$ and $\alpha - \beta$ are roots.

5.1. LEMMA (i) *Let* $\alpha \in \Delta(\mathfrak{l}_c)$. *Then* $(s\rho,\alpha) = (\rho_L,\alpha)$ *if and only if* $s^{-1}\alpha$ *is simple.*

(ii) $\Delta_s = \{\alpha \in \Delta(\mathfrak{l}_c) \mid s^{-1}\alpha$ *and* $s^{-1}w_m w_G\alpha$ *are simple*$\}$.

The assertion (ii) follows from (i) and the definition of Δ_s (see 3.1(1) and 3.2(6)).

Proof of (i): Recall that

(1) $$2(\rho,\beta) = (\beta,\beta) \quad \text{if } \beta \text{ is simple}$$

and similarly

(2) $$2(\rho_L,\beta) = (\beta,\beta) \quad \text{if } \beta \in \Delta(\mathfrak{l}_c).$$

If $s^{-1}\alpha = \beta$ is simple, then

$$2(s\rho,\alpha) = 2(\rho,s^{-1}\alpha) = (\beta,\beta) = (s^{-1}\alpha,s^{-1}\alpha) = (\alpha,\alpha) = 2(\rho_L,\alpha).$$

Assume now $(s\rho,\alpha) = (\rho_L,\alpha)$. Since $s^{-1}\alpha > 0$ we may write $s^{-1}\alpha = \Sigma\, c_\beta\beta$ with β simple, $c_\beta \in \mathbb{N}$ and we have then

(3) $$(s\rho,\alpha) = \Sigma\, c_\beta(\rho,\beta).$$

or equivalently

(4) $$(\alpha,\alpha) = \Sigma\, c_\beta(\beta,\beta).$$

The possible values for the square norms of the roots, suitably normalized, are either 1 or 1 and 2 or 1 and 3. In the first case, (4) shows that $s^{-1}\alpha$ is simple. In the last case, (which occurs only for the Lie type $\mathbf{G}_2$), α should be long, the β with $c_\beta \neq 0$ should be short and $s^{-1}\alpha$ should be a sum of three distinct simple roots, which is absurd since $\mathfrak{g}$ has rank 2. In the second case, if $s^{-1}\alpha$ is not simple, the only possibility is

(5) $$s^{-1}\alpha = \beta_1 + \beta_2, \quad (\beta_1,\beta_2) + (\beta_2,\beta_2) = (\alpha,\alpha).$$

Then β_1,β_2 are short and α long. However if the sum of two simple short roots is a root, it is also short (being the transform of one by the reflection to the other), a contradiction since $s^{-1}\alpha$ is also long.

5.2. We shall write τ for $w_M w_G$ and set $\tau' = -\tau$. Then $s' = \tau.s$. Also, τ' leaves $\Phi_{\mathfrak{n}}$ stable. Both τ and τ' are of order 2. We claim

(1) $$\tau' B_s \cap B_{s'} = \phi.$$

In fact, if $\alpha \in B_s$ then

$$s'^{-1}.\tau'.\alpha = s^{-1}\tau\tau'\alpha = -s^{-1}\alpha > 0.$$

Since $\mathrm{Card}\,\Phi_{\mathfrak{n}} = \ell(s) + \ell(s') = \mathrm{Card}\,B_s + \mathrm{Card}\,B_{s'}$, it follows that $\Phi_{\mathfrak{n}}$ is the disjoint union of $B_{s'}$ and $\tau' B_s$, or also of $B_s, C_s, \tau' B_s$ and that $\tau' C_s = C_s$. We have then the following characterizations of these subsets in terms of the sign of $s^{-1}\alpha$, $s'^{-1}\alpha$ or $s^{-1}\tau'\alpha$:

(2)

$\alpha \in$	$\tau' B_s$	C_s	B_s
$s^{-1}\alpha$	>0	>0	<0
$s'^{-1}\alpha$	>0	<0	<0
$s^{-1}\tau'\alpha$	<0	>0	>0

5.3. LEMMA. (i) *For* $\alpha \in \Phi^+$ *we have the equivalences* $\tau'\alpha = \alpha \Leftrightarrow \alpha|\mathfrak{t} = 0 \Leftrightarrow \alpha \in C_s^{\tau'}$.

(ii) $\Phi(\mathfrak{z}_1) = C_s^{\tau'} \cup (-C_s^{\tau'})$.

(iii) $C_s^{\tau'}$ *contains a basis of* $(\mathfrak{a}_o \oplus \mathfrak{a})^*$.

Proof. The map τ', viewed as transformation of $\mathfrak{h} = \mathfrak{t} \oplus \mathfrak{a}_o \oplus \mathfrak{a}$, is the identity on $\mathfrak{a}_o \oplus \mathfrak{a}$ and -1 on $\mathfrak{t}$. This proves the first equivalence in (i). Since $\mathfrak{t}$ contains regular elements of $\mathfrak{m}$, any positive root identically zero on $\mathfrak{t}$ belongs to Φ_n. Being fixed under τ', such a root must then belong to C_s, since $B_s \cap \tau' B_s = \phi$. This proves (i). The roots of $\mathfrak{z}_1$ are clearly those of $\mathfrak{g}$ which are zero on $\mathfrak{t}$, therefore (i) $\Rightarrow$ (ii). We already pointed out (1.3) that $\mathfrak{z}_1$ is semisimple, and that $\mathfrak{a}_o \oplus \mathfrak{a}$ is a Cartan subalgebra of $\mathfrak{z}_1$. Hence (ii) $\Rightarrow$ (iii).

5.4. LEMMA. *The space* $V_s = \oplus_{\alpha \in C_s} \mathfrak{g}_\alpha$ *is invariant under* $\mathfrak{m}'_s$, *acting by the adjoint representation.*

It suffices to show that $[\mathfrak{g}_\beta, \mathfrak{g}_\alpha] \subset V_s$ if $\beta \in \pm\Delta_s$ and $\alpha \in C_s$. This is clear if $\alpha \perp \beta$. If not, we have to show that if $\alpha + \varepsilon\beta$ is a root, where ε is equal to 1 or to -1, then it belongs to C_s. It is the sum of a positive root α and of $\varepsilon\beta$, with β simple and $\varepsilon = \pm 1$. Therefore it is positive. Its transform

$$(1) \qquad s^{-1}(\alpha+\varepsilon\beta) = s^{-1}\alpha + \varepsilon \,.\, s^{-1}\beta$$

is also positive for the same reason: $s^{-1}\alpha$ is positive since $\alpha \in A_s$ and $s^{-1}\beta$ is simple by 5.1. Similarly, since $\tau'\alpha \in C_s$, we have $s^{-1}\tau'\alpha > 0$ and

$$(2) \qquad s^{-1}\tau'(\alpha+\varepsilon\beta) = s^{-1}\tau'\alpha - \varepsilon s^{-1}\beta^* > 0.$$

But (1), (2) and 5.2(2) imply that $\alpha + \varepsilon\beta \in C_s$.

5.5. LEMMA. *Let* $\Psi = \{\beta \in \Phi^+(\mathfrak{m}'_s) \mid \beta \not\perp C_s^\tau\}$. *Then*

$$2 \text{ Card } (\psi/\langle\tau'\rangle) + \text{Card } C_s^{\tau'} \leq \text{Card } C_s.$$

[By definition, $\Phi^+(\mathfrak{m}'_s)$ is invariant under τ', hence so is Ψ. It therefore makes sense to speak of the set $\Psi/\langle\tau'\rangle$ of orbits on Ψ of the group $\langle\tau'\rangle = \{1,\tau'\}$ generated by τ'.]

Let $\alpha_1,\ldots,\alpha_q$ be the elements of $C_s^{\tau'}$. Set

$$E_1 = \{\beta \in \Psi \mid \beta \not\perp \alpha_1\}$$
$$E_i = \{\beta \in \Psi \mid \beta \perp \alpha_1,\ldots,\beta \perp \alpha_{i-1}, \beta \not\perp \alpha_i\} \quad (2 \leq i \leq q).$$

Then $\psi = \coprod_i E_i$.

If $\beta \in E_i$, then for some $\varepsilon_\beta = \pm 1$, the element $\alpha_i + \varepsilon_\beta\beta$ is a root, and belongs to C_s by 5.4. Moreover, if $\beta = \beta^*$, then both $\alpha_i + \beta$ and $\alpha_i - \beta$ belong to C_s (because if $\alpha_i + \varepsilon_\beta\beta \in C_s$, then $\tau'(\alpha_i+\varepsilon_\beta\beta) = \alpha_i - \varepsilon_\beta\beta \in C_s$.) None of these belongs to $C_s^{\tau'}$, since no root of $\mathfrak{m}_s'$ restricts to zero on $\mathfrak{t}$. It suffices therefore to show that these elements are all distinct, i.e. we are reduced to proving

(*) *Let* $\beta \in E_i$, $\beta' \in E_j$ *and assume that* $\alpha_i + \varepsilon\beta$, $\alpha_j + \varepsilon'\beta' \in C_s$ *and* $\alpha_i + \varepsilon\beta = \alpha_j + \varepsilon'\beta'$. *Then* $i = j$, $\varepsilon = \varepsilon'$ *and* $\beta = \beta'$.

We may assume that $i \leqq j$. Assume first that $i < j$. Then we have

(1) $$\alpha_i - \alpha_j = \varepsilon'\beta' - \varepsilon\beta \neq 0.$$

Being fixed under τ', the roots α_i and α_j are zero on $\mathfrak{t}$, hence so is $\varepsilon'\beta' - \varepsilon\beta$. But no root of $\mathfrak{m}$ is zero on $\mathfrak{t}$ (recall that $\mathfrak{t}$ contains regular elements of $\mathfrak{m}$). Therefore $\varepsilon'\beta - \varepsilon\beta$ and $\alpha_i - \alpha_j$ are not roots, and we have

(2) $$(\alpha_i,\alpha_j) \leqq 0, \quad (\varepsilon'\beta',\varepsilon\beta) \leqq 0.$$

But

$$(\alpha_i+\varepsilon\beta,\alpha_j+\varepsilon'\beta') = (\alpha_i+\varepsilon\beta,\alpha_i+\varepsilon\beta) > 0,$$

therefore

(3) $$(\alpha_i,\alpha_j) + (\alpha_i,\varepsilon'\beta') + (\varepsilon\beta,\alpha_j) + (\varepsilon\beta,\varepsilon'\beta') > 0.$$

By definition $\beta' \perp\!\!\!\perp \alpha_i$, hence a fortiori $(\beta',\alpha_i) = 0$. In view of (2), the relation (3) implies therefore

(4) $$(\varepsilon\beta,\alpha_j) > 0.$$

It follows that $\alpha_j - \varepsilon\beta$ is a root, but then so is $\alpha_i - \varepsilon'\beta'$, which contradicts the definition of β'. This proves (*) when $i < j$.

If now $i = j$. Then $\varepsilon\beta = \varepsilon'\beta'$. Since β and β' are >0, this implies $\varepsilon = \varepsilon'$ and $\beta = \beta'$.

5.6. By 5.3(iii), Card $C_s^{\tau'} \geqq \dim \mathfrak{a}_o + 1$. In order to prove 4.4 it suffices therefore to show

(1) $$2 \text{ Card } (\psi/\langle\tau'\rangle) \geqq \dim X_s'.$$

The dimension of X'_s is the complex dimension of the -1 eigenspace of θ (extended by linearity to $\mathfrak{g}_c$) on $\mathfrak{m}'_{s,c}$. We let σ be the complex conjugation of $\mathfrak{g}_c$ with respect to $\mathfrak{g}$. We have

$$\sigma\beta = -\beta^*, \quad \sigma\beta^* = -\beta \qquad (\beta\in\Phi).$$

Let $\beta \in \Phi^+(\mathfrak{m}'_s)$. We distinguish three cases:

(i) $\beta \neq \beta^*$. Then

$$\mathfrak{g}_\beta + \mathfrak{g}_{\beta^*} + \mathfrak{g}_{-\beta} + \mathfrak{g}_{-\beta^*}$$

is 4-dimensional and σ stable. θ permutes $\mathfrak{g}_\beta$ and $\mathfrak{g}_{\beta^*}$ (resp. $\mathfrak{g}_{-\beta}$ and $\mathfrak{g}_{-\beta^*}$) (recall that $\theta|\mathfrak{h}_c$ is defined by *). Therefore its -1 eigenspace there has dimension 2. But $\beta \neq \beta^*$ implies that $\beta|\mathfrak{a}_o \neq 0$, hence that $\beta \not\perp C_s^{\tau'}$ by 5.3(iii). Therefore β belongs to Ψ and gives a contribution 2 to the left-hand side of (1).

(ii) $\beta = \beta^*$, $\beta \in \Psi$. In this case, $\mathfrak{g}_\beta \oplus \mathfrak{g}_{-\beta}$ is σ-stable and θ-stable. The contribution to the dimension of X'_s is at most two, but $\beta \in \Psi$ adds two to the left-hand side of (1).

(iii) $\beta = \beta^*$, $\beta \notin \Psi$. Then again $\mathfrak{g}_\beta + \mathfrak{g}_{-\beta}$ is invariant under σ and θ. The root β is not in Ψ, hence does not contribute to the left-hand side of (1). Therefore we must show that θ is the identity on $\mathfrak{g}_\beta \oplus \mathfrak{g}_{-\beta}$.

By assumption $\beta \perp C_s^{\tau'}$. Then 5.3(ii) shows that $\mathfrak{g}_\beta$, $\mathfrak{g}_{-\beta}$ centralize $\mathfrak{z}_1$. Assumption 1.3(B) then yields the result.

§6. *The condition* (B).

6.1. If L is compact, condition (B) of 1.3 is automatically fulfilled. In this case $C(L,\mu) = 0$ for all μ's and 2.1(1) amounts to the relation

$$2.\ell(s) < \dim N + 1 \quad \text{if} \quad s\lambda|\mathfrak{a} > 0,$$

which already follows from 2.6 in [C]. As pointed out in [B], it can very easily be checked case by case. A proof is also included in [Z].

6.2. As already remarked at the end of §1, condition (B) is satisfied if $\mathfrak{a}_o$ is maximal $\mathbb{R}$-split in $\mathfrak{l}$, i.e. if

$$\text{(1)} \qquad \operatorname{rk} K(L) + \operatorname{rk}_{\mathbb{R}}\mathfrak{l} = \operatorname{rk} \mathfrak{l}.$$

This equality holds in particular if $\mathfrak{l}$ is a complex simple Lie algebra, viewed as a real Lie algebra, (in which case rank $K(L)$ and $\operatorname{rk}_{\mathbb{R}}(L)$ are equal to half of rk (L)), or if L is locally isomorphic to $\mathbf{SO}(2m+1,1)$ ($m \in \mathbb{N}$), in which case $K(L) = \mathbf{O}(2m+1)$ and $rk(L) = m+1$, $rk_{\mathbb{R}}(L) = 1$, $rk\, K(L) = m$.

6.3. For the following, we refer to [BB: §§1,2]. We now consider the irreducible bounded symmetric domains. Let G be a simple non-compact Lie group such that the symmetric space $X = G/K(G)$ of maximal compact subgroups of G is a bounded symmetric domain. Underlying the construction of V^* (see 2.3), there is first a Satake compactification $\bar{X}$ of X. Its boundary is the union of finitely many orbits of G, each one of which is fibered in the so-called real boundary components. If G is given moreover a $\mathbf{Q}$-structure, then the rational boundary components are among the real ones. A real boundary component is characterized by its normalizer, which is a proper maximal parabolic subgroup of G. Let P be one, $P = NAM$ its Langlands decomposition and X_P the associated boundary component. Then the identity component M^0 of M has an almost direct product decomposition $M^0 = L.G_P$, where L is the greatest connected normal subgroup of M^0 acting trivially on X_P and $X_P = G_P/K(G_P)$. [The notation is the one used so far in this paper, itself borrowed from [BC]. In [BB], F stands for the present X_P, $N(F)$ is the normalizer of F, and $Z(F) \cap M^0$ (resp. $G(F)$) stands for L (resp. G_P).] The space X_P is also a bounded symmetric domain, therefore 1.3(A) is satisfied.

6.4. In the sequel we are only concerned with types of Lie algebras. Equalities between Lie groups are therefore meant to be only local isomorphisms of identity components.

Let t be the $\mathbb{R}$-rank of G and ${}_{\mathbb{R}}\Delta = \{\alpha_1,\ldots,\alpha_t\}$ the set of simple $\mathbb{R}$-roots of G. The Dynkin diagram for the $\mathbb{R}$-roots is of type $\mathbf{C}_t$ or $\mathbf{BC}_t$. We use the canonical numbering of the vertices of [BB: 1.2]. The Satake compactification $\bar{X}$ of X is the one associated to the last simple $\mathbb{R}$-root. The proper maximal parabolic subgroups, up to conjugacy, are indexed by $b \in [1,t]$. Let P_b be the one assigned to b. The simple $\mathbb{R}$-roots of L (resp. G_P) are the α_i ($i<b$) (resp. $i > b$). The $\mathbb{R}$-diagram of L is then of type $\mathbf{A}_{b-1}$. To find out the type of L, one has first to determine the one of $\mathfrak{l}_c$. This amounts essentially to finding the roots in the Dynkin diagram Dyn $\mathfrak{g}_c$ of $\mathfrak{g}_c$ which restrict to the $\mathbb{R}$-roots of L and is easily read off the tables of [T]. Then one has to decide which real form of $\mathfrak{l}_c$ is $\mathfrak{l} = \mathfrak{g} \cap \mathfrak{l}_c$. These are simple computations, which we do not give in detail.

6.5. Up to local isomorphism, the possible G's are, in E. Cartan's notation (see [H: p. 354])

A III, BD I (p=2), C I, D III, E III, E VII .

We now discuss each type separately.

Type A.III. In Tits notation [T: p. 55] this is ${}^2\mathbf{A}_n$. In this case $L = \mathbf{SL}_{b-1}(\mathbf{C})$ and 6.2 applies.

Type BD I. Here $X = \mathbf{SO}(p,2)/S(\mathbf{O}(2)\times\mathbf{O}(p))$, and $t = 2$. We have to consider the case $b = 2$. Then $L = \mathbf{SO}(p-1,1)$, $K(L) = \mathbf{O}(p-1)$, and the $\mathbb{R}$-rank of L is 1.

If p is even, then 6.2(1) is satisfied.

Let now $p = 2m + 1$ be odd. Then $\operatorname{rk} L = \operatorname{rk} K(L) = m$ and 6.2(1) does not hold. Some computation is needed.

We assume the coordinates x_i in $\mathbb{R}^n$ $(n=2m+3)$ so chosen that G is the orthogonal group of the form $-x_1^2 + x_2^2 - x_3^2 + \Sigma_4^n x_i^2$. As a maximal $\mathbb{R}$-split torus, we may take the product of the orthogonal groups of the two hyperbolic planes spanned by the basis vectors (e_1,e_2) and (e_3,e_4) respectively. The group L is the subgroup of G leaving e_1 and e_2 fixed. It has the orthogonal group of the last $n - 3$ variables as a maximal compact subgroup. The Lie algebra $\mathfrak{t}$ is the one of a maximal torus of the latter group. It is a Cartan subalgebra of $\mathfrak{l}$, hence $\mathfrak{a}_0 = 0$. The derived algebra $\mathfrak{z}_1$ of the centralizer of $\mathfrak{t}$ is the Lie algebra of the orthogonal group $\mathbf{SO}(1,2)$ of the first three variables. Its centralizer is the orthogonal group of the last $n - 3$ variables and is indeed compact.

Type C I. Here G is the symplectic group $\mathbf{Sp}_{2b}(\mathbb{R})$, $K(G) = \mathbf{U}(n)$, X is the Siegel upper-half space of genus n and $t = n$. For $P = P_b$, we have $G_P \simeq \mathbf{Sp}_{2n-2b}(\mathbb{R})$ and $L = \mathbf{SL}_b(\mathbb{R})$. After permutation of the coordinates we may assume that L is a block matrix with entries $A, {}^tA^{-1}, I_{2n-2b}$, $(A\in\mathbf{SL}_b(\mathbb{R}))$. Let $c = [b/2]$. Let $Z = \begin{pmatrix} 0 & 1 \\ -1 & 0 \end{pmatrix}$. As $\mathfrak{t}_o$ we may take for b even the block matrices with blocks

$$y_1 Z, \ldots, y_c Z, \ -y_1 Z, \ldots, \ -y_c Z, \ 0_{2n-2b} \qquad (y_i\in\mathbf{R}, i=1,\ldots,c)$$

and $\mathfrak{t}$ is the direct sum of $\mathfrak{t}_o$ and of a compact Cartan subalgebra of $\mathbf{Sp}_{2n-2b}\mathbb{R}$, operating on the last $2n - 2b$ coordinates. To this we should add the 2×2 zero matrix in (e_b,e_{2b}) if b is odd. In Z we have a direct product of c (resp. $c+1$) copies D_i of $\mathbf{SL}_2(\mathbb{R})$ if b is even (resp. odd). For $i \in [1,c]$, D_i is the group of matrices

$$(1) \qquad \begin{pmatrix} aI_2 & bI_2 \\ cI_2 & dI_2 \end{pmatrix} \qquad (ad-bc=1)$$

acting on the 4-plane V_i with basis $(e_{2i-1}, e_{2i}, e_{b+2i-1}, e_{b+2i})$. For b odd, D_{c+1} is the unimodular group of the plane V_{c+1} spanned by (e_b, e_{2b}). The centralizer of Z_1 is contained in the centralizer of the product of the D_i's. It suffices to show that the intersection of the latter with L is compact. This intersection obviously leaves the V_i's invariant. On V_{c+1}, it centralizes $\mathbf{SL}_2(\mathbb{R})$, hence consists of scalar matrices cI_2. But the symplectic condition imposes that $c = c^{-1}$, hence $c = \pm 1$. On V_i $(1 \leq i \leq c)$ we have to find the centralizer of the group of matrices (1). Writing an element of it in 2×2 blocks, we see easily that it consists of matrices $\begin{pmatrix} A & 0 \\ 0 & A \end{pmatrix}$ $(A \in \mathbf{GL}_2(\mathbb{R}))$. But we must have in addition $A = {}^tA^{-1}$, hence A is orthogonal.

Type D III. Here $G = \mathbf{SO}^*_{2n}$ (notation of [H]), $K(G) = \mathbf{U}_n$ and $t = [n/2]$. In this case, $L = \mathbf{SL}_b(\mathbf{H})$, where $\mathbf{H}$ is the field of (Hamilton) quaternions. The group L has rank 2b-1 and $\mathbb{R}$-rank b-1. Moreover K(L) is the unitary group on quaternionic b-dimensional space and has therefore rank b. This shows that 6.2(1) holds.

Type E III. Here G is real form $\mathbf{E}_{6,-14}$ of $\mathbf{E}_6$, K(G) is locally isomorphism to $\mathbf{SO}(10) \times \mathbf{SO}(2)$, and $t = 2$. For $b = 2$, $Z = \mathbf{SO}(7.1)$, and 6.2(1) is satisfied.

Type E VII. Here G is the real form $\mathbf{E}_{7,-25}$ of $\mathbf{E}_7$ and K(G) is locally isomorphic to $\mathbf{E}_6 \times \mathbf{SO}(2)$, (where the compact $\mathbf{E}_6$ is meant). Moreover $t = 3$.

For $b = 2$, we have $L = \mathbf{SO}(9,1)$ and we may again apply 6.2(1).

Let $b = 3$. Then L is the real form $\mathbf{E}_{6,-26}$ of $\mathbf{E}_6$, with maximal compact subgroup of type $\mathbb{F}_4$. It has real rank 2, hence 6.2(1) is again true.

6.6. I have proposed a slight generalization of Zucker's conjecture, in which the underlying symmetric space is not necessarily hermitian, but where all the real boundary components of the Satake compactification to be used satisfy (A). See [Z] for the statement, where the new cases are also enumerated. Apart from some rank 2 groups there are two new series: $\mathbf{SO}(p,q)$ $(1 \leq p \leq q)$, with p+q odd, and the unitary groups $\mathbf{Sp}(p,q)$ $(1 \leq p \leq q)$ of indefinite quaternionic forms.

If $G = \mathbf{SO}(p,q)$, the system of $\mathbb{R}$-roots is of type $\mathbb{B}_p$ and the relevant Satake compactification is associated to the short simple $\mathbb{R}$-root. (Thus, for $p = 2$, it is different from the one considered above.) The $\mathbb{R}$-rank is p, and the types of boundary components are again indexed by $b \in [1,p]$. Given b, the corresponding group G_P is $\mathbf{SO}(p-b, q-b)$ and $L = \mathbf{SL}_b(\mathbb{R})$. The computations to check (B) are the same as in the case C I.

If $G = \mathbf{Sp}(p,q)$, the Satake compactification is associated to the long simple root. Again the $\mathbb{R}$-rank is p and the types of boundary components are indexed by $b \in [1,p]$. For a given b, we have $G_p = \mathbf{Sp}(p-b,q-b)$ and $L = \mathbf{SL}_b(\mathbb{H})$. As we saw above in the case D III, 6.2(1) is satisfied.

6.7. In all this, we have assumed G to be simple. In fact, in the applications, G is the group of real points of a $\mathbf{Q}$-simple algebraic group, hence is not always simple over $\mathbb{R}$ and symmetric space of G is then a product of symmetric spaces of real simple groups. The needed vanishing conditions follow from 2.1 and a suitable Künneth rule.

The Institute for Advanced Study, Princeton, NJ 08540 U.S.A.

References

[BB] W. Baily and A. Borel, *Compactifications of arithmetic quotients of bounded symmetric domains*, Ann. Math., 84, 1966, p. 442-528.

[B] A. Borel, L^2*-cohomology and intersection cohomology of certain arithmetic varieties*, in E. Noether in Bryn Mawr, Springer, 1983, p. 119-131.

[BC1] A. Borel and W. Casselman, L^2*-cohomology of locally symmetric manifolds of finite volume*, Duke Math. J. 50 (1983), p. 625-647.

[BC2] A. Borel et W. Casselman, *Cohomologie d'intersection et* L^2*-cohomologie de variétés arithmétiques de rang rationnel* 2. C. R. Acad. Sci. Paris 301, (1985), p. 369-373.

[C] W. Casselman, L^2*-cohomology for groups of real rank one*, in Representation theory of reductive groups, Progress in Math., 40, Birkhäuser, Boston, 1983, p. 69-82.

[E] T. Enright, *Relative Lie algebra cohomology and unitary representations of complex Lie groups*, Duke Math. J. 47, (1980), p. 1-15.

[H] S. Helgason, Differential Geometry and Symmetric Spaces, Adademic Press 1962.

[K] B. Kostant, *Lie algebra cohomology and the generalized Borel-Weil theorem*, Annals of Math. 74, (1961), p. 329-387.

[T] J. Tits, *Classification of algebraic semisimple groups*, in Algebraic groups and discontinuous subgroups, Proc. Symp. Pure Math. A.M.S. IX, (1966), p. 33-62.

[V] D. Vogan, *Unitarizability of certain series of representations*, Annals of Math., 120, (1984), p. 141-187.

[VZ] D. Vogan and G. Zuckerman, *Unitary representations with non-zero cohomology*, Comp. Math., 53, 1984, p. 51-90.

[Z] S. Zucker, L^2*-cohomology and intersection homology of locally symmetric varieties II*, preprint, 1984.

NILPOTENT ORBITS, PRIMITIVE IDEALS, AND CHARACTERISTIC CLASSES

Walter BORHO

Bergische Universität - GH Wuppertal, Gaußstr. 20, 5600 Wuppertal 1, FR Germany, and Max-Planck-Institut für Mathematik, Gottfried-Claren-Str. 26, 5300 Bonn 3, FR Germany.

CONTENTS

1. INTRODUCTION

1.1 Let G be a semisimple complex Lie group, say linear algebraic, and connected. The present report deals with the following three, a priori fairly unrelated subjects: (i) The geometry of unipotent conjugacy classes of G, or equivalently, of nilpotent orbits in the Lie algebra $\underline{g}$ of G; (ii) the classification of primitive ideals in the universal enveloping algebra $U(\underline{g})$, say with trivial central character for simplicity; and (iii) characteristic classes in $H^*(X)$ of certain bundles on the "flag variety" X, that is the "universal" complete homogeneous space for G. My main point will be to report from recent joint work with J.-L. Brylinski and R. MacPherson [BBM1,2,3] how (iii) can be used as a tool to get insight into both (i) and (ii) simultaneously, and to understand their relation.

For some time, especially in the seventies and early eighties, these subjects developed more or less independently, each beeing studied for its own sake, by its own specific methods. Extensive work has been done by many authors, and remarkable theories have been developed on the three subjects, making each of them individually into a highly cultivated area of mathematical research. Since they have been reviewed individually on various former occasions, I feel free here to focus attention on some of the fascinating relations between these subjects. For more back-ground on the individual subjects, the reader may consult for instance the Lecture Notes by Steinberg [St], Slodowy [Sl], or Spaltenstein [Sp] on (i), the books by Dixmier [Di] and Jantzen [Ja] on (ii), and say [Hi], [Fu] in combination with [BBM1,2,3] concerning (iii).

1.2 The first hint, suggesting that there must be some deep relation between (i) and (ii), became apparent from the fundamental work of T.A. Springer [S] in 1976, resp. A. Joseph [J1,2] a few years later, on (i) resp. (ii): Their results came down to closely relating (i) resp. (ii) to the same kind of objects, namely irreducible Weyl group representations. A careful comparison of Springer's and Joseph's correspondences, ultimately extended by D. Barbasch and D. Vogan [BV1,2] to exhaustive explicit case by case calculations, confirmed some superficial parallelities on one hand, but also exhibited some intriguing discrepancies on the other hand, so that the real relation remained a mystery for some years.

This situation was considered unsatisfactory by some people, including myself. Strictly speaking, in my case, this challenge dates back already to my 1976 exposé at the séminaire Bourbaki [B1], where I first suggested to relate (i) to (ii) via associated varieties, then reported Jantzen's conjectural partial anticipation of Joseph's theory in case $G = SL_n$ (see loc. cit. 2.9. resp. 5.9), and finally learnt from Springer about his new theory; consequently, it was inevitable for me to wonder how these pieces may fit together into a common frame-work. The solution of this puzzle took me not only some time, but also some new advanced methods, as well as two good friends to teach me how to use them. Much of my joint work with Brylinski or MacPherson was stimulated by the challenge of this puzzle, and is finally involved in the solution reported here. We use the intersection homology approach to (i), as developed in joint work with MacPherson [BM1,2], and the $\mathcal{D}$-module approach to (ii), as developed in joint work with Brylinski [BB1,2], and we combine them using equivariant K-theory (on T*X) as a unifying concept, in order to obtain a common frame-work for the simultaneous study of (i) and (ii) in terms of (iii), as elaborated in joint work with both [BBM1,2,3]. Let me also refer at this point to related work of V. Ginsburg [Gi].

1.3 The purpose of this report is two-fold: First, to popularize the "puzzle" mentioned above (section 2), and second to state our solution (section 3 and 4). In section 2, I tried to illustrate the problem in an intelligible way for the non-expert, using $G = SL_n$ as a standard example. Note that this is not only a courtesy to the reader not familiar with semisimple Lie group theory, but it also avoids here almost totally those "intriguing discrepancies" mentioned above; this will make our problem (to explain the coincidences between Springer's and Joseph's correspondence) particularly clear and persuasive. On the other hand, those "discrepancies" and their explanations for the other simple Lie groups are precisely what fascinates the real gourmet in Lie theory most of all. So in some sense, the experts may consider it a loss of good taste, that I have sacrificed such points radically for the sake of popularity; I hope that they will forgive me.

In sections 3 and 4, I tried to formulate an essential part of our results with as little effort as possible. I spent some care in defining important concepts, but essentially no proofs are included here. However, I added here a final section with comments on the general strategy of proof (section 5), offering at least some flavour of equivariant K-theory and its use in this context.

Remark. This report is essentially identical with my address to the International Congress of Mathematicians at Berkeley, except for the augmentation by a fifth section, concerning methods.

2. REVIEW ON SPRINGER'S AND JOSEPH'S WEYL GROUP REPRESENTATIONS

2.1 Basic notation. We fix a Borel subgroup B in G, and a maximal torus T in B. We denote by $\underline{t} \subset \underline{b} \subset \underline{g}$ the Lie algebras of $T \subset B \subset G$, by $W = N_G(T)/T$ the Weyl group, and by $X = G/B$ the flag variety.

2.2 Standard example. To simplify the present review, I shall sometimes restrict (in the present chapter only) to $G = SL_n$ as a standard example. The reader not familiar with general semisimple Lie group theory may anyway prefer to think in terms of this example through out this paper: Then G is the group $SL(n,\mathbb{C})$ of complex n by n matrices $(n \geq 2)$ with determinant 1, and T resp. B are the subgroups of its diagonal resp. upper triangular matrices. The Lie algebra $\underline{g} = \underline{sl}(n,\mathbb{C})$ consists then of all complex n by n matrices of trace 0, with Lie bracket $[x,y] = xy - yx$ the commutator of matrices, and $\underline{t}$ resp. $\underline{b}$ consists of all traceless diagonal resp. upper triangular matrices. The Weyl group W is the symmetric group S_n in this case, acting on T and $\underline{t}$ by permuting the n eigenvalues of a diagonal matrix. The flag variety $X = G/B$ may in this case be defined alternatively in the original sense: Its points $F \in X$ may be thought of as real "flags" in $\mathbb{C}^n$; that is F is an ascending chain of complex subspaces $F_1 \subset F_2 \subset \ldots$ $\ldots \subset F_n$ of $\mathbb{C}^n$ such that F_i has dimension i.

2.3 Nilpotent orbits. Let N denote the "nilpotent cone" in $\underline{g}$, that is the closed subvariety of all ad-nilpotent elements, which is a cone in $\underline{g}$. (A cone in a vector space is a subset closed under homotheties.) Under the adjoint action of G on $\underline{g}$, N decomposes into a finite number of orbits, called "nilpotent orbits".

In case $G = SL_n$, these orbits are just the conjugacy classes of nilpotent complex n by n matrices. Recall that the set N/G of nilpotent orbits is here in bijection to the set $P(n)$ of partitions of n (theory of Jordan normal form), the "parts" being just the sizes of Jordan blocks. For example, the nilpotent orbit $\mathcal{O}$ in $\underline{sl}_6$ generated by the nilpotent matrix

$$X = \left[\begin{array}{ccc|cc|c} 0 & 1 & & & & \\ & 0 & 1 & & & \\ & & 0 & & & \\ \hline & & & 0 & 1 & \\ & & & & 0 & \\ \hline & & & & & 0 \end{array}\right] \qquad \text{(all other entries zero)}$$

corresponds to the partition $\lambda = (3,2,1)$, which is also denoted

$$\lambda = \begin{array}{c}\boxed{}\end{array}$$

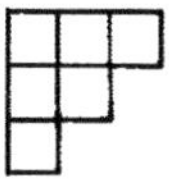

(notation of "Young diagrams").

2.4 Primitive ideals. By definition, a primitive ideal J in $U(\underline{g})$ is the kernel of some irreducible representation of $\underline{g}$, or in other words: the annihilator of some simple (left) $U(\underline{g})$-module L, notation $J = \text{Ann } L$. The center of $U(\underline{g})$ necessarily acts by a character on L, which we assume here to be "trivial", or equivalently, this means $J \subset \underline{g}\, U(\underline{g})$. Let X_0 denote the set of all such primitive ideals. This set is finite. More precisely, one defines a surjection $W \longrightarrow X_0$ as follows: The module L above can always be chosen in a particularly nice way, as a so-called simple highest weight module (theorem of Duflo), and the highest weights in question here come from a single regular W orbit in $\underline{t}^*$ (Harish-Chandra isomorphism), so the modules in question can be suitably indexed by Weyl group elements $w \in W$, and that surjection $W \longrightarrow X_0$ can then be described $w \longmapsto \text{Ann } L_w =: J_w$.

2.5 Combinatorial description. In case $G = SL_n$, the set X_0 of primitive ideals with trivial central character is in bijection to the set $T(n)$ of all <u>tableaux</u> of size n (theory of Joseph's Goldie rank polynomials, see [Ja]).

Here a "tableau" is the combinatorial object also familiar as a "Young standard tableau" from the representation theory of symmetric groups. An example of a tableau of size $m = 6$ is

$$\tau = \begin{array}{|c|c|c|}\hline 1 & 3 & 5 \\\hline 2 & 4 \\\cline{1-2} 6 \\\cline{1-1}\end{array} ;$$

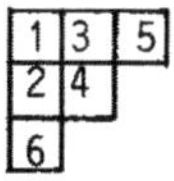

its "<u>shape</u>" is the partition

$$\lambda = (3,2,1) = \begin{array}{c}\boxed{}\end{array} ,$$

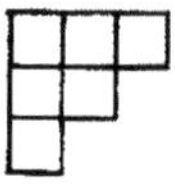

and there are exactly 16 tableaux of this same shape. There is a canonical, easily calculable map $T: W = S_n \longrightarrow T(n)$, producing from each permutation w a tableau $T(w)$ (Robinson-Schensted algorithm). For example, the above tableau τ is produced as $T(w)$ from the permutation $w = 216453$, and from 15 other permutations.

Now the bijection between tableaux and primitive ideals, $T(n) \longrightarrow X_0$,

$\tau \longmapsto J_\tau$, may be described as follows (Joseph): $J_\tau = \mathrm{Ann}\, L_w$ for any permutation w of tableau $\tau = T(w)$.

2.6 Associated varieties. There is also a map $X_0 \longrightarrow N/G$ from primitive ideals to nilpotent orbits. In case $G = SL_n$, in view of the preceding combinatorial descriptions of the two sets $X_0, N/G$, the reader will find it not hard to guess what this map should do in combinatorial terms: It sends the primitive ideal $J = J_\tau$ corresponding to a tableau τ to the nilpotent orbit $\mathcal{O} = \mathcal{O}_\lambda$ corresponding to the shape λ of τ.

But how do we produce directly, and in general, a nilpotent orbit from a primitive ideal J? For this purpose, we identify the associated graded ring of $U(\underline{g})$ with the symmetric algebra $S(\underline{g})$, and interpret it as ring of polynomial functions on $\underline{g}^* = \underline{g}$ (using first the Poincaré-Birkhoff-Witt theorem, and second the Killing form). Then we can define the associated variety of J as the zero set $V(\mathrm{gr}J) \subset \underline{g}$ of the associated graded ideal $\mathrm{gr}\, J \subset S(\underline{g})$. It turns out that this associated variety is irreducible [BB1,2], see also [J3], and contained in N; hence it contains a unique dense orbit $\mathcal{O}$. So the desired map $J \longmapsto \mathcal{O}$ is given by $V(\mathrm{gr}J) = \bar{\mathcal{O}}$. Let me comment here that this relation was suggested [B1] before, but completely proven [BB1] only after Joseph invented his classification theory of primitive ideals (2.9).

2.7 Link to Weyl group representations, illustrated in case $G = SL_n$. Young diagrams, and tableaux, the combinatorial data which we encountered above in the classification of nilpotent orbits resp. primitive ideals, both have a well-known significance in the (Frobenius') theory of representations of the symmetric group: There is a bijection, denoted $\lambda \longmapsto \rho_\lambda$, $P(n) \longrightarrow W^\wedge$, from diagrams λ of size n, to (equivalence classes of) irreducible complex representations of $W = S_n$. Moreover, the tableaux τ of a given shape λ correspond bijectively to a linear basis for the representation ρ_λ. In view of this, the results for $G = SL_n$ reviewed in 2.3, 2.5, may persuade you to make the following guess:

To a nilpotent orbit $\mathcal{O}$, there should correspond an irreducible representation $\rho_{\mathcal{O}}$ of the Weyl group, and to the collection of primitive ideals J with associated variety $\bar{\mathcal{O}}$ should correspond a basis of this same representation. It turns out that this is actually true, not only in case $G = SL_n$, but in general, and the main purpose of my report is to explain how such correspondences can be constructed. I shall next introduce the correspondences of Springer and Joseph, which can be made to do this job, although only with some major effort to verify that the two representations are actually the same. An alternative version, were this difficulty disappears, is then stated in theorems 3.4 and 4.4 below.

2.8 Springer's correspondence. Springer [S] attaches to each nilpotent orbit $\mathcal{O}$ an irreducible representation $\rho_{\mathcal{O}}$ of W as follows: Let $u \in \mathcal{O}$ represent the orbit, and let $X^u \subset X$ be the subvariety of all "flags" respected by u. Then W acts linearly on the homology groups $H_*(X^u)$. (For simplicity, here and below take complex coefficients for (co-)homology groups etc.). For $G = SL_n$, this action is irreducible on the homology group $H_{2d}(X^u)$ of highest degree ($2d = 2\dim X^u$), and this defines $\rho_{\mathcal{O}}$; this is Springer's explicit, geometrical realization of the bijection $N/G \longrightarrow W^\wedge$ described only combinatorially in 2.7.

For G arbitrary, one has to replace $H_{2d}(X^u)$ by its invariants under the action of the isotropy group $G_u \subset G$ of u to define Springer's representation $\rho_{\mathcal{O}}$, and one obtains only an injection $N/G \longrightarrow W^\wedge$ by mapping $\mathcal{O}$ to $\rho_{\mathcal{O}}$.

2.9 Joseph's Goldie rank polynomials. Joseph [J1] attaches to each primitive ideal $J \in X_0$ a polynomial function p_J on the Cartan subalgebra $\underline{t}$. His construction proceeds in two steps. The first is, to replace J by the whole infinite family $(J_\mu)_{\mu \in Z}$ of "translated" primitive ideals J_μ, with variable central character, depending on a parameter μ which varies in a Zariski dense subset Z of $\underline{t}^*$ ("translation principle" of [BJ]). The second step is to consider the Goldie rank rk $U(\underline{g})/J_\mu$ as a function of μ, and to prove that this extends to a polynomial function on $\underline{t}$; this polynomial, by definition, is p_J. (For the reader not familiar with non-commutative Noetherian ring theory, let me also add the definition of Goldie rank: The quotient ring $U(\underline{g})/J_\mu$ has a complete ring of fractions, which is simple Artinian (Goldie's theorem), hence is a matrix ring over some skew fields (Wedderburn-Artin); then rk $U(\underline{g})/J_\mu$ is the rank of this matrix ring.)

2.10 Joseph's correspondence. Next, Joseph attaches to the primitive ideal $J \in X_0$ the W-submodule generated by p_J in $S(\underline{t}^*)$, and proves that the corresponding W-representation, denoted $\sigma(J)$, is irreducible. This defines Joseph's correspondence $X_0 \longrightarrow W^\wedge$, $J \longmapsto \sigma(J)$. Moreover, if J' ranges over all primitive ideals such that $\sigma(J') = \sigma(J)$, then the corresponding Goldie rank polynomials provide a basis for the representation $\sigma(J)$. The reader will find an excellent exposition of this beautiful theory of Joseph in Jantzen's book [Ja].

2.11 Comparison. The intriguing question, how the correspondences of Springer resp. Joseph relate to each other, on which I commented several times before, can be made precise at this point as follows: Does the correspondence from primitive ideals to nilpotent orbits via associated varieties (2.5) combine with Springer's and Joseph's correspondences to a commutative triangle? Or in other words: Is $\sigma(J)$ equivalent to $\rho_{\mathcal{O}}$, if J corresponds to $\mathcal{O}$? In case $G = SL_n$, the explicit combinatorial descriptions given above prove that this is true. Barbasch and Vogan have verified this as a matter of fact for all cases, by an enormous amount of explicit calculations in [BV1,2]. More conceptual reasons are offered by Hotta and Kashiwara [HK] and below.

3. CHARACTERISTIC CLASS APPROACH TO NILPOTENT ORBITS

In these next two sections, I shall now sketch the simultaneous, uniform approach to both subjects, (i) nilpotent orbits, and (ii) primitive ideals, in terms of (iii) characteristic classes on the flag variety X, as suggested in my recent joint work with Brylinski and MacPherson. For the elaboration of more details, I refer to our original papers [BBM1-3].

3.1 Characteristic classes of cone bundles. The concept of a cone bundle K on X generalizes that of a vector bundle: The bundle map $K \longrightarrow X$ is assumed to be a locally trivial fibration of K by cones (in vector spaces). To extend the theory of Chern classes of vector-bundles, Fulton and MacPherson [Fu] introduced the notion of Segre class $s(K)$ of a cone bundle. It may be characterized by two axioms (1) for a vector-bundle K, $s(K) = c(K)^{-1}$ is the inverse of the total Chern class $c(K)$, and (2) $s(K)$ is functorial under proper push-forwards.

Now to define our characteristic class $Q(K)$ in $H^*(X)$, for any cone subbundle K of codimension d in the cotangent bundle T^*X, we multiply its Segre class by the total Chern class of T^*X, and take the lowest (degree 2d) homogeneous term of the product, notationally:

$$Q(K) := [c(T^*X)\,s(K)]^{2d}$$

3.2 Springer's resolution of the nilpotent cone. Another key ingredient for our construction is the famous Springer map $\pi: T^*X \longrightarrow N$, which will allow us to pass from nilpotent orbits to the geometry of the flag variety. So let me recall here that this remarkable map has a very easy, elegant definition (as a Kostant-Souriau momentum map [BB1]): The natural action of $\underline{g}$ by vector-fields on X defines a morphism $\underline{g} \times X \longrightarrow TX$ into the tangent bundle, and the map π of the cotangent bundle T^*X into $\underline{g}^* = \underline{g}$ (Killing form) is then obtained by transposition and projection. The remarkable point about π is then that its image in $\underline{g}$ is the nilpotent cone N, and that it resolves the singularities of N.

3.3 Construction of characteristic classes from a nilpotent orbit. Starting from a nilpotent orbit $\mathcal{O} \subset N$, we first produce a collection of cone bundles on X by taking the preimage $\mathcal{O}$ under Springer's map π and then decomposing the closure $\overline{\pi^{-1}\mathcal{O}}$ into irreducible components $K_1, \ldots, K_r$. These are in fact cone bundles

$K_i \longrightarrow X$, called orbital for $\mathcal{O}$. We know that their codimension d in T^*X depends only on $\mathcal{O}$, more precisely $2d = \mathrm{codim}_N \mathcal{O}$ (work of Spaltenstein and Steinberg).

Next, we take for these "orbital" cone bundles $K_1,\ldots,K_r$ the characteristic classes $Q(K_1),\ldots,Q(K_r)$ as defined in 3.1.

3.4 Theorem.
a) The classes $Q(K_1),\ldots,Q(K_r)$ are linearly independent.
b) They form a W-submodule in $H^{2d}(X)$.
c) This W-representation is equivalent to Springer's $\rho_{\mathcal{O}}$.
d) It transforms the basis $Q(K_1),\ldots,Q(K_r)$ according to the formulae below.

The following formulae were first obtained by Hotta in a slightly different context [H1], [H2], see also [J4].

3.5 Hotta's transformation formulae. For each simple reflection s of W, and for all $i = 1,\ldots,r$, we have either

$$sQ(K_i) = -Q(K_i),$$

or else

$$sQ(K_i) = \sum_{j=1}^{r} n_{ij}(s)\, Q(K_j),$$

for some matrix of non-negative integers $n_{ij}(s)$ with diagonal entries $n_{ii}(s) = 1$. Moreover, there is a geometrical interpretation for these integers, for which I refer to our resp. Hotta's original papers. For example, this says that $n_{ij}(s) = 0$ unless K_j intersects K_i in codimension ≤ 1.

3.6 Algebraic construction of our characteristic classes. For the reader with less inclination for geometrical elegance, but with a preference for abstract algebra, let me also offer here an alternative, more algebraic definition of the classes $Q(K)$ defined geometrically in 3.1. This definition refers to the following ingredients:

- Borel's description of $H^*(X)$ in terms of polynomials on the Cartan subalgebra $\underline{t}$, by a W-equivariant isomorphism $\beta: H^*(X) \xrightarrow{\sim} S(\underline{t}^*)^{\mathcal{H}}$ onto the space of all W-harmonic polynomials on $\underline{t}$.
- The Chern character $ch: K(X) \longrightarrow H^*(X)$, attaching to an (algebraic) vector-bundle on X its total Chern class (an isomorphism here).
- The Grothendieck group $K(X)$ of classes of (algebraic) vector-bundles on X, or equivalently the Grothendieck group of the category of all coherent sheaves of $\mathcal{O}_X$-modules.
- The analogous Grothendieck-group $K(T^*X)$, and its isomorphism σ^* with $K(X)$ induced by the zero-section $\sigma: X \hookrightarrow T^*X$ [Fu].

Given these ingredients, we may attach to any closed subvariety K of codimension d in T^*X a cohomology class $Q(K)$ in $H^{2d}(X)$, which we identify with a degree d homogeneous polynomial on $\underline{t}$ via $\beta(H^{2d}(X) = S^d(\underline{t}^*)^{\natural})$. The construction proceeds as follows: Take the class $[\mathcal{O}_K]$ determined by the structure sheaf of K in $K(T^*X)$, and apply the composition of the maps

$$K(T^*X) \xrightarrow[\sim]{\sigma^*} K(X) \xrightarrow[\sim]{ch} H^*(X) \xrightarrow[\sim]{\beta} S(\underline{t}^*)^{\natural};$$

finally take the lowest degree term. So formally, this alternative, algebraic definition of $Q(K)$ reads:

$$Q(K) := [\beta \text{ ch } \sigma^*[\mathcal{O}_K]]^d.$$

For a cone bundle K it can be shown that this definition coincides with that in terms of Segre class as stated in 3.1.

4. CHARACTERISTIC CLASS OF A PRIMITIVE IDEAL

4.1 Characteristic variety of a primitive ideal. Starting from a primitive ideal $J \in X_0$, we first construct a cone-subbundle in T^*X as follows. Later, we shall define the characteristic class of J as the characteristic class of this cone bundle (4.3).

Consider the quotient $U(\underline{g})/J$ as a left $\underline{g}$-module M, and take the corresponding sheaf of modules

$$\mathcal{M} := \mathcal{D}_X \otimes_{U(\underline{g})} M$$

over the sheaf $\mathcal{D}_X$ of rings of differential operators on X (Beilinson-Bernstein localization of M on X); then the characteristic variety resp. cycle of $\mathcal{M}$ are well-defined notions from the general theory of $\mathcal{D}$-modules, denoted $Ch(\mathcal{M})$ resp. $\mathcal{C}h(\mathcal{M})$ here. By definition, $Ch(\mathcal{M})$ is a closed subvariety in T^*X, and $\mathcal{C}h(\mathcal{M})$ is a formal integer linear combination

$$\mathcal{C}h(\mathcal{M}) = \sum_i m_i [V_i],$$

in which each irreducible component V_i of the characteristic variety occurs with some well-defined positive multiplicity m_i. Now characteristic varieties are (by construction) fibred by cones, and in the present case, this fibration is locally trivial on X (as a consequence of the stability of J under the adjoint G-action, and the G-homogeneity of X). So $Ch(\mathcal{M})$, and hence its components V_i, are actually cone bundles over X.

4.2 Relation to nilpotent orbits. By my work with Brylinski [BB2], these cone bundles are actually orbital for some nilpotent orbit $\mathcal{O}$. In more detail, by loc.cit. Springer's map π maps $Ch(\mathcal{M})$ onto the associated variety $V(\text{gr}J)$, which is the closure of a nilpotent orbit, as reported already in 2.6. In the sequel $\mathcal{O}$ denotes this nilpotent orbit determined by J. It follows that $Ch(M)$ is contained in $\overline{\pi^{-1}\mathcal{O}}$, and now some tricky dimension arguments show that each component V_i is even contained in $\overline{\pi^{-1}\mathcal{O}}$ as one of the irreducible components, hence is one of the orbital cone bundles $K_1,\ldots,K_r$ (notation 3.3). - Hence we may write our characteristic cycle as well as

$$Ch(M) = \sum_{i=1}^{r} z_i\,[K_i]$$

where we admit some of the "multiplicities" z_i to be zero.

4.3 Definition. Now we define the characteristic class of J in $H^*(X)$ as the characteristic class of its characteristic variety by:

$$P(J) := Q(Ch(M)) := \sum_{i=1}^{r} z_i\, Q(K_i).$$

Here $Q(K_i)$ is the characteristic class of a cone bundle as defined in 3.1 (or 3.6).

4.4 Theorem: Let $J_1,\ldots J_r \in X_0$ be the collection of primitive ideals corresponding to a nilpotent orbit $\mathcal{O}$ (that is with associated variety $\overline{\mathcal{O}}$). Then:

a) The characteristic classes $P(J_1),\ldots,P(J_r)$ are linearly independent.

b) They span a W-submodule in $H^{2d}(X)$. (Note $2d = \operatorname{codim}_N \mathcal{O}$.)

c) This W-representation is equivalent to Springer's $\rho_{\mathcal{O}}$.

d) The class $P(J_i)$ is proportional to Joseph's Goldie rank polynomial of J_i, that is $\beta P(J_i) = \gamma p_{J_i}$ for some scalar $0 \neq \gamma \in \mathbb{Q}$.

5. THE EQUIVARIANT K-THEORY SET UP FOR PROOFS

Let me conclude this report with the following comments concerning the general strategy of our uniform proofs of theorems 3.4 and 4.4. The purpose of these comments is to indicate the rôle played by the formalism of equivariant K-theory, and to sketch a few ideas which are crucial in our approach. I refer to my original papers with Brylinski and MacPherson for detailed expositions of proofs, and also for more elaborated statements of these and further results.

5.1 Refinement to the G-equivariant level. Our definition of characteristic classes $Q(K)$ as given in section 3 refers only to the geometrical structure of the cone bundle K. This is fine from the point of view of elegance of results, since it allows purely geometrical interpretations. However, from the point of view of proof of some of our results, it is more convenient to take also account of the additional structure on an orbital cone bundle K provided by the group action. The advantage of our algebraic construction of the characteristic class $Q(K)$ given in 3.6 is that it applies almost word by word to the definition of a more refined notion of "equivariant characteristic class", denoted $Q_G(K)$, as well. To be a little more precise, this is defined for any G-stable closed subvariety K in T^*X as a homogeneous polynomial on $\underline{t}$ as follows: Take the class $[\mathcal{O}_K]$ determined by the structure sheaf on K in the Grothendieck group $K_G(T^*X)$ of the category of G-equivariant coherent sheaves on T^*X, and then apply to it the following chain of homomorphisms

$$K_G(T^*X) \xrightarrow{\sigma_G^*} K_G(X) \xrightarrow{ch_G} H_G^{even}(X) \xrightarrow{\beta_G} \hat{S}(\underline{t}^*),$$

which refines the completely analogous one in subsection 3.6 to the G-equivariant level. Here the "equivariant Chern character" ch_G maps the equivariant K-group of the flag variety X into the even degree part of the equivariant cohomology group of X, which in turn by the "equivariant Borel picture" is identified with the ring $\hat{S}(\underline{t}^*)$ of formal power series on $\underline{t}$. Finally, define $Q_G(K)$ as the lowest degree term of the power series manufactured from K by this procedure.

5.2 Relating the equivariant to the geometric level. Now that we haven taken account of the additional G-structure on K we may look at the process of "forgetting" it again; this provides a canonical "forgetful" homomorphism $K_G(X) \to K(X)$, which corresponds to projecting a power series F on $\underline{t}$ to its W-harmonic part $F^{\natural}$, as is easy to see. It turns out that the lowest degree term $Q_G(K)$ does not project to 0, or has non-zero harmonic part. This fact reflects an equality of the codimension of support of a coherent sheaf on X and the degree of the corresponding element in $K_G(X)$ with respect to Grothendieck's γ-filtration, which is true in this case, though a little delicate to prove. We conclude then that our characteristic class $Q(K)$ can be reobtained from the G-equivariant version $Q_G(K)$ just by taking the W-harmonic part:

$$Q(K) = Q_G(K)^{\natural}.$$

5.3 Reduction to the T-equivariant level. A key technique of our approach consists in switching from G-equivariant K-theory on the flag manifold to T-equivariant K-theory on a vector-space E as follows. Starting from a G-equivariant coherent sheaf on the cotangent bundle T^*X, we first restrict the group action from G to T, and then restrict the sheaf to the single fibre E of T^*X at the base point, which is fixed by T, to obtain a T-equivariant sheaf on E; on the other hand, we restrict the sheaf on T^*X to the zero section X, as we did already in 5.1. These restriction processes give isomorphisms

$$K_G(X) \cong K_G(T^*X) \cong K_T(E). \qquad (*)$$

The point of these manipulations is now that equivariant K-theory of a linear torus action can be carried out very conveniently in terms of calculations with formal characters, as I shall explain a little more precisely below, and that we can reduce our problems from the context of G-equivariant K-theory on X into this more convenient setting. Furthermore, as explained in subsection 5.2, the link to our previous geometrical considerations is made by means of the homomorphisms

$$K_G(X) \longrightarrow K(X) \xrightarrow{ch} H^*(X).$$

In conclusion, this exhibits our technique of translating statements from a context most convenient for computational manipulations (formal characters) into a context most convenient for geometrical interpretation (cohomology of the flag variety), and vice versa. This is one of the crucial ideas underlying the strategy of proofs in [BBM3].

5.4 Formal characters. Since T acts linearly on E, the zero point is T-stable. So the inclusion $\iota: \{0\} \longrightarrow E$ induces a functorial ring homomorphism ι^* of $K_T(E)$ into $K_T(0)$. But $K_T(0)$ is nothing else but R(T), the representation ring of T, which may also be considered as the group algebra of the character group X(T) of T. The map $\iota^*: K_T(E) \longrightarrow R(T)$ turns out to be an isomorphism, which may be described explicitely as follows.

Let F be a T-equivariant coherent sheaf on E. Then $M = \Gamma(E,F)$ is a finitely generated $S(E^*)$-module equipped with an equivariant T-action, and so it decomposes into a direct sum of weight spaces M_χ, where $\chi \in X(T)$. Now one can define the formal character of M as usual as a formal sum

$$ch(M) = \sum_{\chi \in X(T)} (\dim M_\chi)\,[\chi].$$

(Here one has to note that the weight multiplicities $\dim M_\chi$ are all finite because of the positivity of the weights of E.) We may multiply such expressions in an obvious way by elements of R(T), for instance by

$$\Delta := \prod_\chi (1-\chi),$$

where the product is extended over the positive roots (the weights of T in E). Then the desired formula for ι^* reads as follows:

Proposition: $\iota^*[F] = \Delta\, ch(M)$.

In particular, $ch(M)$ may be considered as an element in the fraction field of R(T). It is easy to deduce from this formula the

Corollary: ι^* is an isomorphism of $K_T(E)$ onto R(T).

5.5 A formula for characteristic classes in terms of formal characters. We consider R(T) as a subring of the ring $\hat{S}(\underline{t}^*)$ of formal power series on $\underline{t}$. This is done by making a character χ of T correspond to the exponential of its differential:

$$e^{d\chi} = \sum_{n\geq 0} \frac{1}{n}(d\chi)^n = \chi.$$

If $P \in \hat{S}(\underline{t}^*)$ is any power series, then we denote $[P]^d$ its degree d homogeneous term, which is of course a homogeneous polynomial.

Theorem: Let K be an orbital cone bundle. Let $M = \Gamma(E, \mathcal{O}_K)$, the ring of the regular functions on its fibre over the base point, considered as a T- equivariant $S(E^*)$-module.

Then

a) As a formal power series on $\underline{t}$, $\Delta\, ch(M)$ has its lowest nonzero homogeneous term in degree

$d := \operatorname{codim}_{T^*X} K$.

b) The equivariant characteristic class of K (as a polynomial on $\underline{t}$) is given by the formula

$Q_G(K) = [\Delta\, ch(M)]^d$.

c) The characteristic class of K is given by the formula

$Q(K) = ([\Delta\, ch(M)]^d)^{\natural}$

as the harmonic part of the lowest degree term of $\Delta\, ch(M)$.

5.6 Conclusive remarks on the proof of theorems 3.4 and 4.4. This is the desired explicit expression for our characteristic classes in terms of the formal characters. Since this expression relates our characteristic classes to the "character polynomials" as studied in the previous literature by Joseph, Jantzen, Vogan, and others, it enables us to prove parts d) and c) of theorems 3.4 and 4.4. In case of theorem 4.4, one has to use the work on characteristic varieties of primitive ideals in [BB2] and some $\mathcal{D}$-module theory as an additional ingredient.

To make the identification with Springer's representations, that is to prove part c) of the theorems 3.4 and 4.4, we do not need the equivariant level but we work directly on the geometrical level, using as additional main ingredient the work on intersection homology of closures of nilpotent orbits in [BM1,2].

References:

[BV1] **Barbasch, D. - Vogan, D.**: Primitive ideals and orbital integrals in complex classical groups; Math. Ann. 259 (1982), 153-199.

[BV2] **Barbasch, D. - Vogan, D.**: Primitive ideals and orbital integrals in complex exceptional groups; J. Algebra 80 (1983), 350-382.

[B1] **Borho, W.**: Recent advances in enveloping algebras of semisimple Lie algebras; Séminaire Bourbaki 1976, Springer LNM 677 (1978), exposé 489.

[BJ] **Borho, W. - Jantzen, J.C.**: Über primitive Ideale in der Einhüllenden einer halbeinfachen Lie-Algebra; Invent. Math. 39 (1977), 1-53.

[BB1] **Borho, W. - Brylinski, J.L.**: Differential operators on homogeneous spaces I; Invent. Math. 69 (1982), 437-476.

[BB2] **Borho, W. - Brylinski, J.L.**: Differential operators on homogeneous spaces III; Invent. Math. 80 (1985), 1-68.

[BBM1] **Borho, W. - Brylinski, J.L. - MacPherson, R.**: A note on primitive ideals and characteristic classes; in: Geometry Today, Birkhäuser: Progress in Math. 60 (1985), 11-20.

[BBM2] **Borho, W. - Brylinski, J.L. - MacPherson, R.**: Springer's Weyl group representations through characteristic classes of cone bundles; IHES preprint M/85/70, Dec. 1985.

[BBM3] **Borho, W. - Brylinski, J.L. - MacPherson, R.**: Equivariant K-theory approach to nilpotent orbits; IHES preprint M/86/13, March 1986.

[BM1] **Borho, W. - MacPherson, R.**: Représentations des groupes de Weyl et homologie d'intersection pour les variétés nilpotentes; C.R. Acad. Sci. Paris (A) 292 (1981), 707-710.

[BM2] **Borho, W. - MacPherson, R.**: Partial resolutions of nilpotent varieties; in: Analyse et Topologie sur les Espaces Singuliers, Soc. Math. de France, Astérisque 101 (1983), 23-74.

[Di] **Dixmier, J.**: Algèbres enveloppantes; Paris: Gauthier Villars 1974.

[Fu] **Fulton, W.**: Intersection theory, Springer: Berlin-Heidelberg- New York - Tokio 1984.

[Gi] **Ginsburg, V.**: $\underline{g}$-modules, Springer's representations and bivariant Chern classes; Advances Math. 59 (1986).

[H] **Hiller, H.**: Geometry of Coxeter groups; Res. Notes in Math. 54, Pitman: Boston-London-Melbourne 1982.

[H1] **Hotta, R.**: On Joseph's construction of Weyl group representations; Tohoku Math. J. 36 (1984), 49-74.

[H2] **Hotta, R.**: On Springer's representations, J. Fac. Sci., Univ. of Tokyo, IA 28 (1982), 836-876.

[HK] **Hotta, R. - Kashiwara, M.**: The invariant holonomic system on a semisimple Lie algebra; Invent. Math. 75 (1984), 327-358.

[Ja] **Jantzen, J.C.**: Einhüllende Algebren halbeinfacher Lie-Algebren; Springer: Berlin-Heidelberg-New York-Tokio 1983.

[J1] **Joseph, A.:** Goldie rank in the enveloping algebra of a semisimple Lie algebra I,II; J. of Algebra 65 (1980), 269-306

[J2] **Joseph, A.:** Kostant's problem, Goldie rank, and the Gelfand-Kirillov conjecture; Invent. Math. 56 (1980), 191-213.

[J3] **Joseph, A.:** On the associated variety of a primitive ideal; J. of Algebra 93 (1985), 509-523.

[J4] **Joseph, A.:** On the variety of a highest weight module; J. of Algebra 88 (1984), 238-278.

[Sl] **Slodowy, P.:** Simple singularities and simple algebraic groups; Springer LNM 815 (1980).

[Sp] **Spaltenstein, N.:** Classes unipotentes et sous-groupes de Borel; Springer LNM 946 (1982).

[St] **Steinberg, R.:** Conjugacy classes in algebraic groups; Springer LNM 366 (1974).

[S] **Springer, T.A.:** Trigonometric sums, Green functions of finite groups, and representations of Weyl groups; Invent. Math. 36 (1976), 173-207.

SOME EXAMPLES OF HOCHSCHILD AND CYCLIC HOMOLOGY

Jean-Luc Brylinski*
Brown University
Department of Mathematics
Box 1917, Providence, Rhode Island 02912, U.S.A.

The theory of algebraic groups and their representations has made important progress in the last decade; let us point out two remarkable aspects of this progress.

1) the use of sophisticated (co)homology theories like étale cohomology and intersection cohomology, in the work of Deligne, Kazhdan, Lusztig, Springer, and many others. On the other hand, algebraic groups actions provide most interesting examples and much motivation to experts in intersection cohomology.

2) the geometric importance of non-commutative algebras. It has proven important to consider the Springer resolution, which is the cotangent bundle of the flag variety, as the "shadow" of a non-commutative object, the algebra of differential operators on the flag variety.

In this article, we somehow combine both themes, by looking at the cyclic homology of some interesting non-commutative algebras. We mostly consider two sorts of algebras. One is the convolution algebra $C_c^\infty(G)$, where G is a (real or p-adic) Lie group. Together with P. Blanc, we show that the Hochschild homology of that algebra is

*partially supported by a National Science Foundation grant

equal to the differentiable group homology $H_*^{diff}(G, C_c^\infty(G))$, where G acts on $C_c^\infty(G)$ via the adjoint action. This ties in, in a very interesting way, with the study of orbital integrals on G, and of course with the orbit structure of G itself. Even though cyclic homology is in some sense determined by Hochschild homology, it is not clear what the cyclic homology of $C_c^\infty(G)$ is. One remarkable feature, though, is that cyclic homology does make a difference between the compact conjugacy classes in G and the others. In particular, for p-adic Lie Groups, P. Blanc and I prove an abstract Selberg principle, which says the following: if e is an idempotent of $C_c^\infty(G)$, and if γ is a regular element of G which is not compact, then

$$\int_{G/G_\gamma} e(g\gamma g^{-1})\, d\dot{g} = 0 ,$$

where G_γ is the centralizer of γ . Such an "abstract Selberg principle" was first proven, for G of split-rank 1, by Julg and Valette [18], by very different methods.

The second type of algebras is provided by relative differential operators. The main application is to obtain the Hodge cohomology $\bigoplus_{p,q} H^q(Y, \Omega_Y^p)$ of a smooth projective algebraic variety Y as the Hochschild homology of an algebra A, obtained as follows. According to Jouanolou or Karoubi, there exists a fibration $F : X \longrightarrow Y$ with fibre an affine space A^N and total space X affine. Then A is the algebra of algebraic differential operators on X which only involve differentiations along the fibres of F. This result is joint work with Jean-Benoit Bost and Christophe Soulé; see (2.26) for a precise statement. We conjecture that the cyclic homology of A is equal to the direct sum of hypercohomology groups of truncated de Rham complexes, and that the Hochschild to cyclic spectral sequence is the Hodge to de Rham spectral sequence for X (hence degenerates at E^1).

I am convinced that cyclic homology will prove extremely valuable to geometers and to representation theorists. For example, it will be very useful in the study of group actions on manifolds, where it is appropriate to introduce the "crossed product" algebra; I have computed the Hochschild and cyclic homology of such crossed products for a differentiable action of a compact Lie group on a manifold; the result involves an interesting auxillary space associated to a group action, and will be described elsewhere.

Finally, I will point out that the two sorts of algebras are rather similar in spirit, and hopefully the "equivariant" and "differential" themes may be combined in interesting ways.

It is a pleasure to thank the organizers of this Symposium, which has been very informative and stimulating. In addition, I wish to thank Philippe Blanc, Jean-Benoit Bost and Christophe Soulé, with whom I am presently collaborating. I am also grateful to Joseph Bernstein, Laurent Clozel, Alain Connes, Hélène Esnault, Pierre Julg, David Kazhdan and Alain Valette for useful discussions. In particular, A. Connes pointed out, in the Fall of 1985, that the algebra of relative differential operators on a Jouanolou-Karoubi fibration should have interesting cyclic homology.

§1 Convolution group algebras

All the work described below is joint with Philippe Blanc.

§1.1 Discrete groups

Let k be a commutative ring, with unit, G an abstract group, k[G] the group algebra. For $g \in G$, we denote by g the corresponding element of $k[G]$. A left (resp. right) $k[G]$-module is the same thing as a k-module equipped with a k-linear left (resp. right) action of G. Hence, taking k as a base ring, a $k[G]$-bimodule is a k-module equipped with a k-linear left action of G, and a k-linear right action of G, which commute with each other. If M is a $k[G]$-bimodule, we let M_{ad} be the G-module M, on which $g \in G$ acts by $m \rightarrowtail g\, m\, g^{-1}$. The following proposition is due to Cartan-Eilenberg [10, Chapter 10].

Underline: <u>Proposition 1.1.1</u> For any $k[G]$-bimodule M, one has:

$$H_*(k[G]\,,\, M) = H_*(G\,,\, M_{ad})\,.$$

Here the first group is a Hochschild homology group, the second is group homology. Let us point that the statement is obvious, since both sides are derived functors of the functor

$$M \longrightarrow H_0(k\,[G]\,,\, M) = M/\{g\, m - m\, g\;;\;\; m \in M\,,\, g \in G\}$$

$$= M/\{g\, m\, g^{-1} - m\;;\;\; m \in M\,,\, g \in G\} = H_0(G\,,\, M_{ad})\,.$$

We are interested in the bimodule $M = k[G]$. The G-module $k[G]_{ad}$ is equal to the direct sum, over adjoint orbits $\mathcal{O}$ of G, of $k[\mathcal{O}]_{ad}$; now if $x \in \mathcal{O}$, then $\mathcal{O}$ is isomorphic to G/G_x as a G-set, hence $k[\mathcal{O}]_{ad}$ is isomorphic to $k[G/G_x]$. By Shapiro's lemma [6, Prop. 6.2], $H_*(G\,,\, k[G/G_x]) \cong H_*(G_x\,,\, k)$. Hence if Σ is a system of representatives for the adjoint action of G on itself, we obtain

<u>Corollary 1.1.2</u> $H_*(k[G]) \cong \bigoplus_{x \in \Sigma} H_*(G_x\,,\, k)\,.$

This result is due to Burghelea [9]. However, his method does not show the elementary nature of the computation.

Burghelea also computes the cyclic homology group $H\,C_*(k\,[G])$, for k a $\mathbb{Q}$-algebra. We will state his result as follows. The group $\mathbb{Z}$ acts on G_x, the generator of $\mathbb{Z}$ acting by multiplication by x; hence, up to homotopy, $S^1 = B\mathbb{Z}$ acts on BG_x. The result is

$$HC_*(k[G]) \cong \underset{x \in \Sigma}{\otimes} H_*^{S^1}(BG_x, k).$$

If x is of finite order, $H_*^{S^1}(BG_x, k) \cong H_*(BG_x, k) \otimes H_*(BS^1)$. If x has infinite order, setting $N_x = G_x/x^{\mathbb{Z}}$, we have:

$$H_*^{S^1}(BG_x, k) = H_*(BN_x, k).$$

Up to homotopy, we have a fibration

$$\begin{array}{ccc} S^1 = B\,\mathbb{Z} & \longrightarrow & B\,G_x \\ & & \downarrow \\ & & B\,N_x . \end{array}$$

The Gysin exact sequence for this fibration:

$$H_{*-1}(B\,N_x, k) \longrightarrow H_*(B\,G_x, k) \longrightarrow H_*(B\,N_x, k) \longrightarrow H_{*-2}(B\,N_x, k) \to -$$

is a direct summand of the exact sequence of Connes [13], [24]

$$HC_{*-1}(k[G]) \xrightarrow{B} H_*(k[G], k[G]) \xrightarrow{I} HC_*^{\bullet}(k[G]) \xrightarrow{S} HC_{*-2}(k[G]) \to .$$

In particular, B and S have a clean geometric interpretation.

It is useful to construct an explicit isomorphism between $H_*(k[G], M)$ and $H_*(G, M_{ad})$, for M a $k[G]$-bimodule, and to use it to compute B explicitly. The first group is the homology of the Hochschild complex

$$\longrightarrow M \underset{k}{\otimes} k[G^{i+1}] \xrightarrow{b} M \underset{k}{\otimes} k[G^i] \longrightarrow \ldots$$

where $b(m \otimes (g_0, \ldots, g_i)) = (m\, g_0) \otimes (g_1, \ldots, g_i)$

$$+ \sum_{j=0}^{i-1} (-1)^{j+1}\ m \otimes (g_0\ , \dots , g_j\, g_{j+1}\ , \dots , g_i)$$

$$+ (-1)^{i+1}\ (g_i\, m) \otimes (g_0\ , \dots , g_{i-1})\ .$$

The second group is the homology of the standard complex

$$\cdots \longrightarrow M \underset{k}{\otimes} k[G^{i+1}] \xrightarrow{\ d\ } M \underset{k}{\otimes} k[G^i] \to - \cdot$$

where $d(m \otimes (g_0\ , \dots , g_i)) = (g_0^{-1}\ m\ g_0) \otimes (g_1\ , \dots\dots , g_i)$

$$+ \sum_{j=0}^{i-1} (-1)^{j+1}\ m \otimes (g_0\ , \dots , g_j\, g_{j+1}\ , \dots , g_i)$$

$$+ (-1)^{i+1}\ m \otimes (g_0\ , \dots\dots , g_{i-1})\ .$$

An isomorphism φ from the first complex to the second is constructed as follows:

$$\varphi(m \otimes (g_0\ , \dots , g_i)) = (g_0 \cdots g_i\ m) \otimes (g_0\ , \dots , g_i)\ .$$

Next, for $M = k[G]$, one may compute the operator

$$B' = \varphi \circ B \circ \varphi^{-1} : k[G] \underset{k}{\otimes} k[G^i] \longrightarrow k[G] \underset{k}{\otimes} k[G^{i+1}]\ .$$

One finds, working <u>modulo degenerate cycles</u>, i.e., those

$(g_0 \otimes (g_1\ , \dots , g_i))$ such that $g_j = 1$ for some j with $1 \le j \le i$

$$B'(g_0 \otimes (g_1\ , \dots , g_i))$$

$$= \sum_{j=1}^{i} (-1)^{ij} (g_{i-1}^{-1} \cdots g_1^{-1}\, g_0 g_1 \cdots g_{i-1}) \otimes (g_j, \dots, g_i, g_i^{-1} \cdots g_1^{-1}\, g_0, g_1, \dots, g_{j-1})$$

$$+ (g_i^{-1} \cdots\cdots g_1^{-1}\, g_0\, g_1 \cdots g_i) \otimes (g_i^{-1} \cdots g_1^{-1}\, g_0\ , g_1\ , \dots\dots , g_i)\ .$$

From this formula, it is apparent that for $\mathcal{O} \subset G$ the adjoint orbit of x , B' maps $k[\mathcal{O}] \underset{k}{\otimes} k[G^i]$ into $k[\mathcal{O}] \underset{k}{\otimes} k[G^{i+1}]$; we denote again B'_x the induced map on group homology

$$B'_x : H_i(G, k[\mathcal{O}]) \longrightarrow H_{i+1}(G, k[\mathcal{O}]) .$$

For each $x \in \Sigma$, let $\mathcal{O}$ be its adjoint orbit, and define $B_x : H_i(G_x, k) \longrightarrow H_{i+1}(G_x, k)$ by decreeing the following diagram commutative

$$\begin{array}{ccc} H_i(G_x, k) & \xrightarrow[\sim]{\alpha} & H_i(G, k[G]) \\ B_x \downarrow & & \downarrow B'_x \\ H_{i+1}(G_x, k) & \xrightarrow[\sim]{\alpha} & H_{i+1}(G, k[\mathcal{O}]) \end{array}$$

where α is Shapiro's isomorphism.

To describe explicitly α, recall that if Γ is a group, Δ a subgroup, M a (left) Δ-module, the (produced) Γ-module $P_\Delta^\Gamma(M)$ is the group of all maps $F : \Gamma \longrightarrow M$ such that

(i) $F(x . g) = g^{-1}. F(x)$ for any $x \in \Gamma$, $g \in \Delta$

(ii) F has finite support modulo Δ.

Then the map $\alpha : M \underset{\mathbb{Z}}{\otimes} \mathbb{Z}[\Delta^i] \longrightarrow P_\Delta^\Gamma(M) \underset{\mathbb{Z}}{\otimes} \mathbb{Z}[\Gamma^i]$ is defined by

$\alpha(m \otimes (g_1, \ldots, g_i)) = \alpha(m) \otimes (g_1, \ldots, g_i)$ where $\alpha(m)$ is the function from Γ to M such that

(a) $\alpha(m)(g) = g^{-1} m$ for $g \in \Delta$

(b) $\alpha(m)(g) = 0$ for $g \in \Gamma \setminus \Delta$.

Then α gives a morphism of standard complexes; the induced map on homology is Shapiro's isomorphism.

With the help of this concrete α, one easily verifies the following formula for B_x :

$$B_x(g_1, \ldots, g_i) = \sum_{j=1}^{i} (-1)^{ij}(g_j, \ldots, g_i, g_i^{-1} \ldots g_1^{-1} x, g_1, \ldots, g_{j-1})$$

$$+ (g_i^{-1} \ldots g_1^{-1} x, g_1, \ldots, g_i)$$

Example 1.1.3 For $i = 1$, $B_x(g) = (g^{-1} x , g) - (g , g^{-1} x)$.
For $i = 0$, $B_x(\mathbb{1}) = x \in H_1(G_x)$.

(Once again, all these formulae are given modulo degenerate elements.)

§1.2 Lie groups

Let F be a non-discrete locally compact topological field, whose topology is defined by an absolute value $| \quad |$. A Lie group over F (or F-Lie group) is defined as in [6, §1.1]. We make no assumption on the number of connected components of a $\mathbb{R}$- or $\mathbb{C}$-Lie group.

In [3, 3°], P. Blanc introduced the category $\mathcal{D}_G$ of differentiable G-modules, for G a real or complex Lie group. Let us recall that an object M of $\mathcal{D}_G$ is a complete locally convex topological vector space over $\mathbb{C}$, endowed with a group homomorphism $\rho : G \longrightarrow \mathrm{Aut}(M)$, satisfying the following conditions:

(D_1) if $C \subset G$ is compact, $\rho(C)$ is equicontinuous

(D_2) for any $m \in M$, the function $g \longrightarrow \rho(g) . m$ is differentiable

($\mathcal{D}_3$) the natural map $M \longrightarrow C^\infty(G , M)$, sending m to the function $g \mapsto \rho(g) . m$, is continuous.

The fact that $\mathcal{D}_G$ is not an abelian category complicates the algebraic homology of $\mathcal{D}_G$.

For F non-archimedean, we let $\mathcal{D}_G$ be the category of admissible G-modules, in the sense of Harish-Chandra. An object of $\mathcal{D}_G$ is a complex vector space M , with an action of G , which is such that the stabilizer of any vector of M is open in G . We will also talk of differentiable (instead of admissible) modules. $\mathcal{D}_G$ is now an abelian category.

The definition of the differentiable homology, due to P. Blanc in the real case, may be extended to this more general context as follows: $H_i^{diff}(G, M)$ is the homology in degree i of the complex of [loc. cit., 6°]

$$\ldots \longrightarrow C_c^{\infty}(G^{n+2}, M)_G \xrightarrow{\partial} C_c^{\infty}(G^{n+1}, M)_G \longrightarrow \ldots$$

where $(\partial \psi)(g_0, \ldots, g_n) =$

$$= \sum_{i=0}^{n+1} (-1)^i \int_G \psi(g_0, \ldots, g_{i-1}, g, g_i, \ldots, g_n)\, d\, g .$$

See also [12] for another approach to differentiable homology.

Here $d\,g$ is a fixed left Haar measure on G, C^{∞}-functions are assumed to be locally constant in the non-archimedean case, and N_G is the space of co-invariants of the G-module N; i.e., $N_G = H_0(G, N)$.

One may shift, as usual, from this "homogeneous" complex to the "inhomogeneous" complex. Consider the isomorphism

$$\beta : C_c^{\infty}(G^{n+1}, M)_G \xrightarrow{\sim} C_c^{\infty}(G^n, M)$$

defined by

$$(\beta \psi)(g_1, \ldots, g_n) = \int_G g^{-1} \psi(g, g\, g_1, g\, g_1\, g_2, \ldots, g\, g_1 \cdots g_n)\, d\, g .$$

Using this isomorphism, the previous complex transforms into

$$\ldots \longrightarrow C_c^{\infty}(G^{n+1}, M) \xrightarrow{d} C_c^{\infty}(G^n, M) \longrightarrow \ldots$$

where $(d\,F)(g_1, \ldots, g_n) = \int_G g^{-1} F(g, g_1, \ldots, g_n)\, d\, g$

$$+ \sum_{j=1}^{n} (-1)^j \int_G F(g_1, \ldots, g_{j-1}, g, g^{-1} g_j, g_{j+1}, \ldots, g_n)\, d\, g$$

$$+ (-1)^{n+1} \int_G F(g_1, \ldots, g_n, g)\, d\, g .$$

Now let $C_c^\infty(G)$ be the algebra of compactly supported C^∞ complex-valued functions on G, endowed with the convolution product $(\varphi * \psi)(g) = \int_G \varphi(g') \psi(g'^{-1} g) \, d g'$. In the non-archimedean case, no topology need be put on $C_c^\infty(G)$. In the archimedean case, $C_c^\infty(G)$ is given its usual locally convex topology, for which it is complete. Therefore, when we speak of Hochschild homology of that algebra, we use completed tensor products, as in [13, II, §5], in case F is archimedean.

<u>Proposition 1.2.1</u> For M an object of $\mathcal{D}_{G \times G}$, one has:

$$H_*(C_c^\infty(G), M) \cong H_*^{diff}(G, M_{ad}).$$

Notice that for $M \in \mathcal{D}_{G \times G}$, M is a $C_c^\infty(G)$-bimodule. For instance, the left action of $C_c^\infty(G)$ is: $F . m = \int_G F(g) \, g . m \, d g$. As in the discrete group case, studied in §1.1, the proposition follows from an isomorphism of complexes given by:

$$\varphi : M \otimes C_c^\infty(G^{i+1}) \longrightarrow M \otimes C_c^\infty(G^{i+1})$$

given by $\varphi(F)(g_0, \ldots, g_i) = g_0 \ldots g_i F(g_0, \ldots, g_i)$; i.e., the same formula as in §1.1 (φ is a map from the Hochschild complex to the standard complex of group homology).

Now we concentrate on the bimodule $M = C_c^\infty(G)$. The definition of B in [13, II, §3], [24] is not applicable, since our algebra has no unit, and B involves a homotopy s for the cyclic complex $(C_c^\infty(G^{i+1}), b')$, which uses a unit. However, one may construct another map $\sigma : C_c^\infty(G) \longrightarrow C_c^\infty(G^2)$, from the choice of $u \in C_c^\infty(C)$ with $\int_G u(g)\, d\, g = 1$ (u may be viewed as an "approximate unit") as follows:

$$(\sigma F)(g_1, g_2) = u(g_1)\, F(g_1\, g_2)\,.$$

(It should be possible to extend σ to a homotopy, but we have not written this down.) In any case, in degree 0 , the map

$B : H_0(C_c^\infty(G), C_c^\infty(G)) \longrightarrow H_1(C_c^\infty(G), C_c^\infty(G))$ is then given by $B = (1-t) \circ \sigma$. So we find

$$(BF)(g_1, g_2) = u(g_1)\, F(g_1\, g_2) - u(g_2) F(g_2\, g_1)\,.$$

So the morphism $B' : C_c^\infty(G) \longrightarrow C_c^\infty(G) \hat{\otimes} C_c^\infty(G) = C_c^\infty(G \times G)$ defined by $B' = \varphi \circ B \circ \varphi^{-1} = \varphi \circ B$, is:

$$(B'F)(g_1, g_2) = u(g_2^{-1}\, g_1\, g_2) - u(g_2) F(g_1)\,.$$

Let $C^\infty(G)$ operate on $C_c^\infty(G^{i+1})$, for all i , by multiplication, as function of the 1st variable; i.e.,

$$(F\, \psi)(g_0, \ldots, g_i) = F(g_0)\,.\,\psi(g_0, \ldots g_i)\,.$$

Then the formula for B' shows:

<u>Lemma 1.2.2</u> $B' : C_c^\infty(G) \longrightarrow C_c^\infty(G^2)$ is $C_c^\infty(G)$-linear.

Let us now assume that G is (the group of F-points of) a <u>connected</u> reductive algebraic group over F . Let $\mathcal{O}$ be the G-orbit of a <u>regular semi-simple</u> element $x \in G$, so that $\mathcal{O}$ is closed in G and G_x is a Cartan subgroup of G . Apart from a bit of nuisance coming from stable conjugacy versus conjugacy, Lemma 1.2.2 implies the existence of a factorization

$$\begin{array}{ccc} H_0^{diff}(G, C_c^\infty(G)) & \xrightarrow{B'} & H_1^{diff}(G, C_c^\infty(G)) \\ \downarrow & & \downarrow \\ H_0^{diff}(G, C_c^\infty(\mathcal{O})) & \xrightarrow{B'_x} & H_1^{diff}(G, C_c^\infty(\mathcal{O})) . \end{array}$$

As in §1.1, it is appropriate to introduce Shapiro's isomorphism

$H_i^{diff}(G, C_c^\infty(\mathcal{O})) \xleftarrow[\sim]{\alpha} H_i^{diff}(G_x, \mathbb{C})$ (cf [3, n^0 11]) . We then define

$B_x : H_i^{diff}(G_x, \mathbb{C}) \longrightarrow H_{i+1}^{diff}(G_x, \mathbb{C})$ as $B_x = \alpha^{-1} \circ B'_x \circ \alpha$. Now let $\bar{x}$ be the image in $H_1^{diff}(G_x, \mathbb{C})$ of the canonical element of $H_1(\mathbb{Z}, \mathbb{C})$; let $\mathbb{1}$ be the canonical element of $H_0^{diff}(G_x, \mathbb{C})$.

<u>Lemma 1.2.3</u> $B_x(\mathbb{1}) = \bar{x} \in H_1^{diff}(G_x, \mathbb{C})$.

This extension of Example 1.1.3 to Lie groups requires an explicit description of Shapiro's isomorphism. For this, one uses the method of "effacement": for $M \in \mathcal{D}_G$, one introduces the exact sequence $0 \longrightarrow A \longrightarrow C_c^\infty(G, M) \longrightarrow M \longrightarrow 0$. Then $H_i(G, M)$ identifies as a subgroup of $H_{i-1}(G, A)$; this allows to construct Shapiro's isomorphism inductively. The final formulae involve choices: a measurable decomposition $G = Y \cdot G_x$, a Haar measure on G_x (hence we get a measure on Y) , a compactly-supported C^∞-function χ on Y such that $\int_Y \chi(y)dy = 1$ and a compactly-supported C^∞-function u on G such that $\int_G u(g)dg = 1$. However, one may use these choices and let χ and u tend to "δ-functions"; this in some sense reduces the computation to a maximal discrete subgroup A of G_x such that G_x/A is compact (this reduction is obtained by averaging over G_x/A). So in fact no computation is ever really necessary.

This lemma implies the abstract Selberg principle mentioned in the introduction. To explain this, we first need to relate $H_0^{diff}(G, C_c^\infty(G))$ to orbital integrals.

<u>Lemma 1.2.4</u> For any $x \in G$, $\mathcal{O}$ its adjoint orbit, the composed map $C_c^\infty(G) \longrightarrow H_0^{diff}(G, C_c^\infty(G)) \longrightarrow H_0^{diff}(G, C_c^\infty(\mathcal{O}) \xrightarrow[\alpha^{-1}]{\sim} \mathbb{C}$ is the "orbital integral" functional $F \mapsto \int_{G/G_x} F(g x g^{-1}) d\dot{g}$.

This lemma is a tautology, since the Shapiro isomorphism α^{-1} is given by $\varphi \mapsto \int_{G/G_x} \varphi(\dot{g}) d\dot{g}$ (note that the choice of a Haar measure on G_x appears both in Shapiro isomorphism and in the definition of the orbital integral).

Now Connes has defined a mapping

$$K_0(\mathcal{A}) \longrightarrow H\,C_{2n}(\mathcal{A})$$

for any (topological) algebra $\mathcal{A}$ (see [13,]). If e is an idempotent in $\mathcal{A}$, its image in $HC_0(\mathcal{A}) = H_0(\mathcal{A}, \mathcal{A})$ belongs to the image of $S : H\,C_2(\mathcal{A}) \longrightarrow HC_0(\mathcal{A})$ which equals the kernel of $B : H_0(\mathcal{A}, \mathcal{A}) \longrightarrow H_1(\mathcal{A}, \mathcal{A})$. Here for $\mathcal{A} = C_c^\infty(G)$, it follows from Lemma 1.2.2, 1.2.3 and 1.2.4 that for $x \in G$ regular semi-simple, one has:

$[\int_{G/G_x} e(g x g^{-1}) d\dot{g}] . \bar{x}$ should be 0 in $H_1^{diff}(G_x, \mathbb{C})$. Now assume

x is not compact (i.e., x does not generate a compact subgroup of G). Let A be the group of 1-parameter subgroups (in the algebraic sense) of G_x, defined over F. Then the absolute value morphism

$$G_x \xrightarrow{\log |\ |} A \otimes_{\mathbb{Z}} \mathbb{R}$$

gives an isomorphism on differentiable homology. Since x is not compact, $|x| \neq 0$. On the other hand, $H_1^{diff}(A \otimes_{\mathbb{Z}} \mathbb{R}, \mathbb{R}) \cong A \otimes_{\mathbb{Z}} \mathbb{R}$ and $\bar{x}$ maps to $|x|$ under this isomorphism. Hence $\bar{x}$ is non-zero in $H_1^{diff}(G_x, \mathbb{C})$. It follows that

$\int_{G/G_x} e(g x g^{-1})\, d\dot{g}$ must be 0. Hence we have proved the

<u>Theorem 1.2.5</u> (abstract Selberg principle)

Let G be a connected reductive Lie Group over F, e be an idempotent in $C_c^\infty(G)$, x a regular semi-simple element of G which is not compact. Then $\int_{G/G_x} e(g x g^{-1})\, dg = 0$.

The term "abstract Selberg principle" was coined by Julg and Valette, who established it for groups of F-split rank one [18]. They use analysis on the Bruhat-Tits tree, and Fredholm modules (in the sense of Connes).

The "concrete" Selberg principle is deduced by taking e to be $e(g) = \lambda \langle \pi(g) v, v^* \rangle$ where (π, V) is a cuspidal representation of G (here F is non-archimedean), $v \in V$, $v^* \in V^*$ with $(v, v^*) \neq 0$, and λ is a suitable constant. Indeed for λ well-chosen, e is idempotent in $C_c^\infty(G)$ (cf [11, Théorème 1.1]). For a classical proof of the Selberg principle, see [16].

The exact computation of $H_i^{diff}(G, C_c^\infty(G))$ is an open problem. For $i = 0$, the following statement:

"$F \in C_c^\infty(G)$ has zero image in $H_0^{diff}(G, C_c^\infty(G))$ iff all orbital integrals of F are zero", is conjectured by P. Blanc for F archimedean [4]; the case F non archimedean is due to Harish-Chandra [15].

For F non-archimedean, let $U \subset G$ be the open set of regular semi-simple elements, $Y = G - U$.

Let S be the multiplicative subset of $C^\infty(G)^G$ formed of functions which vanish nowhere on U.

<u>Lemma 1.2.6</u> The $C^\infty(G)^G$-linear map

$$H_i^{diff}(G, C_c^\infty(U)) \longrightarrow H_i^{diff}(G, C_c^\infty(G))$$

becomes an isomorphism after localizing at S.

Indeed, there exists a function -- the function Δ - in $C^\infty(G)^G$ -- which vanishes exactly on Y. Hence Δ kills $C_c^\infty(Y)$ and belongs to S. The lemma follows then from the exact sequence

$$0 \longrightarrow C_c^\infty(U) \longrightarrow C_c^\infty(G) \longrightarrow C_c^\infty(Y) \longrightarrow 0 .$$

The interest of this lemma is that $C_c^\infty(U)$ is computable. Indeed, since

$$U = \coprod_{T \bmod conj} (G \overset{N(T)}{\times} T')$$

where T is a Cartan subgroup of G, $T' \subset T$ is the regular subset, one obtains

$$H_i^{diff}(G, C_c^\infty(U)) = \bigoplus_{T \bmod conj} H_i^{diff}(N(T), C_c^\infty(T'))$$

$$= \bigoplus_{T\,\mathrm{mod\,conj}} [H_i(T, \mathbb{C}) \otimes C_c^\infty(T')]^{W(T)}$$

where $W(T) = N(T) / T$ acts diagonally on $H_i(T, \mathbb{C}) \otimes C_c^\infty(T')$.

The complete computation of $H_i^{\mathrm{diff}}(G, C_c^\infty(G))$ will require at least a clever use of a stratification of the Springer resolution of the nilpotent variety. It appears to be a quite challenging problem.

Proof; The Grothendieck-Cousin fundamental class of $\Delta_X \subset X \times X$ is an element γ of $\underline{H}^n_{\Delta_X}(p_2^* \omega_X)$, hence we have an $\mathcal{O}_{X\times X}$-linear map $p_2^*(\omega_X^{\otimes -1}) \longrightarrow \underline{H}^n_{\Delta_X}(\mathcal{O}_{X\times X})$, which extends to a $\mathcal{D}_{X\times X}$-linear morphism

$$p_2^*(\omega_X^{\otimes -1}) \underset{\mathcal{O}_{X\times X}}{\otimes} \mathcal{D}_{X\times X} \longrightarrow \underline{H}^n_{\Delta_X}(\mathcal{O}_{X\times X})$$

where $\mathcal{D}_{X\times X}$ is viewed as a right module over itself. It remains to show that this morphism factors through the quotient $\mathcal{D}_X^{(2)}$ of $p_2^*(\omega_X^{\otimes -1}) \underset{\mathcal{O}_{X\times X}}{\otimes} \mathcal{D}_{X\times X}$. But γ is killed by the ideal of Δ_X in $\mathcal{O}_{X\times X}$, and if $(x_1, \ldots, x_n)$ are local coordinates on X, we get local coordinates $(x_1, .., x_n ; y_1, \ldots, y_n)$ on $X \times X$ near Δ_X, and the point is that $\frac{\partial}{\partial xi}$ and $\frac{\partial}{\partial yi}$ have the same action on

$\gamma = \dfrac{dy_1 \wedge --- \wedge dy_n}{(y_1-x_1) ---- (y_n-x_n)}$ (notice $\frac{\partial}{\partial yi}$ acts on γ by the opposite of Lie derivation). Since $\mathcal{D}_X^{(2)}$ and $\underline{H}^n_{\Delta_X}(\mathcal{O}_{X\times X})$ are both simple holonomic $\mathcal{D}_{X\times X}$-modules, the map between them, which is non-zero, is an isomorphism, q.e.d.

The crucial lemma is the following

Lemma 2.1.6 The dual $(\mathcal{D}_X^{(1)})^*$ of $\mathcal{D}_X^{(1)}$ is isomorphic to $\underline{H}^n_{\Delta_X}(\mathcal{O}_{X\times X})\,[-2n]$.

Proof: Since locally $\mathcal{D}_X^{(1)}$ is isomorphic to $\mathcal{D}_X^{(2)}$ (the monkey business of left and right may be arranged locally), it is a simple holonomic right $\mathcal{D}_{X\times X}$-module. Hence $(\mathcal{D}_X^{(1)})^*$ $[-2n]$ is a simple holonomic left $\mathcal{D}_{X\times X}$-module (see [20] or [2] for the duality on holonomic modules). Its characteristic variety is $T^*_{\Delta_X}(X\times X)$. Hence it is locally (on $X \cong \Delta_X$) isomorphic to $H^n_{\Delta_X}(\mathcal{O}_{X\times X})$. The sheaf of germ of automorphisms of the $\mathcal{D}_{X\times X}$-module $\underline{H}^n_{\Delta_X}(\mathcal{O}_{X\times X})$ is constant, equal to k^*. The lemma then follows from $H^1(X, k^*) = 0$; recall a constant sheaf is flasque, for the Zariski topology.

Now use the Japanese notation $\mathcal{B}_{\Delta_X/X\times X} = \underline{H}^n_{\Delta_X}(\mathcal{O}_{X\times X})$. We have obtained so far:

$$\mathcal{D}_X \overset{\mathbb{L}}{\underset{\mathcal{D}_X \boxtimes \mathcal{D}_X^0}{\otimes}} \mathcal{D}_X \cong \mathcal{D}_X^{(1)} \otimes \mathcal{D}_X^2 \cong$$

$$\cong R\mathcal{H}om_{\mathcal{D}_{X\times X}}(\mathcal{B}_{\Delta_X/X\times X}, \mathcal{B}_{\Delta_X/X\times X})[2n].$$

Recall the following lemma, where for Y a smooth variety, $\mathcal{M}$ a left $\mathcal{D}_Y$-module, $DR(\mathcal{M})$ denotes the de Rham complex:

$$\begin{array}{ccc} \mathcal{M} \xrightarrow{d} \Omega^1_Y \underset{\mathcal{O}_Y}{\otimes} \mathcal{M} \longrightarrow \dots \xrightarrow{d} & \Omega^{\dim Y}_Y \underset{\mathcal{O}_Y}{\otimes} \mathcal{M} \\ | & | \\ d^0(-\dim Y) & d^0 0 \end{array}$$

<u>Lemma 2.1.7</u> If $Z \overset{i}{\hookrightarrow} Y$ is a closed immersion of smooth varieties over k, $DR(\mathcal{B}_{Z/Y})$ is quasi-isomorphic to $i_*(\Omega^{\bullet}_Z)[\dim Z]$.

This lemma may be extracted from [20, §4].

We thus obtain a morphism of complexes of sheaves:

$$\mathcal{D}_X \overset{\mathbb{L}}{\underset{\mathcal{D}_X \boxtimes \mathcal{D}_X^0}{\otimes}} \mathcal{D}_X \to \mathbb{R}\mathrm{Hom}_{X\times X}(i_* \Omega_X^{\bullet}, i_* \Omega_X^{\bullet})[2n].$$

Using the inclusion $k \hookrightarrow \Omega_X^{\bullet}$, we get a map

$$\mathcal{D}_X \overset{\mathbb{L}}{\underset{\mathcal{D}_X \boxtimes \mathcal{D}_X^0}{\otimes}} \mathcal{D}_X \to \mathbb{R}\mathrm{Hom}_{X\times X}(i_* k, \Omega_X^{\bullet})[2n\} = \Omega_X^{\bullet}[2n].$$

<u>Proposition 2.1.8</u> $\mathcal{D}_X \overset{\mathbb{L}}{\underset{\mathcal{D}_X \boxtimes \mathcal{D}_X^0}{\otimes}} \mathcal{D}_X \to \Omega_X^{\bullet}[2n]$

is a quasi-isomorphism.

<u>Remark</u>: This may be viewed as a refinement of Theorem 2.1.1.

<u>Proof</u> (sketch): The question being local on X, one may assume there exists an étale morphism $\varphi : X \longrightarrow \mathbb{A}_k^n$. The diagonal $\Delta_X \subset X \times X$ is open and closed in $\varphi^{-1}(\Delta_{\mathbb{A}^n})$. In what follows, we consider sheaves which are supported on $\varphi^{-1}(\Delta_{\mathbb{A}^n})$, and we restrict them to a neighborhood of X in $X\times X$, so that we may behave as if Δ_X was equal to $\varphi^{-1}(\Delta_{\mathbb{A}^n})$.

With these considerations in mind, we let $(x_1, \ldots, x_n)$ be the standard coordinates on $\mathbb{A}^n$, hence on X. On $X\times X$, we let $(x_1, \ldots, x_n)$ (resp. $(y_1, \ldots, y_2)$) be the coordinates on the first (resp. second) copy of X. The $\mathcal{D}_X \boxtimes \mathcal{D}_X^{\circ}$-right module $\mathcal{D}_X$ is the quotient of $\mathcal{D}_X \boxtimes \mathcal{D}_X^{\circ}$ the right ideal generated by

$$(x_1-y_1 , \ldots, x_n-y_n ; \frac{\partial}{\partial x_1} - \frac{\partial}{\partial y_1} , \ldots, \frac{\partial}{\partial x_n} - \frac{\partial}{\partial y_n}) .$$

These elements commute with each other; to see this, notice that if a and b belong to $\mathcal{D}_X$, we have: $[a , b]_0 = -[a , b]$, where $[a , b]$ is their commutator in $\mathcal{D}_X$, $[a , b]_0$ their commutator in $\mathcal{D}^\circ_X$; hence, e.g.,

$$[x_1-y_1 , \frac{\partial}{\partial x_1} - \frac{\partial}{\partial y_1}]_0 = [x_1 , \frac{\partial}{\partial x_1}] + [y_1 , \frac{\partial}{\partial y_1}]_0$$

$$= [x_1 , \frac{\partial}{\partial x_1}] - [y_1 , \frac{\partial}{\partial y_1}] = -1 + 1 = 0 .$$

Hence one may introduce the Koszul complex

$$K(\mathcal{D}_X \boxtimes \mathcal{D}^\circ_X ; x_1-y_1 , \ldots, x_n-y_n , \frac{\partial}{\partial x_1} - \frac{\partial}{\partial y_1} , \ldots, \frac{\partial}{\partial x_n} - \frac{\partial}{\partial y_n}) :$$

this is a complex of right $\mathcal{D}_X \boxtimes \mathcal{D}^\circ_X$-modules.

<u>Lemma 2.1.9</u> $K^\cdot(\mathcal{D}_X \boxtimes \mathcal{D}^\circ_X ; x_1-y_1 , \ldots, \frac{\partial}{\partial x_n} - \frac{\partial}{\partial y_n})$ is a resolution of $\mathcal{D}_X$. Since this Koszul complex consists of projective modules, $\mathcal{D}_X \overset{\mathbb{L}}{\underset{\mathcal{D}_X \boxtimes \mathcal{D}^\circ_X}{\otimes}} \mathcal{D}_X$ is realized by the Koszul complex

$$K^\cdot(\mathcal{D}_X ; x_1-y_1 , \ldots, x_n-y_n ; \frac{\partial}{\partial x_1} - \frac{\partial}{\partial y_1}, \ldots, \frac{\partial}{\partial x_n} - \frac{\partial}{\partial y_n})$$

for the left $\mathcal{D}_X \boxtimes \mathcal{D}^\circ_X$-module $\mathcal{D}_X$. We will analyze this Koszul complex in two steps; dividing the sequence

$(x_1-y_1 , \ldots, \frac{\partial}{\partial x_n} - \frac{\partial}{\partial y_n})$ into the subsequences

$(x_1-y_1, \ldots, x_n-y_n)$ and $(\frac{\partial}{\partial x_n} - \frac{\partial}{\partial y_1} , \ldots, \frac{\partial}{\partial x_n} - \frac{\partial}{\partial y_n})$.

Let $L^{\cdot} = K^{\cdot}(\mathcal{D}_X ; \; x_1 - y_1 , \ldots , x_n - y_n)$; then our Koszul complex is quasi-isomorphic to the simple complex deduced from the double complex

$$K^{\cdot}(L^{\cdot} ; \frac{\partial}{\partial x_1} - \frac{\partial}{\partial y_1} , \ldots, \frac{\partial}{\partial x_n} - \frac{\partial}{\partial y_n}) .$$

So first we study $L^{\cdot}$ which may be described as the complex

$$\begin{array}{ccc} \ldots \longrightarrow \Lambda^{n-i}(k^n) \underset{k}{\otimes} \mathcal{D}_X & \xrightarrow{\delta} & \Lambda^{n-i+1}(k^n) \underset{k}{\otimes} \mathcal{D}_X \longrightarrow \ldots \\ | & & | \\ d^0 - i & & d^0(-i+1) \end{array}$$

where δ is described as follows. Let $(e_1 , \ldots , e_n)$ be the canonical basis of K^n ; then we have:

$$\delta((e_k \wedge \ldots \wedge e_l) \otimes P) = \sum_{j=1}^{n} ((e_j \wedge e_k \wedge \ldots \wedge e_l) \otimes [x_j , P]) .$$

There is a map of complexes $\mathcal{O}_X[n] \longrightarrow L^{\cdot}$. which maps $F \in \mathcal{O}_X$ to $F \in \mathcal{D}_X$ (note that $[x_j , F] = 0$ for $F \in \mathcal{O}_X$)

<u>Lemma 2.1.10</u> $\mathcal{O}_X[n] \longrightarrow L^{\cdot}$ is a quasi-isomorphism.

<u>Proof</u>: On $T^* X$, introduce coordinates $(x_1 , \ldots , x_n ; \; \xi_1 , \ldots , \xi_n)$, where $(\xi_1 , \ldots , \xi_n)$ are dual to $(x_1 , \ldots , x_n)$. Filter $L^{\cdot}$ by the subcomplexes

$$\begin{array}{ccc} L^{\cdot}(m) : \ldots \longrightarrow \Lambda^{n-i}(k^n) \otimes \mathcal{D}_X(m-i) & \longrightarrow & \Lambda^{n-i+1}(k^n) \otimes \mathcal{D}_X(m-i+1) \\ | & & | \\ d^0 - i & & d^0(-i+1) \end{array}$$

Filter $\mathcal{O}_X$ in the stupid way: $\mathcal{O}_X[n](0) = \mathcal{O}_X[n]$, $\mathcal{O}_X[n](-1) = 0$. Then $\mathcal{O}_X[n] \longrightarrow L^\cdot$ is a morphism of filtered complexes. We identify $gr(L^\cdot) = \bigoplus_m L^\cdot(m) / L^\cdot(m-1)$ with the complex $\Omega^\bullet_{T^*X/X}[n]$ of relative differential forms on T^*X, which is graded by the degree of homogeneity along the fibers of $T^*X \longrightarrow X$ Precisely, we map the element

$(e_1 \wedge \ldots \wedge e_{n-i}) \otimes P$ of $\Lambda^{n-i}(k^n) \otimes \mathcal{D}_X(m-i)$ to the (relative) differential form $\sigma_{m-i}(P)\, d\,\xi_1 \wedge \ldots \wedge d\,\xi_{n-i}$. To show that this is compatible with the differentials, we note that

$$\sigma_{m-i-1}([x_j, P]) = -\frac{\partial}{\partial \xi_j}\sigma_{m-i}(P).$$

Since $\mathcal{O}_X \longrightarrow \Omega^\bullet_{T^*X/X}$ is a quasi-isomorphism, $\mathcal{O}_X[n] \longrightarrow L^\cdot$ is a filtered quasi-isomorphism.

From Lemma 2.1.10, we deduce that $K^\cdot(L^\cdot; \frac{\partial}{\partial x_1} - \frac{\partial}{\partial y_1}, \ldots, \frac{\partial}{\partial x_n} - \frac{\partial}{\partial y_n})$ is quasi-isomorphic to the Koszul complex $K^\cdot(\mathcal{O}_X; \frac{\partial}{\partial x_1} - \frac{\partial}{\partial y_1}, \ldots, \frac{\partial}{\partial x_n} - \frac{\partial}{\partial y_n})$.

Since $\frac{\partial}{\partial x_i} - \frac{\partial}{\partial y_i}$ acts on $\mathcal{D}_X$ by $P \mapsto [\frac{\partial}{\partial x_i}, P]$, it acts the same way on $\mathcal{O}_X \subset \mathcal{D}_X$; for $F \in \mathcal{O}_X$, $[\frac{\partial}{\partial x_i}, F] = \frac{\partial F}{\partial x_i} \in \mathcal{O}_X$.

Hence this complex is just $\Omega^\bullet_X[2n]$.

Hence we have found, locally on X, a quasi-isomorphism $\mathcal{D}_X \overset{\mathbb{L}}{\underset{\mathcal{D}_X \boxtimes \mathcal{D}_X^\cdot}{\otimes}} \mathcal{D}_X \to \Omega^\bullet_X(2n)$; it is still necessary, in order to prove lative differential operators

ıis section, we describe some joint work with J-B. Bost and

e last section, we analyzed the complex of sheaves

$\mathcal{D}_X$, for X a smooth algebraic variety over k .

$\mathcal{D}^{\bullet}_X$

nsider the algebra $\mathcal{D}_{X/Y}$ of relative differential operators

to a smooth morphism $F : X \longrightarrow Y$ of smooth varieties over

y define the sheaf of algebras $\mathcal{D}_{X/Y} \subset \mathcal{D}_X$ in either of the

ays:

$_{X/Y}$ is the sub-algebra of $\mathcal{D}_X$, formed all elements which

ommute with $F^{-1}(\mathcal{O}_Y) \subset \mathcal{O}_X$.

$_{X/Y}$ is the sub-algebra of $\mathcal{D}_X$ generated by $\mathcal{O}_X$ and by

$\subset T_X$, the sheaf of germs of <u>vertical</u> vector fields on X .

$_{X/Y}$ is filtered as sub-ring of $\mathcal{D}_X$ and we have

$$gr(\mathcal{D}_{X/Y}) = \mathcal{O}_{T^*(X/Y)} ,$$

(X/Y) is the relative cotangent space. $\mathcal{D}_{X/Y}$ is a

heaf of algebras, in the sense of [21].

w consider the complex of sheaves

$$\overset{\mathbb{L}}{\otimes}_{\mathcal{D}^{\bullet}_{X/Y}} \mathcal{D}_{X/Y} ;$$

lex of sheaves on X x X , concentrated on the diagonal

as in §2.1, we may view it as a complex of sheaves on

By the analogy of Lemma 2.1.2, if X is <u>affine</u>, the

omology of $D(X/Y) = \Gamma(X, \mathcal{D}_{X/Y})$ is computed from the

(2.18), to compare this quasi-isomorphism with before (2.1.8); we will neglect here to do that.

§2.2 R

In t

C. Soulé.

In t

$\mathcal{D}_X \overset{\mathbb{L}}{\otimes}$

$\mathcal{D}_X \boxtimes$

Here we c

associated

k . We m

following v

(i)

c

(ii)

T

where T

noetherian

We n

$\mathcal{D}_{X/Y}$ $\mathcal{D}_{X/}$

'it is a com

Δ_X . Henc

$\Delta_X \cong X$.

Hochschild

complex

§2.2 Relative differential operators

In this section, we describe some joint work with J-B. Bost and C. Soulé.

In the last section, we analyzed the complex of sheaves

$$\mathcal{D}_X \overset{\mathbb{L}}{\underset{\mathcal{D}_X \boxtimes \mathcal{D}_X^\circ}{\otimes}} \mathcal{D}_X \,, \text{ for } X \text{ a smooth algebraic variety over } k\,.$$

Here we consider the algebra $\mathcal{D}_{X/Y}$ of relative differential operators associated to a smooth morphism $F : X \longrightarrow Y$ of smooth varieties over k. We may define the sheaf of algebras $\mathcal{D}_{X/Y} \subset \mathcal{D}_X$ in either of the following ways:

(i) $\mathcal{D}_{X/Y}$ is the sub-algebra of $\mathcal{D}_X$, formed all elements which commute with $F^{-1}(\mathcal{O}_Y) \subset \mathcal{O}_X$.

(ii) $\mathcal{D}_{X/Y}$ is the sub-algebra of $\mathcal{D}_X$ generated by $\mathcal{O}_X$ and by $T_F \subset T_X$, the sheaf of germs of vertical vector fields on X.

$\mathcal{D}_{X/Y}$ is filtered as sub-ring of $\mathcal{D}_X$ and we have

$$\mathrm{gr}(\mathcal{D}_{X/Y}) = \mathcal{O}_{T^*(X/Y)} \,,$$

where $T^*(X/Y)$ is the relative cotangent space. $\mathcal{D}_{X/Y}$ is a noetherian sheaf of algebras, in the sense of [21].

We now consider the complex of sheaves

$$\mathcal{D}_{X/Y} \overset{\mathbb{L}}{\underset{\mathcal{D}_{X/Y} \boxtimes \mathcal{D}_{X/Y}^\circ}{\otimes}} \mathcal{D}_{X/Y} \,;$$

it is a complex of sheaves on $X \times X$, concentrated on the diagonal Δ_X. Hence, as in §2.1, we may view it as a complex of sheaves on $\Delta_X \cong X$. By the analogy of Lemma 2.1.2, if X is affine, the Hochschild homology of $D(X/Y) = \Gamma(X, \mathcal{D}_{X/Y})$ is computed from the complex

(2.18), to compare this quasi-isomorphism with the morphism defined before (2.1.8); we will neglect here to do that.

$$\mathbb{R}\Gamma(X, \mathcal{D}_{X/Y} \overset{\mathbb{L}}{\underset{\mathcal{D}_{X/Y} \boxtimes \mathcal{D}^{\circ}_{X/Y}}{\otimes}} \mathcal{D}_{X/Y})$$

We will set $d = \dim(Y)$, $n = \dim(X) - \dim(Y)$. Following the same strategy as in §2.1, we will produce a morphism of complexes:

$$\mathcal{D}_{X/Y} \overset{\mathbb{L}}{\underset{\mathcal{D}_{X/Y} \otimes \mathcal{D}^{\circ}_{X/Y}}{\otimes}} \mathcal{D}_{X/Y} \to \mathbb{R}\mathrm{Hom}_{\mathcal{O}_{S\times S}}(\omega_S^{\otimes -1}, \mathcal{O}_S) \underset{k}{\otimes} \Omega_{X/S}[2n+d]$$

and then justify it is a quasi-isomorphism by local computation. First, we introduce the relative dualizing sheaf $\omega_{X/Y} = \Omega^n_{X/Y}$. $\omega_{X/Y}$ may be used to transform left $\mathcal{D}_{X/Y}$-modules into right $\mathcal{D}_{X/Y}$-modules. In particular:

$\mathcal{D}^{(1)}_{X/Y} = p_2^* \omega_{X/Y} \underset{p_2^* \mathcal{O}_X}{\otimes} \mathcal{D}_{X/Y}$ is a right $\mathcal{D}_{(X \times X)/(Y \times Y)}$-module ;

and

$\mathcal{D}^{(2)}_{X/Y} = p_2^*(\omega_{X/Y}^{\otimes -1}) \underset{p_2^* \mathcal{O}_X}{\otimes} \mathcal{D}_{X/Y}$ is a left $\mathcal{D}_{(X \times X)/(Y \times Y)}$-module .

<u>Lemma 2.2.1</u> $\mathcal{D}_{X/Y}{}^{(2)}$ is canonically isomorphic, as a left $\mathcal{D}_{(X \times X)/(Y \times Y)}$-module, to $\underline{H}^n_{\Delta_X}(\mathcal{O}_{X \underset{Y}{\times} X})$.

This is proven just like Lemma 2.1.5, using the <u>relative fundamental class</u> of Angéniol-Elzein [1], for the inclusion $\Delta_X \hookrightarrow X \underset{Y}{\times} X$ of schemes over Y. This fundamental class belongs to $\underline{H}^n_{\Delta_X}(X \underset{Y}{\times} X, p_2^* \omega_{X/Y})$.

<u>Lemma 2.2.2</u> The dual $(\mathcal{D}_{X/Y}{}^{(1)})^*$ of $\mathcal{D}_{X/Y}{}^{(1)}$ is (maybe non-canonically) isomorphic to $\underline{H}^n_{\Delta_X}(\mathcal{O}_{X \times_Y X} \otimes_{\mathcal{O}_Y} \omega_Y^{\otimes -1})[-d-2n]$.

<u>Idea of Proof</u>: Locally on X, $(\mathcal{D}_{X/Y}{}^{(1)})^*[+d+2n]$ is isomorphic, as a left $\mathcal{D}_{(X \times X)/(Y \times Y)}$-module, to $\underline{H}^n_{\Delta_X}(\mathcal{O}_{X \times_Y X})$. The sheaf of germs of automorphisms of that module is equal to $\mathcal{O}_Y^*$ (recall that $\mathcal{O}_Y$ is central in $\mathcal{D}_{X/Y}$), or more precisely to the inverse image $F^{-1}\mathcal{O}_Y^*$. Therefore we have:

$$(\mathcal{D}_{X/Y}{}^{(1)})^* \cong \underline{H}^n_{\Delta_X}(\mathcal{O}_{X \times_Y X}) \underset{F^{-1}\mathcal{O}_Y}{\otimes} F^{-1}\mathcal{L}[-d-2n]$$

for some invertible sheaf $\mathcal{L}$ on Y. It remains to identify $\mathcal{L}$. For this purpose, we endow the two sides of this isomorphism with their natural filtrations; as a simple holonomic $\mathcal{D}_{X \times_Y X}$-module, $\underline{H}^n_{\Delta_X}(\mathcal{O}_{X \times_Y X})$ has a natural "depth of layer" filtration (under the isomorphism 2.2.1), this corresponds to the order filtration on $\mathcal{D}_{X/Y}{}^{(2)}$. Hence $(\mathcal{D}_{X/Y}{}^{(1)})^*[d+2n]$ also has a filtration. We will show that $\mathcal{L} \cong \omega_Y^{\otimes -1}$ by comparing the graded modules for these filtrations. There is a natural filtration on the dual of a filtered complex. For $\mathcal{D}_{X/Y}{}^{(1)}$, the "duality spectral sequence" of [2, Chapter 2] degenerates, since $\mathrm{gr}\,\mathcal{D}_{X/Y}{}^{(1)}$ is Cohen-Macaulay, and we obtain

$$\mathrm{gr}(\mathcal{D}_{X/Y}{}^{(1)*}) \cong \mathrm{gr}(\mathcal{D}_{X/Y}{}^{(1)})^*$$

(on the right-hand side, $*$ denotes duality for complexes of $\mathcal{O}_{T^*(X/Y) \times T^*(X/Y)}$-modules). Now, let Λ be the <u>(relative) co-normal bundle</u> for the immersion $X \hookrightarrow X \times X$ of $Y \times Y$-schemes;

i.e., Λ is a vector bundle over X, which is the quotient of the usual co-normal bundle $T^*_X(X \times X)$ by its horizontal part (in the case where F is a trivial fibration so that $X = Y \times Z$, we have $\Lambda = Y \times T^*_Z(Z \times Z))$.

Then $\mathrm{gr}\, \mathcal{D}_{X/Y}{}^{(1)} \cong \mathcal{O}_\Lambda \otimes \omega_{X/Y}$.

Now recall the following simple case of Grothendieck duality (see [14]).

<u>Lemma 2.2.3</u> Let $S \overset{i}{\hookrightarrow} T$ be a closed immersion of smooth k-varieties, then the dual $(\mathcal{O}_S)^*$ of the $\mathcal{O}_T$-module $\mathcal{O}_S$ is equal to $\omega_S \otimes i^*(\omega_T^{\otimes -1})[-d]$, where d is the codimension of Y in X.

We apply this lemma to the embedding

$$\Lambda \subset T^*(X/Y) \times T^*(X/Y) \ .$$

We get $(\mathcal{O}_\Lambda)^* = \omega_\Lambda \otimes \omega_{T^*(X/Y) \times T^*(X/Y)}{}^{\otimes(-1)}[-2n - d]$. . We have a smooth morphism $T^*(X/Y) \longrightarrow Y$, the fibers of which are the symplectic manifolds $T^*(X_y)$; since the canonical sheaf of a symplectic manifold is trivial, we have $\omega_{T^*(X/Y)} \cong \omega_Y$; hence we get $(\mathcal{O}_\Lambda)^* = \omega_\Lambda \otimes \omega_Y^{\otimes -2}[-2n - d]$. Similarly, the fibers of the smooth morphism $\Lambda \longrightarrow Y$ are the symplectic manifolds

$$T^*_{X_y}(X_y \times X_y) \cong T^*(X_y) \ ;$$

therefore $\omega_\Lambda = \omega_Y$ and $(\mathcal{O}_\Lambda)^* = \omega_Y^{\otimes -1}[-2n - d]$, so we get:

$$[\mathrm{gr}\, \mathcal{D}^{(1)}_{X/Y}]^* \cong \omega_{X/Y}^{\otimes -1} \otimes (\mathcal{O}_\Lambda)^* \cong \omega_{X/Y}^{\otimes -1} \otimes \omega_Y^{\otimes -1}[-2n - d] \ .$$

On the other hand, using Lemma 2.2.1, we find:

$$\mathrm{gr}(\mathbb{H}^n_{\Delta_X}(\mathcal{O}_{X\times_Y X})) = \mathrm{gr}(\mathcal{D}_{X/Y^{(2)}}) = \omega_{X/Y} \otimes \mathrm{gr}(\mathcal{D}_{X/Y}).$$

It follows that $\mathcal{L}$ must be equal to $\omega_Y^{\otimes -1}$. This includes the proof of Lemma 2.2.2.

Now, just in §2.1, we have:

$$\mathcal{D}_{X/Y} \overset{\mathbb{L}}{\underset{\mathcal{D}_{X/Y} \boxtimes \mathcal{D}^\circ_{X/Y}}{\otimes}} \mathcal{D}_{X/Y} = \mathbb{R}\mathrm{Hom}_{\mathcal{D}_{(X\times X)/(Y\times Y)}}((\mathcal{D}_{X/Y^{(1)}})^*, \mathcal{D}_{X/Y^{(2)}})$$

$$= \mathbb{R}\mathrm{Hom}_{\mathcal{D}_{(X\times X)/(Y\times Y)}}(\mathcal{B}_{\Delta_X/X\times_Y X} \otimes \omega_X^{\otimes -1}, \mathcal{B}_{\Delta_X/X\times_Y X})[2n+d].$$

Now we need the following:

<u>Definition 2.2.4</u> For $\mathcal{M}^\cdot$ a complex of left $\mathcal{D}_{X/Y}$-module (where $F : X \longrightarrow Y$ is a smooth morphism between smooth algebraic varieties over k), we let $DR(\mathcal{M}^\cdot) = DR_{X/Y}(\mathcal{M}^\cdot)$ be the simple complex associated to the double complex (where $\Omega^i_{X/Y} \underset{\mathcal{O}_X}{\otimes} \mathcal{M}^j$ is in degree $i+j-n$)

$$\begin{array}{ccc} \to \Omega^i_{X/Y} \underset{\mathcal{O}_X}{\otimes} \mathcal{M}^{j+1} & \xrightarrow{d} & \Omega^{i+1}_{X/Y} \underset{\mathcal{O}_X}{\otimes} \mathcal{M}^{j+1} \to \cdots \\ \uparrow & & \uparrow \\ \to \Omega^i_{X/Y} \underset{\mathcal{O}_X}{\otimes} \mathcal{M}^{j} & \xrightarrow{d} & \Omega^{i+1}_{X/Y} \underset{\mathcal{O}_X}{\otimes} \mathcal{M}^{j} \to \cdots \end{array}$$

where the horizontal differential uses the action on $\mathcal{M}^\cdot$ of the vertical vector fields on X.

Notice $DR(\mathcal{M}^\cdot)$ is a complex of $F^{-1}(\mathcal{O}_Y)$-modules. We apply this construction to $F \times F : X \times X \to Y \times Y$, and $\mathcal{M} = \mathcal{B}_{\Delta_X / X \times_Y X}$.

<u>Lemma 2.2.5</u> $DR_F(\mathcal{B}_{\Delta_X / X \times_Y X})$ is isomorphic to $i_*(\Omega^{\bullet}_{X/Y})[n]$,

where $i : \Delta_X \hookrightarrow X \times X$ is the inclusion.

(This is a generalization of Lemma 2.1.7.)

Hence we obtain a morphism of complexes

$$\mathcal{D}_{X/Y} \overset{\mathbb{L}}{\underset{\mathcal{D}_{X/Y} \boxtimes \mathcal{D}^{\circ}_{X/Y}}{\otimes}} \mathcal{D}_{X/Y} \to \mathbb{R}\mathrm{Hom}_{\mathcal{O}_{Y \times Y}}(i_*(\Omega^{\bullet}_{X/Y}) \otimes$$

$$\otimes\, \omega_Y^{\otimes -1}, i_*(\Omega^{\bullet}_{X/Y}))[2n + d].$$

There is a $\mathcal{O}_{Y \times Y}$-linear injection $i_* \mathcal{O}_Y \hookrightarrow \Omega^{\bullet}_{X/Y}$ of complexes, hence we obtain a morphism of complexes

$$\mathcal{D}_{X/Y} \overset{\mathbb{L}}{\underset{\mathcal{D}_{X/Y} \boxtimes \mathcal{D}^{\circ}_{X/Y}}{\otimes}} \mathcal{D}_{X/Y} \to \mathbb{R}\mathrm{Hom}_{\mathcal{O}_{Y \times Y}}(i_*(\Omega^{\bullet}_{X/Y}) \otimes$$

$$\otimes\, \omega_Y^{\otimes -1}, i_*(\Omega^{\bullet}_{X/Y}))[2n + d].$$

Now notice that the action of $\mathcal{O}_{Y \times Y}$ on $i_* \Omega^{\bullet}_{X/Y}$ factors through the quotient $\mathcal{O}_Y = \mathcal{O}_{\Delta_Y}$ of $\mathcal{O}_{Y \times Y}$. It follows that the last complex is equal to $\mathbb{R}\mathrm{Hom}_{\mathcal{O}_{Y \times Y}}(\omega_Y^{\otimes -1}, \mathcal{O}_Y) \overset{\mathbb{L}}{\underset{\mathcal{O}_Y}{\otimes}} \Omega^{\bullet}_{X/Y}[2n + d]$.

Now we want to prove that this morphism is a quasi-isomorphism. Just as in §2.1, we may assume that X is étale over $\mathbb{A}^n \times Y$; then essentially we are in a product situation and we have:

$$\mathcal{D}_{X/Y} \cong \mathcal{D}_{\mathbb{A}^n} \boxtimes \mathcal{O}_Y .$$

Then our morphism is a tensored product of the morphism

$$\mathcal{D}_{\mathbb{A}^n} \overset{\mathbb{L}}{\underset{\mathcal{D}_{\mathbb{A}^n} \boxtimes \mathcal{D}^\circ_{\mathbb{A}^n}}{\otimes}} \mathcal{D}_{\mathbb{A}^n} \to \Omega^{\bullet}_{\mathbb{A}^n}[2n]$$

which we know by §2.1 to be a quasi-isomorph... ...d of a morphism

$$\mathcal{O}_Y \overset{\mathbb{L}}{\underset{\mathcal{O}_{Y \times Y}}{\otimes}} \mathcal{O}_Y \to \mathbb{R}\mathrm{Hom}_{\mathcal{O}_{Y \times Y}}(\omega_Y^{\otimes -1}, \mathcal{O}_Y)[d]$$

which is a special case of the isomorphism (2.1.3). Hence we have found a quasi-isomorphism between

$$\mathcal{D}_{X/Y} \overset{\mathbb{L}}{\underset{\mathcal{D}_{X/Y} \boxtimes \mathcal{D}^\circ_{X/Y}}{\otimes}} \mathcal{D}_{X/Y}$$

and

$$\mathbb{R}\mathrm{Hom}_{\mathcal{O}_{Y \times Y}}(\omega_Y^{\otimes -1}, \mathcal{O}_Y) \overset{\mathbb{L}}{\otimes} \Omega^{\bullet}_{X/Y}[2n + d]$$

which is the same as

$$(\mathcal{O}_Y \overset{\mathbb{L}}{\underset{\mathcal{O}_{Y \times Y}}{\otimes}} \mathcal{O}_Y) \overset{\mathbb{L}}{\otimes} \Omega^{\bullet}_{X/Y}[2n] .$$

Now Loday and Quillen [24] prove an isomorphism

$$\mathcal{O}_Y \overset{\mathbb{L}}{\underset{\mathcal{O}_{Y \times Y}}{\otimes}} \mathcal{O}_Y \Sigma \cong \bigoplus_i \Omega^{\bullet}_Y[i]$$

(this is a little stronger than the classical computation of Hochschild-Kostant-Rosenberg [17]). Hence we have proved the following quasi-isomorphic to $\underset{i}{\otimes}(\Omega^i_Y \underset{\mathcal{O}_Y}{\otimes} \Omega^{\bullet}_{X/Y})[i+2n]$.

<u>Theorem 2.2.6</u> The complex of sheaves

$$\mathcal{D}_{X/Y} \overset{\mathbb{L}}{\underset{\mathcal{D}_{X/Y} \boxtimes \mathcal{D}^{\circ}_{X/Y}}{\otimes}} \mathcal{D}_{X/Y} \text{ on } X$$

is quasi-isomorphic to $\underset{i}{\oplus}(\Omega^i_Y \underset{\mathcal{O}_Y}{\otimes} \Omega^{\bullet}_{X/Y})[i+2n]$.

The operator B has a nice interpretation on the direct image $\underset{i}{\oplus} \mathbb{R}F_*(\Omega^i_Y \underset{\mathcal{O}_Y}{\otimes} \Omega^{\bullet}_{X/Y})[i+2n]$; then $\mathbb{R}F_*(\Omega^{\bullet}_{X/Y})$ is a quasi-coherent complex of $\mathcal{O}_Y$-modules, endowed with a connection ∇, called the <u>Gauss-Manin</u> connection [23].

<u>Proposition 2.2.7</u> B maps $\Omega^i_Y \underset{\mathcal{O}_Y}{\otimes} \mathbb{R}F_*(\Omega^{\bullet}_{X/Y})[i+2n]$ to

$\Omega^{i+1}_Y \underset{\mathcal{O}_Y}{\otimes} \mathbb{R}F_*(\Omega^{\bullet}_{X/Y})[i+1+2n]$,

and is induced by the Gauss-Manin connection.

Details will be presented elsewhere.

Now for X <u>affine</u>, the Hochschild homology of $D(X/Y)$ is computed from the complex $\mathbb{R}\Gamma(X, \mathcal{D}_{X/Y} \overset{\mathbb{L}}{\underset{\mathcal{D}_{X/Y} \otimes \mathcal{D}^{\bullet}_{X/Y}}{\otimes}} \mathcal{D}_{X/Y})$.

We want to explicit this for $F : X \longrightarrow Y$ a <u>Jouanolou-Karoubi</u>

resolution of a smooth projective variety Y, i.e., F is supposed to satisfy

(i) X affine;

(ii) F is a locally trivial fibration with fibre $\mathbb{A}^n$. For any smooth projective Y, the existence of such $F : X \longrightarrow Y$ was proved by Jouanolou and Karoubi (see e.g., [25,]).

We first apply $\mathbb{R}F_*$ to the Hochschild complex of $\mathcal{D}_{X/Y}$, and we obtain $\bigoplus_i (\Omega^i_Y \otimes_{\mathcal{O}_Y} \mathbb{R}F_*(\Omega^{\cdot}_{X/Y}))[i+2n]$. Because of property (ii), $\mathbb{R}F_*(\Omega^{\cdot}_{X/Y})$ is quasi-isomorphic to $\mathcal{O}_Y$, we obtain $\bigoplus_i \Omega^i_Y[i+2n]$.

Hence we have proved

Theorem 2.2.8 If $F : X \longrightarrow Y$ is a Jouanolou-Karoubi resolution, then $H_k(D(X/Y), D(X/Y)) \cong \bigoplus_i H^{2n+i-k}(Y, \Omega^i_X)$.

We have not verified the following description. of the cyclic homology of $D(X/Y)$, which in any case is extremely likely (as the Connes spectral sequence should degenerate). We should have: $HC_k(D(X/Y)) \cong \bigoplus_i \mathbb{H}^{2n-k+2i}(Y, \Omega^i_Y \to \ldots \to \Omega^i_Y)$ a direct sum of hypercohomology groups of truncated de Rham complexes. The Connes spectral sequence then would become, essentially the Hodge to de Rham spectral sequence, and then degenerate at E^1. Conversely, of course, if the Connes spectral sequence was shown to degenerate by some cyclic homology wizzard, the degeneration of the Hodge to de Rham spectral sequence would follow.

The higher algebraic K-theory of $D(X/Y)$ is easily determined

Proposition 2.2.9 $K_j(D(X/Y)) = K_j(Y)$.

Proof: $D(X/Y)$ is filtered by order, hence by [25,],
$K_j(D(X/Y)) = K_j(\mathcal{O}(X)) = K_j(X)$ and by [25,], $K_j(X) = K_j(Y)$.

Karoubi defines [19, §3.25] Chern characters

$$\tilde{Ch}_j^m : K_j(A) \longrightarrow HC_{j+2m}(A)$$

for any algebra A , with the property that

$$\tilde{Ch}_j^{m-1} = S_\circ \tilde{Ch}_j^m.$$

In our case, it appears that the interesting Chern character is $\tilde{Ch}_j^n$ (for j variable). Indeed this will give for each i and j a Chern character $K_j(Y) = K_j(D(X/Y)) \longrightarrow \mathbb{H}^{2i-j}(Y, \mathcal{O}_Y \longrightarrow \ldots \longrightarrow \Omega_Y^i)$ (using a conjectural description of $HC_*(D(X/Y))$. Presumably this Chern character factors through $H^{i-j}(Y, \Omega_Y^i)$; this would follow from the vanishing of $\tilde{Ch}_j^{n-1}$. Of course, this sort of Chern character should coincide with those obtained by Karoubi, in de Rham cohomology.

We hope that some modification of the algebra $D(X/Y)$, probably some sort of algebra of Toeplitz operators, will have closer relations with the Beilinson-Deligne cohomology theory.

Bibliography

[1] Angéniol, B. and Elzein, F.; *La classe fondamentale relative d'un cycle*; Bull. Soc. Math. Mémoire n°58 (1978); pp. 67-93.

[2] Björk, E.; *Rings of differential operators*; North Holland (1982).

[3] Blanc, P.; *Cohomologie différentiable et changement de groupes*; Astérisque 124-125 (1985); pp. 113-30.

[4] Blanc, P.; *Sur les fonctions d'intégrales orbitales nulles sur un groupe réductif*; preprint Ecole Polytechnique (1985).

[5] Blanc, P. and Wigner, D.; *Homology of Lie groups and Poincaré duality*; Letters in Math. Physics 7 (1983); pp. 259-270.

[6] Bourbaki, N.; *Groupes et algèbres de Lie*; *chapitre III*; Diffusion C.C.L.S., Paris.

[7] Brown, K.S.; *Cohomology of groups*; Graduate texts in mathematics n°87, Springer Verlag (1982).

[8] Brylinski, J.-L.; *A differential complex for Poisson manifolds*; preprint I.H.E.S./M/86/12 (1986).

[9] Burghelea, D.; *The cyclic homology of the group rings*; preprint Ohio State University (1984).

[10] Cartan, H. and Eilenberg, S.; *Homological algebra*; Annals of math. studies; Princeton University Press n°19n (1956).

[11] Cartier, P.; *Representations of p-adic groups*; Proc of Symp. in Pure Math. vol. 33 (1979); pp. 111-155.

[12] Casselman, W.; *A new non-unitarity argument for p-adic representations*; Journal of the Faculty of Science, University of Tokyo 28 (1982); pp. 907-928.

[13] Connes, A.; *Non-commutative differential geometry*; Publ. Math. I.H.E.S. 62 (1986); pp. 257-360.

[14] Grothendieck, A.; *Cohomologie locale des faiseaux cohérents et théorémes de Letchetz locaux et globaux*; (SGA); North Holland.

[15] Harish-Chandra; *Admissible distributions on reductive p-adic groups*; Queen's papers 48 (1978); pp. 281-348.

[16] Harish-Chandra and van Dijk, G.; *Harmonic analysis on reductive p-adic groups*; Lecture Notes in Math.

[17] Hochschild, G., Kostant, B. and Rosenberg, A.; *Differential forms on regular affine algebras*; Trans. Amer. Math. Soc. 102 (1962); pp. 383-408.

[18] Julg, P. and Valette, A.; *Twisted coboundary operators and the Selberg principle*; preprint (1986); to appear in J. Oper. Theory.

[19] Karoubi, M.; *Homologie cyclique et K-theorie I*; preprint; Paris (1985).

[20] Kashiwara, M.; *On the holonomic systems of linear differential equations II*; Invent. Math. 49 (1978); pp. 121-135.

[21] Kashiwara, M. and Kawai, T.; *On the holonomic systems of linear differential equations (systems with regular singularities) III*; Publ. R.I.M.S./Kyoto University 17 (1981); pp. 813-979.

[22] Kassel, C. and Mitschi, C.; *Algèbres d'opérateurs différentiels et 'cohomologie de de Rham*; in preparation.

[23] Katz, N. and Oda, T.; *On the differentiation of de Rham cohomology classes with respect to parameters*; J. Math. Kyoto University 8-2 (1968); pp. 199-213.

[24] Loday, J.-L. and Quillen, D.; *Cyclic homology and the Lie algebra homology of matrices*; Comment. Math. Helv. 59 (1984); pp. 565-591.

25] Quillen, D.; *Higher algebraic K-theory*; Springer Lecture Notes in Math 341 (1973); pp. 85-147.

On the Topology of Algebraic Torus Actions

by

M. Goresky[1] and R. MacPherson[2]

To T. A. Springer

On his sixtieth birthday

§ 1. Introduction.

Suppose a compact complex algebraic variety X has an action of an algebraic torus $(\mathbb{C}^*)^n$. As a Lie group, the algebraic torus is the product of two topological subgroups: the compact torus $(S^1)^n \subset (\mathbb{C}^*)^n$ and $(\mathbb{R}^+)^n \subset (\mathbb{C}^*)^n$, where $\mathbb{R}^+$ is the positive reals. In this note, we determine the following information:

1. the topology of the orbit space $B = X/(S^1)^n$, and
2. the topological structure of the $(\mathbb{R}^+)^n$ action on B.

By the topological structure of the $(\mathbb{R}^+)^n$ action, we mean the orbits and the stabilizer subgroups $\mathrm{Stab}_{(\mathbb{R}^+)^n}(b) \subset (\mathbb{R}^+)^n$. Knowledge of this information goes a long way towards reconstructing X topologically, as explained in § 8.

---

1. Partially supported by N.S.F. grant # DMS 860-1161

2. Partially supported by N.S.F. grant # DMS 850-2422

We express this information in terms of certain "torus action data" which can be associated to the $(\mathbb{C}^*)^n$ action on X. Torus action data is of two types: The first is a collection of polyhedra in Euclidean n-space. The second is a collection of algebraic varieties Z_F and some algebraic maps $\zeta_{FG}: Z_F \longrightarrow Z_G$ between them. If X had dimension k and the $(\mathbb{C}^*)^n$ action is effective, then the varieties Z_F have dimension at most k-n.

The association of polyhedra to the torus action is done by the moment map ([A1], [GS], [MW], [K]). The varieties Z_F are the various symplectic quotients associated to different points in the image of the moment map. These may also be identified with the geometric invariant theory quotients of various subvarieties of semistable points in X. All of this is standard. The new ingredient here is the collection of algebraic maps ζ_{FG}: $Z_F \longrightarrow Z_G$ and the role that they play in reconstructing the quotient space B.

This is a largely expository paper. The results are easy consequences of what are by now standard techniques. Our main contribution consists of an efficient presentation of the rather complicated picture of the orbit structure of a torus action.

§ 2. Definitions. In this section we give some elementary topological definitions which will be used throughout the paper.

A piecification of a topological space X is a partially ordered set $\mathcal{F}$ (with partial ordering denoted $\leq$) and a choice for each $F \in \mathcal{F}$ of a subset (or "piece") $X^F \subset X$ such that

(a) If $F \neq G$ then $X^F \cap X^G = \phi$

(b) $\cup \{X^F \mid F \in \mathcal{F}\} = X$

(c) If $X^G \cap \overline{X^F} \neq \phi$ then $G \leq F$

__Remarks__. A stratification is a piecification, however a piecification is more general: the pieces may be singular, and a piecification does not necessarily satisfy the axiom of the frontier (i.e. the closure of a piece is not necessarily a union of pieces). We allow the possibility that $X^F = \phi$. The partial ordering axiom (c) implies that each piece is locally closed.

__Definition__. A __space-valued cofunctor__ $\mathscr{Z}$ on a partially ordered set $\mathscr{F}$ is a collection of topological spaces Z_F (indexed by the elements of $\mathscr{F}$) together with continuous maps

$$\varsigma_{FG} : Z_F \longrightarrow Z_G$$

whenever $G \le F$, with the property that if $H \le G \le F$ then $\varsigma_{FH} = \varsigma_{GH}\varsigma_{FG}$ and ς_{FF} is the identity.

__Definition__. Suppose $\mathscr{F}$ is a partially ordered set, X is a piecified space with pieces indexed by $\mathscr{F}$, and $\mathscr{Z}$ is a space-valued cofunctor on $\mathscr{F}$. The __realization__ $R(\mathscr{Z})$ __over X__ of the triple $(\mathscr{Z}, X, \mathscr{F})$ is the topological space

$$R(\mathscr{Z}) = \coprod_{F\in\mathscr{F}} Z_F \times \overline{X^F} / \sim$$

where $\sim$ identifies a point $(z,x) \in Z_F \times \overline{X^F}$ with $(\varsigma_{FG}(z), x)$ whenever $x \in X^G \cap \overline{X^F}$.

__Example of a realization__: The mapping cylinder. Suppose the partially ordered set $\mathscr{F}$ consists of two elements $G \le F$. Let $X = [0,1]$ with piecification $X^G = \{0\}$ and $X^F = (0,1]$. A cofunctor $\mathscr{Z}$ over $\mathscr{F}$ is a pair of spaces Z_F, Z_G, together with a continuous map $\varsigma_{FG}: Z_F \longrightarrow Z_G$. The realization $R(\mathscr{Z})$ over X is the mapping cylinder of ς_{FG}.

__Remarks__. The realization is canonically piecified with pieces

$$R(\mathscr{Z})^F = Z_F \times X^F$$

The realization comes with an obvious projection $\pi: R(\mathscr{Z}) \longrightarrow X$, which is proper if and only if each of the spaces Z_F is compact. $R(\mathscr{Z})$ is Hausdorff if X is Hausdorff and each Z_F is Hausdorff (this uses the commutation relations). $R(\mathscr{Z})$ is locally compact if X is locally compact, each Z_F is locally compact and each ζ_{FG} is proper.

§ 3. Torus Action Data

In this section we define a collection of data which can be obtained from any projective variety X with an action of an algebraic torus. In § 5 we will show how to reconstruct the topological space $X/(S^1)^n$ from this data.

Recall that a convex polyhedron $C \subset \mathbb{R}^n$ is the convex hull of a finite set of points. Its affine hull is the smallest affine subspace A containing C. The interior C^o of C is the topological interior of C, viewed as a subspace of A. The interior of a point is itself. The span of C is the Euclidean subspace span(C) which is obtained by translating the affine hull A so that it passes through the origin.

Definition. TA Data consists of the following four ingredients: TAD1, TAD2, TAD3, TAD4 :

TAD1 is a finite collection $\mathscr{C}$ of (closed) convex polyhedra (of various dimensions, possibly overlapping, possibly sharing interior points) in $\mathbb{R}^n$ such that

(a) If $C \in \mathscr{C}$ then each face D of C is also an element of $\mathscr{C}$.

(b) Each $C \in \mathscr{C}$ is rational, i.e. the Euclidean subspace span(C) $\subset \mathbb{R}^n$ has a basis consisting of integral points $b_1, \ldots, b_r \in \mathbb{Z}^n$.

Remarks: We obtain a partial order on $\mathscr{C}$ by defining

$$D \leq C \iff D \text{ is a face of } C$$

<u>Define</u> $$P = \cup \mathscr{C}$$

to be the compact subset of $\mathbb{R}^n$ which is the union of all these polyhedra. There is a natural (coarsest) piecification $\mathscr{F}$ of the topological space P with the property that each $C \in \mathscr{C}$ is a union of pieces: two points $x,y \in P$ are in the same piece of P if and only if they are contained in <u>exactly</u> the same convex polyhedra $C \in \mathscr{C}$. Thus $\mathscr{F}$ is the set of subsets of $\mathscr{C}$ and is partially ordered by inclusion. The pieces of P are then given by

$$P^F = \cap \{C \mid C \in F\} - \cup \{C \mid C \in \mathscr{C}-F\}$$

for each subset $F \subset \mathscr{C}$.

We remark that this piecification of P is in fact a Whitney stratification and in particular it satisfies the axiom of the frontier: $P^F \cap \overline{P^G} \neq \phi \iff P^F \subset \overline{P^G} \iff F \subset G$

<u>TAD 2</u> is a cofunctor $\mathscr{R}$ of complex (not necessarily compact) algebraic varieties over the partially ordered set $\mathscr{C}$, i.e. for each $C \in \mathscr{C}$ an algebraic variety R^C, and for each face $D \leq C$ an algebraic map $\rho_{CD}\colon R^C \longrightarrow R^D$.

<u>TAD 3</u> is a cofunctor $\mathscr{Z}$ of complex algebraic varieties over $\mathscr{F}$, i.e. for each $F \in \mathscr{F}$ an algebraic variety Z_F and for each relation $G \leq F$ an algebraic map $\varsigma_{FG}\colon Z_F \longrightarrow Z_G$.

<u>TAD 4</u> is a choice, for each $F \in \mathscr{F}$ and for each $C \in \mathscr{C}$ such that $P^F \subset C^o$, of an inclusion

$$\iota_F^C\colon R^C \longrightarrow Z_F$$

We shall denote the image $\iota_F^C(R^C)$ by Z_F^C . These data are furthermore assumed to satisfy the following axioms:

<u>Axiom 1</u>. Each Z_F is piecified by the images $Z_F^C = \iota_F^C(R^C)$, where C is allowed to vary over the partially ordered set

$$\mathscr{C}_F = \{ C \in \mathscr{C} \mid C^o \supset P^F \}$$

(which is partially ordered by containment, i.e.

$$D \leq C \Leftrightarrow D \subset C)$$

<u>Axiom 2</u>. If $G \leq F \in \mathcal{F}$, and if $C \in \mathcal{C}$, with $P^F \subset C^o$ and $P^G \subset C^o$ then

$$i_F^C \circ \varsigma_{FG} = i_G^C$$

<u>Axiom 3</u>. If $G \leq F \in \mathcal{F}$ and if $D \leq C \in \mathcal{C}$ with $P^G \subset D^o$ and $P^F \subset C^o$ then the following diagram commutes:

$$\begin{array}{ccc} R^C & \xrightarrow{\rho_{CD}} & R^D \\ {\scriptstyle i_F^C}\downarrow & & \downarrow{\scriptstyle i_G^D} \\ Z_F & \xrightarrow{\rho_{FG}} & Z_G \end{array}$$

<u>§ 4: A Torus Action Gives Rise to TA Data.</u>

Suppose X is a projective complex algebraic variety with an action of the algebraic torus $(\mathbb{C}^*)^n$. We assume the torus action extends to a linear action on the ambient projective space $\mathbb{P}^N$. Choose a Kaehler metric on $\mathbb{P}^N$ which is invariant under the compact torus $(S^1)^n \subset (\mathbb{C}^*)^n$ and let $\mu: X \longrightarrow \mathbb{R}^n$ be the (restriction to X of the) associated moment map ([K], [A1], [MS], [A2], [GS]). This map factors

$$X \xrightarrow{\alpha} B \xrightarrow{\beta} \mathbb{R}^n$$

through the quotient space $B = X/(S^1)^n$.

<u>TAD 1</u>: We define a collection $\mathcal{C}$ of convex polyhedra in $\mathbb{R}^n$ as follows: the closure in X

$$\bar{T} = \overline{(\mathbb{C}^*)^n \cdot x}$$

of each torus orbit projects to a polyhedron $C = \mu(\bar{T})$ and the torus orbit itself $(\mathbb{C}^*)^n \cdot x$ projects to the interior C^o of C ([A1], [GS], [K])

<u>Proposition.</u> The polyhedra obtained in this manner constitute TAD 1, i.e. they satisfy the following hypotheses:

1. Each $C = \mu(\bar{T})$ is a <u>rational</u> convex polyhedron
2. Only finitely many polyhedra ppear in the collection $\mathcal{C}$
3. If $C = \mu(\bar{T})$ is a polyhedron in this collection then each face of C is also the μ-image of a torus orbit closure.

This collection $\mathcal{C}$ of polyhedra indexes a canonical piecification of X as follows:

<u>Definition.</u> Let $C \in \mathcal{C}$. A point $x \in X$ is in the piece X^C if and only if the convex polyhedron corresponding to the orbit through x is equal to C, i.e.

$$X^C = \{x \in X \mid \mu(\overline{(\mathbb{C}^*)^n \cdot x}) = C\}$$

<u>TAD 2</u>: Define a space-valued cofunctor $\mathcal{R}$ of complex algebraic varieties over $\mathcal{C}$ as follows: for each convex polyhedron $C \in \mathcal{C}$ let

$$R^C = X^C / (\mathbb{C}^*)^n$$

If $D \leq C$ then there is a unique map $\rho_{CD} : R^C \longrightarrow R^D$ which can be characterized as follows: suppose $x \in X^C$ is a lift of $\bar{x} \in R^C$ and suppose $y \in X^D$ is a lift of $\bar{y} \in R^D$. Then $\rho_{CD}(\bar{x}) = \bar{y}$ if and only if $y \in \overline{(\mathbb{C}^*)^n \cdot x}$.

<u>Proposition</u>. Each map ρ_{CD} is well defined and algebraic.

Now let $P = \mu(X)$ denote the union of the convex polyhedra defined above and let $\mathcal{F}$ be the natural piecification of P as described in § 2.

<u>TAD 3</u>. Define a space-valued cofunctor $\mathcal{Z}$ over $\mathcal{F}$ as follows: given $F \in \mathcal{F}$ choose any $p \in P^F$ and set

$$Z_F = \beta^{-1}(p) = \mu^{-1}(p)/(S^1)^n$$

(This is the "symplectic quotient" which is identified with a particular "geometric invariant theory" algebro-geometric quotient [K], [M], [A2]. It is known that the symplectic quotient does not depend on the point $p \in P^F$).

If $G < F \in \mathcal{F}$ then we obtain a map

$$\zeta_{FG}: Z_F \longrightarrow Z_G$$

which is characterized in the following way: Let $p' \in Z_F$. This means that $p = \beta(p') \in P^F$. Choose a lift $\tilde{p} \in \alpha^{-1}(p')$. Choose $q \in P^G$. Then the closure of the torus orbit $\overline{(\mathbb{C}^*)^n.\tilde{p}}$ intersects $\mu^{-1}(q)$ in a single $(S^1)^n$ orbit, thus determining a single point

$$q' \in \mu^{-1}(q)/(S^1)^n = Z_G.$$

We define $\zeta_{FG}(p') = q'$.

<u>Proposition.</u> The preceding choices of Z_F and ζ_{FG} are well defined and satisfy the criteria of TAD 2, i.e. each Z_F is an algebraic variety and each ζ_{FG} is an algebraic map.

<u>TAD 4</u>. If $P^F \subset C^o$ then there is an inclusion $\iota_F^C: R^C \longrightarrow Z_F$ because (by [K]) for any choice of point $p \in P^F \subset C^o$ there is a natural identification

$$R^C = X^C / (\mathbb{C}^*)^n \cong (\mu^{-1}(p) \cap X^C) / (S^1)^n$$

and it is clear that this second description of R^C is a subset of

$$Z_F = \mu^{-1}(p) / (S^1)^n.$$

<u>Proposition</u>. The data TAD1, TAD2, TAD3, TAD4 defined here satisfy the axioms AX1, AX2, AX3.

<u>Sketch of proof</u>. For each $F \in \mathcal{F}$ there is a $(\mathbb{C}^*)^n$ - invariant set of semistable points,

$$X_F^{ss} = \cup \{X^C \mid P^F \subset C\}$$

consisting of the union of those pieces X^C such that the closure of the μ-image of X^C contains the stratum P^F. Although the topological quotient $X_F^{ss}/(\mathbb{C}^*)^n$ may not even be Hausdorff, there is a categorical quotient ([M]), i.e. an algebraic variety (which we still denote by $X_F^{ss}/(\mathbb{C}^*)^n$) with the property that whenever $f\colon X_F^{ss} \longrightarrow Y$ is an algebraic map which takes each torus orbit to a single point, then f factors through an algebraic map

$$g\colon X_F^{ss}/(\mathbb{C}^*)^n \longrightarrow Y.$$

By [K], the categorical quotient can be identified with

$$Z_F = \mu^{-1}(p)/(S^1)^n$$

for any $p \in P^F$. (Neither Mumford nor Kirwan emphasize the fact that X^{ss} and Z vary with F. In Mumford's language, [M] p.148, a choice of lift of the action of $G = (\mathbb{C}^*)^n$ to the invertible sheaf L must be made, while Kirwan chooses a basepoint p = {0}, or equivalently, an embedding [K] p. 102 of G into PGL(n+1). Kirwan's choice of basepoint p does not necessarily correspond to Mumford's choice of X^{ss}.)

If $G \leq F$ then $X_F^{ss} \subset X_G^{ss}$ so we obtain an algebraic map

$$\varsigma_{FG}\colon \quad X_F^{ss}/(\mathbb{C}^*)^n \longrightarrow X_G^{ss}/(\mathbb{C}^*)^n$$

This agrees with the map ς_{FG} as defined above because Mumford's categorical quotient is homeomorphic to the universal Hausdorff quotient.

§ 5: <u>Construction of the space B from TA Data</u>

Suppose we are given a collection of TA data, i.e.

<u>TAD1:</u> a finite collection $\mathscr{C}$ of convex polyhedra in $\mathbb{R}^n$ with union P = ∪ $\mathscr{C}$ which is piecified by the decomposition $\mathscr{F}$,

<u>TAD2</u>: a space-valued cofunctor $\mathscr{X}$ of algebraic varieites defined over $\mathscr{C}$,

<u>TAD3:</u> a space-valued cofunctor $\mathcal{Z}$ of algebraic varieties defined over $\mathcal{F}$,

<u>TAD4</u>: a system of inclusions $i_F^C : R^C \longrightarrow Z_F$ which piecify Z_F into pieces indexed by the partially ordered set

$$\mathcal{C}_F = \{ C \in \mathcal{C} \mid P^F \subset C^o \}$$

<u>Construction 1.</u> Define a topological space B to be the realization (over P) $B = R(\mathcal{F})$ of the cofunctor triple $(\mathcal{Z}, P, \mathcal{F})$.

<u>Construction 2</u>. Construct a piecification of B indexed by $\mathcal{C}$ as follows: for each $C \in \mathcal{C}$ define a partially ordered set

$$\mathcal{F}_C = \{F \in \mathcal{F} \mid P^F \subset C^o\}.$$

This set indexes the pieces in the piecification of C^o and admits a cofunctor of spaces, $\mathcal{Z}_C$ which associates to any $F \in \mathcal{F}_C$ the algebraic subvariety

$$Z_F^C \subset Z_F$$

<u>Definition</u>. The piece B^C in the piecification of B is the realization $R(\mathcal{Z}_C)$ of the cofunctor triple $(\mathcal{Z}_C, C^o, \mathcal{F}_C)$

<u>Remark</u>. Since each Z_F^C is identified with R^C, and since C^o is a cell, there is a canonical homeomorphism

$$B^C \cong C^o \times R^C$$

and so B^C is foliated by subsets $C^o \times \{\text{point}\}$.

<u>Construction 3</u>. For each $C \in \mathcal{C}$ we associate a subgroup

$$St_C = \exp\ (\iota\ \mathrm{Ann}(\mathrm{span}(C))) \subset (\mathbb{R}^+)^n$$

as follows: Span(C) is a subspace of $\mathbb{R}^n$ which [A1], [GS] has been identified with the dual of the Lie algebra of $(S^1)^n$. Therefore its annihilator lies in the Lie algebra of $(S^1)^n$, and multiplication by $\iota = \sqrt{(-1)}$ identifies this with the Lie algebra of $(\mathbb{R}^+)^n$. The exponential map

$$\exp : \mathrm{Lie}\ (\mathbb{R}^+)^n \longrightarrow (\mathbb{R}^+)^n$$

is an isomorphism. This is summarized in the diagram

$$\begin{array}{ccccccc} (\mathbb{R}^n)^* & \cong & \text{Lie } (S^1)^n & \xrightarrow[\iota]{\cong} & \text{Lie } (\mathbb{R}^+)^n & \xrightarrow[\exp]{\cong} & (\mathbb{R}^+)^n \\ \cup & & & & & & \cup \\ \text{Ann(span(C))} & & & & & & St_C \end{array}$$

__Theorem__. Suppose X is a projective algebraic variety with an action of the algebraic torus $(\mathbb{C}^*)^n$. Extract the corresponding TA Data, TAD1...TAD4. Let B be the space obtained from construction 1 as applied to this TAD, let B^C be the pieces obtained from construction 2 and let St_C be the subgroups obtained from construction 3. Then:

(1) there is a canonical homeomorphism $h : B \longrightarrow X / (S^1)^n$ such that, for each $C \in \mathscr{C}$ we have:

(2) h takes B^C homeomorphically to $X^C / (S^1)^n$

(3) h takes each leaf $C^o \times \{\text{point}\} \subset B^C$ homeomorphically to a single $(\mathbb{R}^+)^n$ orbit in $X/(S^1)^n$

(4) for each $x \in X^C/(S^1)^n$, the isotropy subgroup $\text{Stab}_{(\mathbb{R}^+)^n}(x)$ is precisely the subgroup St_C.

__Example__. See [GGMS] for a family of examples where the polyhedra C are explicitly described and are in one to one correspondance with matroids of rank k on n elements, and where X is the Grassmann manifold $G_{n-k}(\mathbb{C}^n)$ with the usual action of the torus $(\mathbb{C}^*)^n$. The nonempty pieces in the piecification of the Grassmannian correspond to matroids which are representable over the complex numbers.

__§6 An example__.

Suppose that $(\mathbb{C}^*)^2$ acts on $X = \mathbb{CP}^3$, complex projective three space with homogeneous coordinates $(z_1 : z_2 : z_3 : z_4)$, by the formula

$$(s, t) \bullet (z_1 : z_2 : z_3 : z_4) = (z_1 : sz_2 : tz_3 : stz_4)$$

The moment map is then given by

$$\mu([z_1:z_2:z_3:z_4]) = \frac{(|z_2|^2 + |z_4|^2 , |z_3|^2 + |z_4|^2)}{|z_1|^2 + |z_2|^2 + |z_3|^2 + |z_4|^2}$$

The image P is the square

$$\{(x,y) \in \mathbb{R}^2 |\ 0 \le x \le 1,\ 0 \le y \le 1\ \}$$

in $\mathbb{R}^2$. The various polygons C in $\mathscr{C}$ are listed below, along with the part of X which projects to their interior (where we make the convention that no coordinate typed z_i is zero):

TYPE I:

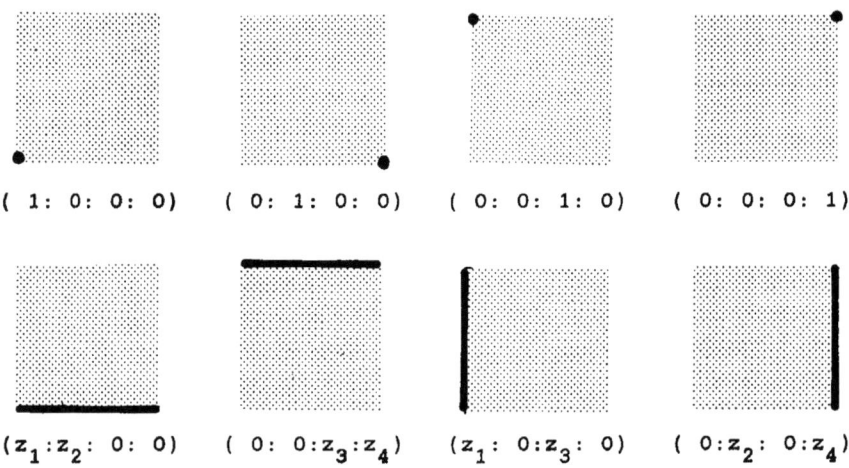

(1: 0: 0: 0)

(0: 1: 0: 0)

(0: 0: 1: 0)

(0: 0: 0: 1)

$(z_1:z_2:\ 0:\ 0)$

$(\ 0:\ 0:z_3:z_4)$

$(z_1:\ 0:z_3:\ 0)$

$(\ 0:z_2:\ 0:z_4)$

TYPE II

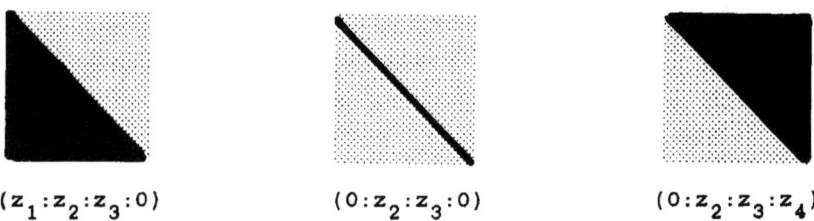

$(z_1:z_2:z_3:0)$

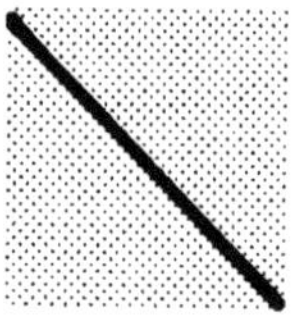

$(0:z_2:z_3:0)$

$(0:z_2:z_3:z_4)$

TYPE III

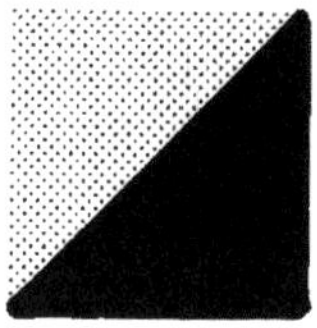

$(z_1:z_2:0:z_4)$

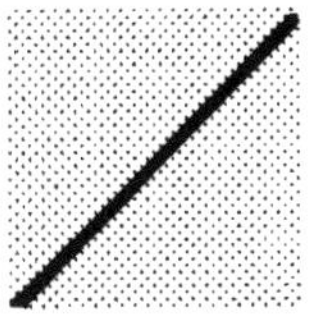

$(0:z_2:0:z_4)$

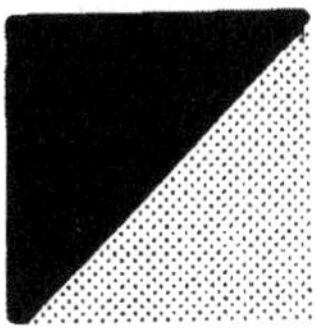

$(z_1:0:z_3:z_4)$

TYPE IV

$(z_1:\ z_2:\ z_3:\ z_4)$

The R^C is a point for C of TYPES I, II, or III, and is $\mathbb{C}^*$ for TYPE IV. The piecification $\mathscr{C}$ of P is like this:

Over each piece F on the edge of the square, Z_F is a point. For each piece F which is contained in the interior of the square, Z_F is a complex projective line, which we may identify with the standard complex projective line (with homogeneous coordinates $[y_1:y_2]$) and we may take each of the maps ζ_{FG} to be the identity. The inclusions $i_F^C\colon R^C \longrightarrow Z_F$ have as their image $(1:0)$ if C is of TYPE II, $(0:1)$ if C is of TYPE III, and the rest of $\mathbb{C}P^1$ for the C of TYPE IV.

It is easy to see that the realization of $\mathscr{F}$ is the four sphere S^4, so it follows from the theorem that the orbit space B is S^4.

<u>§7 Sketch of the proof</u>. First we consider a lemma in pure topology. Suppose that we have:

1. a compact Hausdorff space B mapping to a piecewise linear subset of $\mathbb{R}^n$, $\beta: B \longrightarrow P$

2. a piecification of P (indexed by a partially ordered set $\mathcal{F}$) into finitely many piecewise linear subsets P^F, and which satisfies the axiom of the frontier: the closure $\overline{P^F}$ of any piece is a union of pieces,

3. a disjoint decomposition of B into (possibly uncountably many) topological ("open") balls of various dimensions,

such that:

a. the map β takes each open ball homeomorphically onto a union of pieces F^o of P

b. the closure of each open ball is a "closed" ball which β takes homeomorphically to a union of pieces of P.

For each piece P^F of P, choose a point $p \in P^F$ and let Z_F be the fiber $\beta^{-1}(p)$ over p. Whenever G is a face of F let ζ_{FG}: $Z_F \longrightarrow Z_G$ be defined by the condition that $\zeta_{FG}(z)$ lies in the closure of the open ball through z. This forms space valued cofunctor on $\mathcal{F}$ which we call $\mathcal{Z}$.

<u>Lemma</u> Under these hypotheses, the space B is canonically homeomorphic to the realization $R(\mathcal{Z})$ over P of the cofunctor triple $(\mathcal{Z}, P, \mathcal{F})$.

For example, in the following picture, B is a subset of the plane, P is a subset of the line, β is vertical projection, and the open balls in B and pieces of P are sketched in.

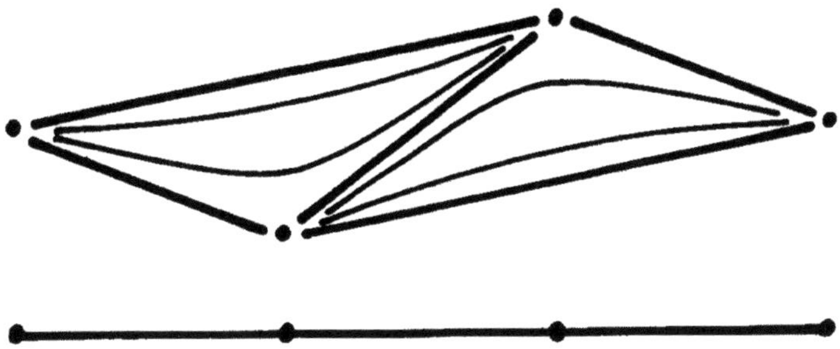

In this example, the Z_F and the maps ζ_{FG} are as follows:

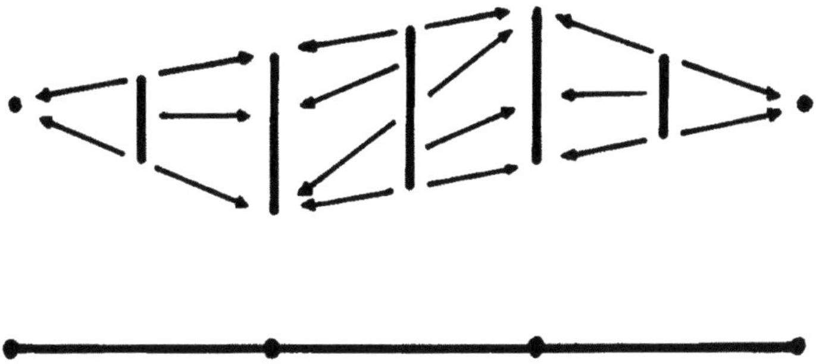

The proof of the lemma is straightforward: fix a stratum P^F of P and a point $p \in P^F$. By (3a) and (3c) there exists a unique homeomorphism

$$h_F\colon P^F \times \beta^{-1}(p) \longrightarrow \beta^{-1}(P^F) \subset B$$

which commutes with the projection to P^F and such that each $h_F(P^F \times \{\text{point}\})$ is a leaf of the foliation (i.e. lies in a single ball). Furthermore, by (3b), h_F extends to a continuous map

$$\overline{h_F}\colon \overline{P^F} \times \beta^{-1}(p) \longrightarrow \beta^{-1}(\overline{P^F}) \subset B$$

and it is easy to check that this is compatible with the relations defining the realization of $\mathcal{Z}$.

To apply this topological lemma to the theorem of §5, we take the decomposition of B into open balls to be the decomposition by $(\mathbb{R}^+)^n$ orbits. We claim that these satisfy the conditions of the topological lemma. This follows from the following facts about the composition

$$X \xrightarrow{\alpha} B \xrightarrow{\beta} \mathbb{R}^n$$

1. A single $(\mathbb{C}^*)^n$ orbit O in X projects to a topological open disk which is a single $(\mathbb{R}^+)^n$ orbit O in B, which projects homeomorphically to the interior C^o of a convex polyhedron C in P.

2. The closure of O in X consists of finitely many $(\mathbb{C}^*)^n$ orbits. It projects to the closure of $\bar{O}$ in B, which is a topological closed disk consisting of finitely many $(\mathbb{R}^+)^n$ orbits, each of which projects to the interior of a face of C in P.

To prove these facts, we observe that the moment map for the closure of O is the restriction to the closure of O of the moment map for X. However the closure of O is a toric variety, and these facts are standard for toric varieties.

§ 8. Reconstructing X.

In §5, we constructed the topology of $B = X/(S^1)^n$ and the stabilizer subgroups $\mathrm{Stab}_{(\mathbb{R}^+)^n}(b)$ from TAD. To what extent can the topology of X itself be reconstructed from this information?

The first remark is that the stabilizer subgroup $\mathrm{Stab}_{(S^1)^n}(x)$ of any point $x \in X$ projecting to b is determined by $\mathrm{Stab}_{(\mathbb{R}^+)^n}(b)$. This is because $\mathrm{Stab}_{(\mathbb{C}^*)^n}(x)$ is determined by $\mathrm{Stab}_{(\mathbb{R}^+)^n}(x) = \mathrm{Stab}_{(\mathbb{R}^+)^n}(b)$ since the $(\mathbb{C}^*)^n$ action is algebraic. In terms of TAD, the group $(S^1)^n$ identifies with $(\mathbb{R}^n)^*/(\mathbb{Z}^n)^*$, the space of linear functionals on $\mathbb{R}^n$ modulo those that take integral values on integral points. If x projects to $b \in B^C$, then $\mathrm{Stab}_{(S^1)^n}(x)$ is the subtorus

$$\mathrm{Stab}_{(S^1)^n}(x) = \mathrm{Ann}(\mathrm{span}(C))/\mathrm{Ann}(\mathrm{span}(C)) \quad (\mathbb{Z}^n)^*.$$

We call X a "piecified torus bundle" over B. The preimage of each piece B^C in B fibers over B^C with fiber the quotient torus $(S^1)^n/\mathrm{Stab}_{(S^1)^n}(x) = \mathrm{span}(C)^* / \Lambda_C$, where a covector in $\mathrm{span}(C)^*$ is in Λ iff it has some extension to $\mathbb{R}^n$ which takes integral values on $\mathbb{Z}^n$. The cohomology of X can be computed from the Leray spectral sequence for the projection from X to B. The above remarks imply that the E_2 term of this Leray spectral sequence can be computed from TAD alone.

The question of topologically reconstructing X from TAD reduces to the purely topological question of classifying "piecified torus bundles". For example, if the torus is a circle and all of the stabilizer subgroups are the identity, then X is a principal S^1 bundle over B, and its topology is determined by the first Chern class. It would be interesting to have a theory of first Chern classes classifying such "bundles" in general.

In case that the map from X to B admits a section, there is no twisting in the "piecified torus bundle" and the topology of X is determined by the TAD alone. This is the case, for example, when X is a toric variety.

Bibliography

[A1] M. F. Atiyah, Convexity and commuting Hamiltonians, Bull. London Math. Soc. 14, 1-15 (1982)

[A2] M. F. Atiyah, Angular momentum, convex polyhedra, and algebraic geometry. Proc. Edinburgh Math. Soc. 26 (1983), 121-138

[BBS] A. Bialynicki-Birula and J. Swiecicka, Complete quotients by algebraic torus actions. in Group Actions and Vectorfields (J.B. Carrell, ed). Springer lecture notes in mathematics # 956, Springer-Verlag, New York (1982)

[BBSo] A. Bialynicki-Birula and A. J. Sommese, Quotients by $\mathbb{C}^*$ and SL(2,$\mathbb{C}$) actions. Trans. Amer. Math. Soc., 1982.

[D] V. I. Danilov, The geometry of toric varieites, Uspekhi Mat. Nauk. 33 (1978), 85-134

[GS] V. Guillemin and S. Sternberg, Convexity properties of the moment mapping, Inv. Math. 67 (1982), 491-513

[GGMS] I.M. Gelfand, M. Goresky, R. MacPherson, and V. Serganova, Combinatorial geometries, convex polyhedra, and Schubert cells. to appear in Advances in Mathematics.

[K] F. C. Kirwan, Cohomology of Quotients in Symplectic and Algebraic Geometry. Mathematical Notes # 31, Princeton University Press, Princeton N.J. (1984)

[MW] J. Marsden and A. Weinstein, Reduction of symplectic manifolds with symmetry, Reports on Math. Phys. 5 (1974), 121-130

[M] D. Mumford and J. Fogarty, Geometric Invariant Theory, (second edition) Springer Verlag, New York (1982)

RESTRICTED LIE ALGEBRA COHOMOLOGY

J.C. Jantzen
Mathematisches Seminar
Universität Hamburg
Bundesstr. 55
D - 2000 Hamburg 13

The cohomology of restricted Lie algebras was first defined by Hochschild in 1954, cf.[11]. It was however only recently that one could get more precise information about these cohomology groups in non-trivial cases. The most fascinating result is still the theorem (proved by Friedlander and Parshall) that for large p the cohomology ring of the Lie algebra of a reductive algebraic group is the ring of regular functions on the nilpotent cone in this Lie algebra. It is the purpose of this article to give a survey of recent developments in this theory.

Throughout this paper let k be an algebraically closed field with $\mathrm{char}(k) = p \neq 0$.

1. General Theory

1.1 The Lie algebra $\mathrm{Lie}(G)$ of an algebraic group G over k has an additional structure: the p-th power map. We can regard $\mathrm{Lie}(G)$ as the set of all (k-linear) derivations of $k[G]$ (the algebra of all regular functions on G) which commute with the left regular representation of G on $k[G]$. For any such derivation x also x^p is a derivation (because of $\mathrm{char}(k) = p$) and commutes again with G, hence belongs to $\mathrm{Lie}(G)$. We get thus a map $x \mapsto x^p$ from $\mathrm{Lie}(G)$ to itself which is usually denoted by $x \mapsto x^{[p]}$ in order to distinguish it from the p-th power in the universal enveloping algebra of $\mathrm{Lie}(G)$.

1.2 The situation as in 1.1 has lead to the general definition of a p-Lie algebra (also called restricted Lie algebra). A p-Lie algebra over k is a Lie algebra $\underline{g}$ over k together with a map $x \mapsto x^{[p]}$ from $\underline{g}$ to itself which has to satisfy some conditions which one can look up e.g. in [12],V.7 or [4],II,§ 7,3.3. For us it will be enough to know the following facts. Any Lie(G) as in 1.1 is a p-Lie algebra. If one takes $G = GL(V)$ for some vector space V over k ($\dim V < \infty$) then one gets $\mathrm{Lie}(G) = \underline{g\ell}(V) = \mathrm{End}_k(V)$ and $x^{[p]}$ is (for any $x \in \underline{g\ell}(V)$) just the p-th power of x as an endomorphism of V. Any Lie subalgebra of a p-Lie algebra which is closed under $x \mapsto x^{[p]}$ is again a p-Lie algebra. Any finite dimensional p-Lie algebra can be embedded into a suitable $\underline{g\ell}(V)$ with $\dim(V) < \infty$ as a p-Lie subalgebra.

1.3 The simplest examples have dimension one. One can make $\underline{g} = k$ into a p-Lie algebra $\underline{g}_1$ by taking $x^{[p]} = x^p$ for all $x \in k$. This is just the structure we get from 1.1 when regarding k as the Lie algebra of the multiplicative group.

On the other hand we can make $\underline{g} = k$ into a p-Lie algebra $\underline{g}_0$ by taking $x^{[p]} = 0$ for all $x \in k$. This is just the p-Lie algebra of the additive group.

1.4 From now on let $\underline{g}$ be a p-Lie algebra over k.

A representation $\rho : \underline{g} \to \underline{g\ell}(V)$ of $\underline{g}$ is called a representation as a p-Lie algebra if $\rho(x^{[p]}) = \rho(x)^p$ for all $x \in \underline{g}$, i.e. if ρ is a homomorphism of p-Lie algebras (taking the standard p-th power map on $\underline{g\ell}(V)$ as in 1.2).

A $\underline{g}$-module is a vector space with a fixed representation of $\underline{g}$ on it. We call it a <u>restricted</u> $\underline{g}$-module if we are dealing with a representation as a p-Lie algebra.

In case $\underline{g} = \mathrm{Lie}(G)$ for some algebraic group G any representation $G \to GL(V)$ of G as an algebraic group (i.e. any "G-module") leads by differentiating to a representation $\underline{g} \to \underline{g\ell}(V)$ of $\underline{g}$ as a p-Lie algebra (i.e. to a restricted $\underline{g}$-module).

1.5 Let $U(\underline{g})$ be the universal enveloping algebra of $\underline{g}$. Then the category of all $\underline{g}$-modules is identified in a natural way with the category of all $U(\underline{g})$-modules. This construction identifies the full subcategory of all restricted $\underline{g}$-modules with that of all $U(\underline{g})$-modules annihilated by all $x^p - x^{[p]}$ with $x \in \underline{g}$. Set $U^{[p]}(\underline{g}) = U(\underline{g})/I$ where I is the ideal generated by all these $x^p - x^{[p]}$. This algebra is called the <u>restricted enveloping algebra</u> of $\underline{g}$. By the remarks above, the category of all restricted $\underline{g}$-modules is identified with that of all $U^{[p]}(\underline{g})$-modules.

Let us look at this construction in the two simple cases from 1.3. In both cases $U(\underline{g})$ can be identified with the polynomial ring $k[X]$ in one indeterminate X. (The embedding $\underline{g} = k \to U(\underline{g})$ maps 1 to X.)

One sees easily $U^{[p]}(\underline{g}_0) = k[X]/(X^p)$ and $U^{[p]}(\underline{g}_1) = k[X]/(X^p-X) \simeq \prod_{a \in \mathbb{F}_p} k[X]/(X-a)$. So $U^{[p]}(\underline{g}_1)$ is isomorphic to a direct product of p copies of k, it is a semi-simple algebra admitting p different simple modules (all of dimension one) and each restricted $\underline{g}_1$-module is semi-simple.

On the other side, there is (up to isomorphism) only one simple restricted $\underline{g}_0$-module (the trivial one). For each i $(1 \le i \le p)$ there is an indecomposable module of dimension i, namely $k[X]/(X^i)$. Any restricted $\underline{g}_0$-module is isomorphic to a direct sum of such $k[X]/(X^i)$. It is a projective (or injective) restricted $\underline{g}_0$-module if and only if all indecomposable summands have dimension p, i.e. are isomorphic to $U^{[p]}(\underline{g}_0)$.

1.6 Both categories, that of all $\underline{g}$-modules and that of all restricted $\underline{g}$-Modules, have been identified with categories of all modules over some ring. So we have injective and projective resolutions in these categories and can use them to compute derived functors.

Let us look at the fixed point functor

$$M \longmapsto M^{\underline{g}} = \{m \in M \mid xm = 0 \quad \text{for all} \quad x \in \underline{g}\}.$$

It is left exact. When regarding it as a functor on all $\underline{g}$-modules (resp. on all restricted $\underline{g}$-modules) we get derived functors which we shall denote by $H^i(\underline{g},?)$ resp. by $H^i_*(\underline{g},?)$. The $H^i(\underline{g},M)$ are called the Lie algebra cohomology groups of M, and the $H^i_*(\underline{g},M)$ for restricted M are the restricted Lie algebra cohomology groups of M. The notation H^i_* has been taken from [11] where these groups were introduced for the first time.

One can interpret the $H^i(\underline{g},M)$ also as $\mathrm{Ext}^i_{\underline{g}}(k,M)$ and similarly the $H^i_*(\underline{g},M)$ as the extension groups of the trivial $\underline{g}$-module k and of M in the category of all restricted $\underline{g}$-modules. We can therefore compute the $H^i(\underline{g},M)$ not only using an injective resolution of M, but also using a projective resolution of k.

The cup product makes $H^\bullet_*(\underline{g},k) = \bigoplus_{i \ge 0} H^i_*(\underline{g},k)$ into a graded associative algebra and any $H^\bullet_*(g,M)$ into an $H^\bullet_*(\underline{g},k)$-module. The group of all automorphisms of $\underline{g}$ as a p-Lie algebra acts on $H^\bullet_*(\underline{g},k)$ through graded algebra automorphisms. If $\underline{g} = \mathrm{Lie}(G)$ for some algebraic group G and if M is a G-module, then G acts on each $H^i_*(\underline{g},M)$. The action of G on $H^\bullet_*(\underline{g},M)$ is compatible with that on $H^\bullet_*(\underline{g},k)$ and with the action of $H^\bullet_*(\underline{g},k)$ on $H^\bullet_*(g,M)$. (There are similar results for the ordinary Lie algebra cohomology groups which we do not mention.)

1.7 The functors $H^i(\underline{g},?)$ and $H^i_*(\underline{g},?)$ have quite different properties. One can observe this already in the simple examples from 1.3. Let us use the notation from 1.5.

Obviously $0 \to k[X] \to k[X] \to k \to 0$ is a free resolution of the trivial $U(\underline{g}_0) = U(\underline{g}_1)$-module k. (The map $k[X] \to k[X]$ is the multiplication by X.) One gets easily

(1) $H^i(\underline{g}_0,k) = H^i(\underline{g}_1,k) \simeq \begin{cases} k & \text{for } i = 0,1, \\ 0 & \text{for } i > 1. \end{cases}$

Any restricted $\underline{g}_1$-module is semi-simple. Therefore the fixed point functor is exact on the category of all restricted $\underline{g}_1$-modules. This implies:

(2) $H^i_*(\underline{g}_1,k) \simeq \begin{cases} k & \text{for } i = 0, \\ 0 & \text{for } i > 0. \end{cases}$

A minimal projective resolution of k as a restricted $\underline{g}_0$-module has the form

$$\ldots \to U^{[p]}(\underline{g}_0) \to U^{[p]}(\underline{g}_0) \to U^{[p]}(\underline{g}_0) \to k \to 0,$$

where the maps from $U^{[p]}(\underline{g}_0) = k[X]/(X^p)$ to itself are alternatingly induced by the multiplication by X and by X^{p-1}. Then an easy computation yields:

(3) $H^i_*(\underline{g}_0,k) \simeq k$ for all $i \geq 0$.

Let us generalize the last result as follows: Consider a finite dimensional vector space V over k as a commutative Lie algebra and make it into a p-Lie algebra with $x^{[p]} = 0$ for all $x \in V$. So V is just a direct product of $\dim(V)$ copies of $\underline{g}_0$. One can get the $H^i_*(V,k)$ from (3) using a Künneth formula. When formulating the result below we want to take into account the algebra structure on $H^\bullet_*(\underline{g},k)$ and the operation of $GL(V)$ on this algebra. (Of course, $GL(V)$ is the group of all automorphisms of the p-Lie algebra V.) One gets (cf.[1],1.6. or [14],I.4.27):

(4) $H^\bullet_*(V,k) \simeq \begin{cases} S(V^*) & \text{for } p = 2, \\ \Lambda(V^*) \otimes S'(V^*)^{(1)} & \text{for } p > 2. \end{cases}$

Here we use the usual grading on the symmetric resp. exterior algebra when writing $S(V^*)$ resp. $\Lambda(V^*)$. The notation $S'(V^*)$ means that we take $S(V^*)$ but with a grading such that each $S^i(V^*)$ appears in degree $2i$. (We shall use this convention also for other vector spaces but V^*.) Furthermore for any vector space M over k and any $r \in \mathbb{Z}$ we write $M^{(r)}$ for the vector space over k which coincides with M as a group and where any $a \in k$ acts as $a^{p^{-r}}$ does on M. (Note that we do not change the action of some $g \in GL(V)$ on $S'(V^*)$ when taking $S'(V^*)^{(1)}$.)

1.8 Let us assume from now on that $\dim(\underline{g}) < \infty$.

The map $x \mapsto 1 \otimes x + x \otimes 1$ induces an algebra homomorphism $U^{[p]}(\underline{g}) \to U^{[p]}(\underline{g}) \otimes U^{[p]}(\underline{g})$. It makes $U^{[p]}(\underline{g})$ into a (co-commutative) Hopf algebra and by dualising $U^{[p]}(\underline{g})^*$ into a (commutative) Hopf algebra. Therefore one can regard $U^{[p]}(\underline{g})^*)$ as the ring of regular functions on some infinitesimal group scheme $\underline{\underline{G}}$ over k, cf.[4],II,§ 7,n°3. Then the category of all $\underline{\underline{G}}$-modules is the same as that of all restricted $\underline{g}$-modules.

Consider the special case $\underline{g} = \mathrm{Lie}(G)$ for some algebraic group G defined over the prime field. Let F be the corresponding Frobenius endomorphism on G. So $F^* : k[G] \to k[G]$ maps any $f \in \mathbb{F}_p[G]$ to f^p. Then $\underline{\underline{G}}$ is the kernel of F (taken in the category of group schemes), cf.[14],I.9.7. Let me mention that F induces an isomorphism $G/\ker(F) \xrightarrow{\sim} G$, cf.[14],I.9.5.

In this situation ($\underline{g} = \mathrm{Lie}(G)$ with F as above) we set $M^{[r]}$ for any G-module M and any $r \in \mathbb{N}$ equal to the G-module which coincides with M as a vector space and where any $g \in G$ acts as $F^r(g)$ does on M. Obviously $\ker(F)$, hence also $\underline{g}$ acts trivially on each $M^{[r]}$ with $r > 0$. Conversely, because of $G/\ker(F) \simeq G$: any G-module M on which $\underline{g}$ acts trivially has the form $M = M_1^{[1]}$ for some other G-module M_1. Then M_1 is uniquely determined by M and denoted by $M^{[-1]}$.

If a G-module M is defined over $\mathbb{F}_p$, then $M^{[r]}$ is isomorphic to $M^{(r)}$ for all $r \in \mathbb{N}$. Indeed, one can choose a basis over $\mathbb{F}_p$. Then there are $f_{ij} \in \mathbb{F}_p[G]$ such that any $g \in G$ has matrix $(f_{ij}(g))$ with respect to this basis. It is also a basis of $M^{(r)}$ and of $M^{[r]}$ and the matrix of g is $(f_{ij}^{p^r}(g))$ in the case of $M^{(r)}$ and $(f_{ij}(F^r(g))$ in the case of $M^{[r]}$. As $F^*(f_{ij}) = f_{ij}^p$ these matrices coincide. One has similarly $M^{(-1)} \simeq M^{[-1]}$ in case the latter module is defined.

1.9 The category of restricted $\underline{g}$-modules is equal to that of all $\underline{\underline{G}}$-modules and the functor $M \mapsto M^{\underline{g}}$ coincides with taking $\underline{\underline{G}}$-fixed points. Therefore the $H_*^i(\underline{g},M)$ are equal to the Hochschild cohomology groups $H^i(\underline{\underline{G}},M)$. These can be computed the "Hochschild complex", cf.[4],II, § 3 or [14],I.4.14 - 4.16. One can construct a natural filtration of this complex. The associated graded complex turns out to be the Hochschild complex which we get when we regard the vector space $\underline{g}$ as a commutative p-Lie algebra with trivial p-th power map and when we regard M as a trivial module over this p-Lie algebra. So the cohomology of the graded complex is given by 1.7(4) with V replaced by $\underline{g}$. It is not difficult to compute the grading. We get thus the E_1-term of a spectral sequence converging to $H^\bullet_*(\underline{g},M)$. There are many terms equal to zero which enables you to simplify the results.

In case $p = 2$ one gets finally $H^\bullet_*(\underline{g},M)$ as the cohomology of a complex of the form

$$(1) \quad 0 \to M \to M \otimes \underline{g}^* \to M \otimes S^2\underline{g}^* \to M \otimes S^3\underline{g}^* \to \dots$$

The differentials can be written down explicitly.

In case $p \neq 2$ one re-indexes the spectral sequence making old $E_{(p-2)r+1}$ terms into E_r terms for the new spectral sequence. One gets then E_0-terms equal to

$$(2) \quad E_0^{i,j}(M) = S^i(\underline{g}^*)^{(1)} \otimes \Lambda^{j-i}(\underline{g}^*) \otimes M.$$

One has enough information about the differentials to be able to compute the E_1-terms:

$$(3) \quad E_1^{i,j}(M) \simeq S^i(\underline{g}^*)^{(1)} \otimes H^{j-i}(\underline{g},M).$$

We get obviously: If $E_r^{i,j}(M) \neq 0$, then $i \le j \le i + \dim(\underline{g})$. If $d_r : E_r^{i,j} \to E_r^{i+r,j+1-r}(M)$ is non-zero, then also $i+r \le j+1-r \le i+r+\dim(\underline{g})$, hence $2r-1 \le j-i \le \dim(\underline{g})$. So for $2r \ge \dim(\underline{g})$ all differentials are zero. This implies:

$$(4) \quad E_\infty(M) = E_r(M) \qquad \text{for all } r \ge \dim(\underline{g})/2.$$

Here we use the notation $E_r(M) = \bigoplus_{i,j} E_r^{i,j}(M)$.

If $\underline{g} = \mathrm{Lie}(G)$ for some algebraic group G and if M is a G-module, then each $E_r^{i,j}(M)$ is a G-module, all differentials are homomorphisms of G-modules, and the filtration of $H^\bullet_*(\underline{g},M)$ with factors of the type $E_\infty^{i,j}(M)$ is compatible with the action of G.

The results mentioned above were proved in [5],5.1/2, [1],1.8, [7],1.1, [14],I,9.10 - 9.19.

1.11 Assume for the moment $p \neq 2$. The cup product yields for all r,i,j,ℓ,m bilinear maps $E_r^{i,j}(k) \times E_r^{\ell,m}(M) \to E_r^{i+\ell,j+m}(M)$, which can be put together to bilinear maps $E_r(k) \times E_r(M) \to E_r(M)$. In the case $M = k$ we get thus a structure as an algebra on $E_r(k)$, in general we get on $E_r(M)$ a structure as an $E_r(k)$-module. The differentials d_r have then a derivation property. One gets the structures on $E_\infty(k)$ and $E_\infty(M)$ by taking the graded structure associated to the filtration of $H^\bullet_*(\underline{g},k)$ resp. $H^\bullet_*(\underline{g},M)$ which has the E_∞-terms as factors.

Any d_r with $r \ge 1$ vanishes on $E_r^{i,i}(k)$ as $E_r^{i+r,i+1-r}(k) = 0$. So $E_{r+1}^{i,i}(k)$ is a homomorphic image of $E_r^{i,i}(k)$. By induction

$\bigoplus_{i\geq 0} E_r^{i,i}(k)$ is a homomorphic image of $\bigoplus_{i\geq 0} E_1^{i,i}(k) = S(\underline{g}^*)^{(1)}$. On the other hand $\bigoplus_{i\geq 0} E_\infty^{i,i}(k)$ is a subalgebra of $H_*^\bullet(\underline{g},k)$, in fact of the commutative subalgebra $H_*^{ev}(\underline{g},k) = \bigoplus_{i\geq 0} H_*^{2i}(\underline{g},k)$. So we have constructed a homomorphism $\kappa : S(\underline{g}^*)^{(1)} \to H_*^{ev}(\underline{g},k)$ of commutative algebras.

For any restricted $\underline{g}$-module M the structure on $E_r(M)$ as an $E_r(k)$-module yields by restriction to $\bigoplus_{i\geq 0} E_r^{i,i}(k)$ a structure as an $S(\underline{g}^*)^{(1)}$-module. The differentials are then homomorphisms of $S(\underline{g}^*)^{(1)}$-modules and $E_{r+1}(M)$ is a subquotient of $E_r(M)$ as an $S(\underline{g})^*$-module. If $\dim(M) < \infty$, then each $H^i(\underline{g},M)$ is finite dimensional, hence $E_1(M)$ is finitely generated over $S(\underline{g}^*)^{(1)}$ by 1.9(3). Now induction and 1.9(4) imply that $E_\infty(M)$ is finitely generated over $S(\underline{g}^*)^{(1)}$.

We have via κ also on $H_*^\bullet(\underline{g},M)$ a structure as an $S(\underline{g}^*)^{(1)}$-module. The structure on $E_\infty(M)$ is just the associated graded one. If the associated graded module is finitely generated, so is the original one. So we have shown at least for $p \neq 2$ what was first proved in [7],1.4:

<u>Proposition</u>: <u>If M is a finite dimensional $\underline{g}$-module, then $H_*^\bullet(\underline{g},M)$ is a finitely generated module over $S(\underline{g}^*)^{(1)}$</u>.

The special case $M = k$ yields immediately:

<u>Corollary</u>: <u>$H_*^{ev}(\underline{g},k)$ is a finitely generated commutative k-algebra</u>.

In the case $p = 2$ one gets $\kappa : S(\underline{g}^*)^{(1)} \to H_*^{ev}(\underline{g},k)$ by mapping any $f \in S^i(\underline{g}^*)^{(1)}$ to $f^2 \in S^{2i}(\underline{g}^*)$. This is the 2i-th term in the complex 1.9(1) for $M = k$. The derivation property of the differential implies that f^2 is a cocycle, hence defines a class in $H_*^{2i}(\underline{g},k)$. As $S(\underline{g}^*)\otimes M$ is a finitely generated module over $\{f^2 \mid f \in S(\underline{g}^*)\} \simeq S(\underline{g}^*)^{(1)}$, the proposition and its corollary follow easily also for $p = 2$.

The homomorphism $\kappa : S(\underline{g}^*)^{(1)} \to H_*^{ev}(\underline{g},k)$ is (for any p) determined by the linear map $\underline{g}^{*(1)} \to H_*^2(\underline{g},k)$. Using the interpretation of $H_*^2(\underline{g},k)$ as set of equivalence classes of central extensions $0 \to \underline{g}_0 \to \tilde{\underline{g}} \to \underline{g} \to 0$, cf.[4],III,§ 6,8.5, one can give a direct construction of this map: One associates to $\varphi \in \underline{g}^*$ the direct product $\tilde{\underline{g}} = \underline{g} \times k$ with p-th power map $(x,a)^{[p]} = (x^{[p]},\varphi(x)^p)$ for all $x \in \underline{g}$ and $a \in k = \underline{g}_0$. One can show that this gives the same map as before, cf.[9],1.1.

2. <u>The Reductive Case</u>

<u>2.1</u> In this part let G be a connected and reductive algebraic group over k, let $T \subset B \subset G$ be a maximal torus and a Borel subgroup of G.

Denote the unipotent radical of B by U. Let us assume $G \neq T$. Set $\underline{g} = \mathrm{Lie}(G)$, $\underline{b} = \mathrm{Lie}(B)$, $\underline{t} = \mathrm{Lie}(T)$, $\underline{u} = \mathrm{Lie}(U)$. We want to assume that all these groups are defined and split over $\mathbb{F}_p$. Let F be the corresponding Frobenius endomorphism. Denote by $X(T)$ the group of all characters of T and by h the maximum of the Coxeter numbers of all components of G.

2.2 As $\underline{u}$ is an ideal in $\underline{b}$ with $\underline{b} = \underline{u} \oplus \underline{t}$ one has $M^{\underline{b}} = (M^{\underline{u}})^{\underline{t}}$ for any $\underline{b}$-module M. The algebraic group T is a direct product of multiplicative groups, so $\underline{t}$ is a direct product of p-Lie algebras isomorphic to $\underline{g}_1$ (as in 1.3). Therefore (cf.1.5) any restricted $\underline{t}$-module is semi-simple and $M \mapsto M^{\underline{t}}$ is exact on restricted $\underline{t}$-modules. This implies easily (for all restricted $\underline{b}$-modules M and all $i \in \mathbb{N}$):

(1) $\qquad H^i_*(\underline{b},M) \simeq H^i_*(\underline{u},M)^{\underline{t}}$

We can apply the construction of a spectral sequence as in 1.10(2) to the Lie algebra $\underline{u}$ and a restricted $\underline{b}$-module M. Then all terms will be restricted $\underline{t}$-modules and all differentials will commute with the operation of $\underline{t}$. Therefore taking $\underline{t}$-fixed points yields a spectral sequence converging to $H^\bullet_*(\underline{b},M)$ because of (1):

(2) $\qquad E_0^{i,j} = (S^i(\underline{u}^*)^{(1)} \otimes \Lambda^{j-i}(\underline{u}^*) \otimes M)^{\underline{t}} \Rightarrow H^{i+j}_*(\underline{b},M)$.

2.3 Any T-module M is the direct sum of its weight spaces, and $M^{\underline{t}}$ is the direct sum of all weight spaces corresponding to a weight in $pX(T)$. Looking at the weights of $\Lambda(\underline{u}^*)$ an elementary argument (cf.[1], 2.2 or [14],II,12.10) yields:

(1) $\qquad$ <u>If</u> $p > h$, <u>then</u> $\Lambda^i(\underline{u}^*)^{\underline{t}} = 0$ <u>for all</u> $i > 0$.

On the other hand, as all groups are defined over $\mathbb{F}_p$, so is the adjoint representation of T on $\underline{u}^*$ and on $S(\underline{u}^*)$. So $S^i(\underline{u}^*)^{(1)} \simeq S^i(\underline{u}^*)^{[1]}$ and $\underline{t}$ acts trivially on this module, cf. 1.8. Therefore the spectral sequence in 2.2(2) degenerates for $p > h$ and $M = k$. We get, using the same convention as in 1.7(4):

(2) $\qquad$ <u>If</u> $p > h$, <u>then</u> $H^\bullet_*(\underline{b},k) \simeq S^\bullet(\underline{u}^*)^{(1)}$.

This was first proved in [1],2.3. There are in [1],2.9(2) also results on $H^\bullet_*(\underline{b},M)$ for other restricted $\underline{b}$-modules M with $\dim(M) = 1$.

2.4 For any B-module M the induced G-module $\mathrm{ind}_B^G M$ is defined as

(1) $\qquad \mathrm{ind}_B^G M = \{f : G \to M \text{ regular} \mid f(gb) = b^{-1}f(g) \text{ for all } g \in G, b \in B\}$

with G acting by left translation. (Here "regular" means that there is a finite dimensional subspace $M' \subset M$ depending on f such that

$f(G) \subset M'$ and such that f is regular in the usual sense as a map $G \to M'$.) Then $\varepsilon_M : \mathrm{ind}_B^G M \to M$, $f \mapsto f(1)$ is a homomorphism of B-modules and $\varphi \mapsto \varepsilon_M \circ \varphi$ induces an isomorphism $\mathrm{Hom}_G(V, \mathrm{ind}_B^G M) \xrightarrow{\sim} \mathrm{Hom}_B(V,M)$ for any G-module V. The induction functor ind_B^G is left exact and right adjoint to the (exact) forgetful functor from G-modules to B-modules. It has right derived functors which we denote by $R^n \mathrm{ind}_B^G$. (All this is true in greater generality, cf.e.g. [14],chapter I 3.)

One can associate to any B-module M a locally free sheaf $\mathcal{L}(M)$ of $\mathcal{O}_{G/B}$-modules on G/B. It has the property that $R^n \mathrm{ind}_B^G M \simeq H^n(G/B, \mathcal{L}(M))$ for all $n \in \mathbb{N}$, cf.e.g.[14],I,5.12. In the case of the trivial B-module k one gets just the structure sheaf $\mathcal{O}_{G/B}$ as $\mathcal{L}(k)$. So Kempf's vanishing theorem implies (cf.e.g.[14],II 4.6):

$$(2) \qquad R^n \mathrm{ind}_B^G k \simeq \begin{cases} k & \text{for } n = 0, \\ 0 & \text{for } n > 0. \end{cases}$$

More generally one has for any G-module M:

$$(3) \qquad R^n \mathrm{ind}_B^G M \simeq \begin{cases} M & \text{for } n = 0, \\ 0 & \text{for } n > 0, \end{cases}$$

because of the "generalized tensor identity", cf.e.g.[14],I.4.8.

2.5 For all G-modules V, V' any homomorphism $\varphi : V'^{[1]} \to V$ of G-modules takes values in $V^{\underline{g}}$ as $\underline{g}$ acts trivially on $V'^{[1]}$. So one has natural isomorphisms, cf.1.8:

$$\mathrm{Hom}_G(V'^{[1]}, V) \simeq \mathrm{Hom}_G(V'^{[1]}, V^{\underline{g}}) \simeq \mathrm{Hom}_G(V', (V^{\underline{g}})^{[-1]})$$

This implies that $V \mapsto (V^{\underline{g}})^{[-1]}$ is a (left exact) functor from {G-modules} to itself which is right adjoint to the exact functor $V \mapsto V^{[1]}$. Its derived functors can be identified with $H^i_*(\underline{g}, ?)^{[-1]}$.

This construction generalizes obviously to all algebraic groups defined over $\mathbb{F}_p$, especially to B.

The composition of the functor $V \mapsto V^{[1]}$ on G-modules with the forgetful functor from G-modules to B-modules is obviously isomorphic to the composition of at first this forgetful functor and then the functor $M \mapsto M^{[1]}$ on B-modules. Therefore also the adjoint functors $M \mapsto ((\mathrm{ind}_B^G M)^{\underline{g}})^{[-1]}$ and $M \mapsto \mathrm{ind}_B^G((M^{\underline{b}})^{[-1]})$ are isomorphic. These functors map injective objects to injective objects, being right adjoint to exact functors. Therefore we get two Grothendieck spectral sequences (first constructed in [1],3.1.) converging to the "same" abutment with E_2-terms

$$(1) \qquad E_2^{i,j} = H^i_*(\underline{g}, R^j \mathrm{ind}_B^G M)^{[-1]}$$

respectively

(2) $\quad E_2^{i,j} = R^i\mathrm{ind}_B^G(H_*^j(\underline{b},M)^{[-1]})$.

2.6 Because of 2.4(2) the spectral sequence 2.5(1) degenerates for $M = k$. As 2.5(2) has the same abutment as 2.5(1) we get therefore a spectral sequence with

(1) $\quad E_2^{i,j} = R^i\mathrm{ind}_B^G(H_*^j(\underline{b},k)^{[-1]}) \Rightarrow H_*^{i+j}(\underline{g},k)^{[-1]}$.

Suppose now $p > h$. Then $H_*^\bullet(\underline{b},k)$ is known by 2.3(2) and we have to compute all $R^i\mathrm{ind}_B^G(S^j\underline{u}^*)$. One can show (cf.[1],3.4 - 3.6) for any p and all $j \in \mathbf{N}$:

(2) $\quad R^i\mathrm{ind}_B^G(S^j\underline{u}^*) = 0 \quad$ for all $i > 0$.

Combining this with (1) we get for $p > h$:

(3) $$H_*^i(\underline{g},k)^{[-1]} \simeq \begin{cases} \mathrm{ind}_B^G S^{i/2}(\underline{u}^*) & \text{for } i \text{ even,} \\ 0 & \text{for } i \text{ odd.} \end{cases}$$

There is a natural homomorphism $S(\underline{g}^*) \to \mathrm{ind}_B^G S(\underline{u}^*) = \bigoplus_{i\geq 0} \mathrm{ind}_B^G S^i(\underline{u}^*)$: Map any $f \in S(\underline{g}^*)$ to the function $G \to S(\underline{u}^*)$, $g \longmapsto f \circ \mathrm{Ad}(g)^{-1}|_{\underline{u}}$. The kernel consists of all functions vanishing on $\mathrm{Ad}(G)\underline{u} = \underline{\underline{N}}$, the variety of nilpotent elements in $\underline{g}$. We get thus an embedding of the algebra $k[\underline{\underline{N}}]$ of regular functions on $\underline{\underline{N}}$ into $\mathrm{ind}_B^G S(\underline{u}^*)$. Comparing the dimensions of the homogeneous pieces one gets, if p does not divide the order $|W|$ of the Weyl group W of G:

(4) $\quad k[\underline{\underline{N}}] \xrightarrow{\sim} \mathrm{ind}_B^G S(\underline{u}^*)$,

cf.[1],3.9. If $p > h$, then p does not divide $|W|$. Therefore:

Theorem: If $p > h$, then $H_*^{ev}(\underline{g},k)^{[-1]}$ is isomorphic to $k[\underline{\underline{N}}]$ and $H_*^{2i+1}(\underline{g},k) = 0$ for all i.

This was first proved (for $p \geq 3h - 1$) in [6]. The approach described here is taken from [1].

2.7 Some results in 2.6 can be interpreted in a different way. Set $Y = \{(gB,x) \in G/B \times \underline{\underline{N}} \mid \mathrm{Ad}(g)^{-1}x \in \underline{u}\}$. Let $\pi : Y \to G/B$ and $\tau : Y \to \underline{\underline{N}}$ be the two projections. The map π is locally trivial with all fibres isomorphic to $\underline{u}$, it is especially affine. The sheaf $\mathcal{L}(S\underline{u}^*)$ mentioned in 2.4 can be checked to be equal to $\pi_*\mathcal{O}_Y$. Hence $R^i\mathrm{ind}_B^G(S\underline{u}^*) \simeq H^i(G/B,\pi_*\mathcal{O}_Y)$ for all i. As π is affine, one has $H^i(G/B,\pi_*\mathcal{O}_Y) \simeq H^i(Y,\mathcal{O}_Y)$. The spectral sequence $H^i(\underline{\underline{N}},R^j\tau_*\mathcal{O}_Y) \Rightarrow H^{i+j}(Y,\mathcal{O}_Y)$ degenerates ($\underline{\underline{N}}$ being an affine variety) and yields isomorphisms

$H^i(Y,\mathcal{O}_Y) \simeq (R^i\tau_*\mathcal{O}_Y)(\underline{\underline{N}})$. So 2.6(2),(4) imply

(1) $R^i\tau_*\mathcal{O}_Y = 0$ for all $i > 0$,

and

(2) If $(p,|W|) = 1$, then $\tau_*\mathcal{O}_Y \simeq \mathcal{O}_{\underline{\underline{N}}}$.

The map τ is Springer's resolution of the nilpotent variety (from [17]). It is proper and at least for "very good" p birational. (Any p not dividing $|W|$ is very good.) As Y is smooth, (2) implies that $\underline{\underline{N}}$ is normal for $(p,|W|) = 1$. These results generalise analogous theorems over $\mathbb{C}$ due to Kostant (the normality of $\underline{\underline{N}}$) resp. to Hesselink (the vanishing as in (1)), cf. [16], [10].

2.8 It is impossible to avoid a bound on p in Theorem 2.6. Indeed, the theorem gets definitely false for $p < h$ as will follow immediately from 3.3(2). There are also some examples in [1], 6.5 - 6.20.

For $p = h$ the results depend on the isogeny class of G. For $G = PGL_p$ and also for $G = GL_p$ one has still $H_*^\bullet(\underline{g},k)^{[-1]} \simeq \mathrm{ind}_B^G S'(\underline{u}^*)$, but this is no longer true for $G = SL_p$, cf. [1], 6.3.

For G-modules of the type $\mathrm{ind}_B^G M$ with $\dim(M) = 1$ there are results on $H_*^\bullet(\underline{g},\mathrm{ind}_B^G M)^{[-1]}$ in [1], 3.7, 5.5 They give rather complete information in case $p > h$ for G of classical type.

One can also ask for the Hochschild cohomology groups $H^i(\ker(F^r),k)$ with $r > 1$. (Recall that $H_*^i(\underline{g},?) = H_*^i(\ker(F),?)$, cf. 1.8/9.) Here one has to expect more complicated results, cf. [1], 2.4. In general one does not even know whether $H^{ev}(\ker(F^r),k)$ is a finitely generated algebra, except for the case $r = 2$ dealt with in [7], 1.11.

3. Support Varieties

Let $\underline{g}$ be a finite dimensional p-Lie algebra over k. All restricted $\underline{g}$-modules are assumed to have finite dimension.

3.1 The algebra $H_*^{ev}(\underline{g},k)^{(-1)}$ is finitely generated over k, so its maximal spectrum $\underline{\underline{X}}_{\underline{g}}$ is a finite dimensional variety over k. It is called the cohomology variety of $\underline{g}$.

For any restricted $\underline{g}$-module M the annihilator of $H_*^\bullet(\underline{g},M \otimes M^*)^{(-1)}$ in $H_*^{ev}(\underline{g},k)^{(-1)}$ defines a closed subvariety $\underline{\underline{X}}_{\underline{g}}(M)$ of $\underline{\underline{X}}_{\underline{g}}$. It is called the support variety of M. Obviously $\underline{\underline{X}}_{\underline{g}} = \underline{\underline{X}}_{\underline{g}}(k)$.

Such varieties have been studied extensively for finite groups (instead of p-Lie algebras), cf.the survey in [2],2.24 - 2.27. Some results proved there carry over to our situation (cf.[8],1.5,3.2), for example:

(1) $\dim \underline{\underline{X}}_{\underline{g}}(M) = 0 \Leftrightarrow M$ is an injective restricted module.

(One has to know that in the category of all restricted $\underline{g}$-modules injective objects are projective and vice versa. This is however a general theorem about finite dimensional Hopf algebras like $U^{[p]}(\underline{g})$.) The complexity of M is the smallest integer c such that there are a constant b and a projective resolution $0 \leftarrow M \leftarrow P_0 \leftarrow P_1 \leftarrow P_2 \leftarrow \cdots$ of M in the category of restricted $\underline{g}$-modules with $\dim(P_n) \le b\, n^{c-1}$ for all n. Then:

(2) $\dim \underline{\underline{X}}_{\underline{g}}(M)$ is the complexity of M.

3.2 Consider the homomorphism $\kappa : S(\underline{g}^*) \to H_*^{ev}(\underline{g},k)^{(-1)}$ as in 1.11. It defines a morphism $f : \underline{\underline{X}}_{\underline{g}} \to \underline{g}$. Via κ any $H_*^{\bullet}(\underline{g}, M \otimes M^*)^{(-1)}$ is made into an $S(\underline{g}^*)$-module. Let $\underline{\underline{V}}_{\underline{g}}(M)$ be the variety in $\underline{g}$ defined by the annihilator of $H_*^{\bullet}(\underline{g}, M \otimes M^*)^{(-1)}$ in $S(\underline{g}^*)$. Proposition 1.11 implies that f is a finite map onto $\underline{\underline{V}}_{\underline{g}}(k)$ and that $f(\underline{\underline{X}}_{\underline{g}}(M)) = \underline{\underline{V}}_{\underline{g}}(M)$ for any restricted $\underline{g}$-module M. So also

(1) $\dim \underline{\underline{V}}_{\underline{g}}(M) = \dim \underline{\underline{X}}_{\underline{g}}(M)$

is the complexity and

(2) $\underline{\underline{V}}_{\underline{g}}(M) = \{0\} \Leftrightarrow M$ is an injective restricted $\underline{g}$-module.

So one may hope not to lose too much information when working not with the $\underline{\underline{X}}_{\underline{g}}(M)$, but with the $\underline{\underline{V}}_{\underline{g}}(M)$ which seem to be more accessible than the others. Any variety $\underline{\underline{V}}_{\underline{g}}(M)$ is homogeneous (for the scalar multiplication), and any homogeneous and closed subvariety of $\underline{\underline{V}}_{\underline{g}}(k)$ can occur as $\underline{\underline{V}}_{\underline{g}}(M)$ for a suitable M. If M is indecomposable, then the image of $\underline{\underline{V}}_{\underline{g}}(M) - \{0\}$ in the projective space $\mathbb{P}(\underline{g})$ is connected, cf.[9], 2.2.

The only homogeneous subvarieties of $\underline{g}_0 = k$ as in 1.3 are $\underline{g}_0$ and $\{0\}$. So (2) implies for any restricted $\underline{g}_0$-module:

(3) If M is not injective as a restricted $\underline{g}_0$-module, then $\underline{\underline{V}}_{\underline{g}_0}(M) = \underline{g}_0$.

3.3 The construction of the $\underline{\underline{V}}_{\underline{g}}(M)$ is natural. If $\varphi : \underline{g}' \to \underline{g}$ is a homomorphism of p-Lie algebras and if M is a restricted $\underline{g}$-module, then

(1) $\varphi \underline{\underline{V}}_{\underline{g}'}(M) \subset \underline{\underline{V}}_{\underline{g}}(M)$

where we regard M as a restricted $\underline{g}'$-module via φ.

For any $x \in \underline{g}$, $x \neq 0$ with $x^{[p]} = 0$ the subspace kx is a p-Lie subalgebra of $\underline{g}$ isomorphic to $\underline{g}_0$ (as above). So (1) and 3.2(3) yield one inclusion ("$\supset$") in the formulas (2),(3) below.

Proposition: One has

(2) $\underline{\underline{V}}_{\underline{g}}(k) = \{x \in \underline{g} \mid x^{[p]} = 0\}$

and for any restricted $\underline{g}$-module M

(3) $\underline{\underline{V}}_{\underline{g}}(M) = \{0\} \cup \{x \in \underline{\underline{V}}_{\underline{g}}(k) \mid x \neq 0$, M is not injective as a restricted kx-module$\}$.

Formula (2) is proved in [13], formula (3) in [9]. This second result had been known before in special cases (for unipotent p-Lie algebras by [8], for $p = 2$ by [13]). In order to get the only inclusion still missing ("$\subset$") one wants to apply (1) to some embedding $\underline{g} \to \underline{g\ell}(V)$ for some finite dimensional vector space V resp. to the given representation $\underline{g} \to \underline{g\ell}(M)$. Thus one has to prove only for any finite dimensional vector space V that $\underline{\underline{V}}_{\underline{g\ell}(V)}(k) \subset \{x \in \underline{g\ell}(V) \mid x^p = 0\}$ and that $\underline{\underline{V}}_{\underline{g\ell}(V)}(V) \subset \{x \in \underline{g\ell}(V) \mid x^p = 0,\ x \neq 0,\ V$ not injective for $kx\} \cup \{0\}$. This problem can be reduced to a similar one for the Lie algebra $\underline{b}$ of a Borel subgroup of $GL(V)$ using the spectral sequences from 2.5(1),(2). The proofs for $\underline{b}$ are still rather complicated.

3.4 For all restricted $\underline{g}$-modules M_1, M_2 one gets immediately from 3.3(3):

(1) $\underline{\underline{V}}_{\underline{g}}(M_1 \oplus M_2) = \underline{\underline{V}}_{\underline{g}}(M_1) \cup \underline{\underline{V}}_{\underline{g}}(M_2)$,

(2) $\underline{\underline{V}}_{\underline{g}}(M_1 \otimes M_2) = \underline{\underline{V}}_{\underline{g}}(M_1) \cap \underline{\underline{V}}_{\underline{g}}(M_2)$,

and for each p-Lie subalgebra $\underline{h}$ of $\underline{g}$:

(3) $\underline{\underline{V}}_{\underline{h}}(M_1) = \underline{\underline{V}}_{\underline{g}}(M_1) \cap \underline{h}$.

Any restricted $\underline{g}_0$-module (as in 1.3) is injective if and only if all its indecomposable direct summands have dimension equal to p. So 3.3(3) implies for any restricted $\underline{g}$-module M:

(4) If $(p, \dim(M)) = 1$, then $\underline{\underline{V}}_{\underline{g}}(M) = \underline{\underline{V}}_{\underline{g}}(k)$.

3.5 Let us assume from now on that we are in the situation of 2.1 and let us introduce some additional notations. For any T-module V and any $\lambda \in X(T)$ let V_λ be the weight space of weight λ. Denote the root system of G with respect to T by R. Let S be the set of simple roots in R such that the weights of T in $\underline{u}$ are the negative roots with respect to S.

For any subset $I \subset S$ let G_I be the (connected and reductive) closed subgroup of G generated by T and the root subgroups for all roots in $\mathbb{Z}I$. Then $\underline{g}_I = \mathrm{Lie}(G_I)$ is the direct sum of $\underline{t}$ and of all $\underline{g}_\alpha$ with $\alpha \in R_I = R \cap \mathbb{Z}I$. For any G-module M and any class $\Lambda \in X(T)/\mathbb{Z}I$ the sum

(1) $$M_\Lambda = \bigoplus_{\lambda\in\Lambda} M_\lambda$$

is a G_I-submodule of M and M is the direct sum of all M_Λ. The same argument as for 3.4(4) yields:

(2) <u>If there is</u> $\Lambda \in X(T)/\mathbb{Z}I$ <u>with</u> $(p,\dim(M_\Lambda)) = 1$, <u>then</u> $\{x \in \underline{g}_I \mid x^{[p]} = 0\} \subset \underline{\underline{V}}_{\underline{\underline{g}}}(M)$.

3.6 Let $(e(\lambda) \mid \lambda \in X(T))$ be the canonical basis of the group ring $\mathbb{Z}[X(T)]$ and let $(e(\Lambda) \mid \Lambda \in X(T)/\mathbb{Z}I)$ be the canonical basis of the group ring $\mathbb{Z}[X(T)/\mathbb{Z}I]$. So $e(\lambda)\,e(\mu) = e(\lambda+\mu)$ for all $\lambda,\mu \in X(T)$ and there is a ring homomorphism $r_I : \mathbb{Z}[X(T)] \to \mathbb{Z}[X(T)/\mathbb{Z}I]$ with $r_I(e(\lambda)) = e(\lambda + \mathbb{Z}I)$ for all $\lambda \in X(T)$.

For any finite dimensional G-module M its formal character $\mathrm{ch}(M)$ is defined as

$$\mathrm{ch}(M) = \sum_{\lambda\in X(T)} \dim(M_\lambda)\, e(\lambda) \in \mathbb{Z}[X(T)].$$

Then

(1) $$r_I\,\mathrm{ch}(M) = \sum_{\Lambda\in X(T)/\mathbb{Z}I} \dim(M_\Lambda)\, e(\Lambda).$$

Let $\rho \in X(T)\otimes\mathbb{Q}$ be half the sum of all positive roots. So $\langle\rho,\alpha^\vee\rangle = 1$ for all $\alpha \in S$ where we write $\beta^\vee$ for the dual root of any $\beta \in R$. Fix some $\lambda \in X(T)$ dominant, i.e. with $\langle\lambda,\alpha^\vee\rangle \geq 0$ for all $\alpha \in S$. Set $H^0(\lambda) = \mathrm{ind}_B^G(k_\lambda)$ where k_λ is k regarded as a B-module with $tu \in TU = B$ operating as $\lambda(t)$. Then Weyl's character formula holds, cf.[14],II,5.10:

(2) $$\mathrm{ch}(H^0(\lambda)) = \prod_{\alpha>0} (1-e(-\alpha))^{-1} \sum_{w\in W} \det(w)e(w(\lambda+\rho) - \rho).$$

Let W_I be the subgroup generated by all simple reflections s_β with $\beta \in I$ and let W^I be the set of all $w \in W$ with $w^{-1}(\beta) > 0$ for all $\beta \in I$. Then each $w \in W$ can be written uniquely in the form $w = w_1w_2$ with $w_1 \in W_I$ and $w_2 \in W^I$. Furthermore we can write $\prod_{\alpha>0} (1-e(-\alpha)) = \chi_1\chi_2$ where χ_1 (resp. χ_2) is the product of all $1-e(-\alpha)$ with $\alpha > 0$ and $\alpha \notin \mathbb{Z}I$ (resp. with $\alpha \in \mathbb{Z}I$). Then

$$\chi_1\,\mathrm{ch}\,H^0(\lambda) = \sum_{w\in W^I} \det(w) \sum_{w'\in W_I} \det(w')e(w'(w(\lambda+\rho))-\rho)\chi_2^{-1}.$$

Now $\sum_{w'\in W_I} \det(w')e(w'(w(\lambda+\rho))-\rho)\chi_2^{-1}$ is the formal character of the G_I-module $H_I^0(w(\lambda+\rho)-\rho)$ induced from $B\cap G_I$ in an analogous way to $H^0(\lambda)$. Let us use the abbreviation $w_\bullet\lambda = w(\lambda+\rho)-\rho$. If one applies r_I to the character of $H_I^0(w(\lambda+\rho)-\rho)$ one gets $e(w_\bullet\lambda+\mathbb{Z}I)\dim(H_I^0(w_\bullet\lambda))$. Hence:

(3) $r_I(\chi_1 \operatorname{ch} H^0(\lambda)) = \sum_{w\in W^I} \det(w)\dim(H_I^0(w_\bullet\lambda))\, e(w_\bullet\lambda+\mathbb{Z}I)$.

So, if there is $w\in W^I$ with $(p,\dim H_I^0(w_\bullet\lambda)) = 1$ and with $w_\bullet\lambda - w'_\bullet\lambda \notin \mathbb{Z}I$ for all $w'\in W^I, w'\neq w$ with $(p,\dim H_I^0(w'_\bullet\lambda)) = 1$, then $r_I(\operatorname{ch}H^0(\lambda))$ is not divisible by p, so in that case $\{x\in \underline{g}_I \mid x^{[p]} = 0\} \subset \underline{\underline{V}}_{\underline{g}}(H^0(\lambda))$ by (1) and 3.5(2).

The dimension of $H_I^0(w_\bullet\lambda)$ is given by Weyl's dimension formula

$$\dim H_I^0(w_\bullet\lambda) = \prod_\alpha \frac{\langle w(\lambda+\rho),\alpha^\vee\rangle}{\langle\rho,\alpha^\vee\rangle}$$

where the product is over all $\alpha\in R_I$, $\alpha>0$. Set $R_\mu = \{\alpha\in R \mid \langle\mu+\rho,\alpha^\vee\rangle \in \mathbb{Z}p\}$ for any $\mu\in X(T)$. Obviously $R_{w_\bullet\mu} = w(R_\mu)$ for all w,μ. Hence:

(4) <u>If</u> $R_I\cap w(R_\lambda) = \emptyset$, <u>then</u> $(p,\dim H_I^0(w_\bullet\lambda)) = 1$.

The converse holds if $p > \langle\rho,\alpha^\vee\rangle$ for all $\alpha\in R_I$, e.g. for $p\geq h$.

<u>3.7</u> Assume now that R is of type A_n. Let us use the notations from [3], planche I. Then there is a partition $\pi(\lambda) = (m_1\geq m_2\geq \ldots \geq m_r > 0)$ of $n+1$ such that R_λ is of type $A_{m_1-1}\times A_{m_2-1}\times\ldots\times A_{m_r-1}$. Let ${}^t\pi(\lambda) = (\ell_1\geq\ell_2\geq\ldots\geq\ell_s>0)$ be the dual partition, so $\ell_1 = r$ and $s = m_1$.

We claim that $\ell_1\leq p$. Indeed, we can write $\lambda+\rho = \sum_{i=1}^{n+1}(a_i+a)\varepsilon_i$ with $a\in\mathbb{Q}$ and $a_i\in\mathbb{Z}$. Set $L_j = \{i \mid 1\leq i\leq n+1, a_i\equiv j \bmod p\}$. Then $\{1,\ldots,n+1\}$ is the disjoint union of all L_j with $0\leq j<p$. A root $\alpha = \varepsilon_i-\varepsilon_j$ is in R if and only if there is some h with $i,j\in L_h$. If $L_h = \{i_{h1}<i_{h2}<\ldots<i_{ht_h}\}$, then L_h contributes to R_λ a component of type A_{t_h-1} with basis $\varepsilon_{i_{h1}}-\varepsilon_{i_{h2}}, \varepsilon_{i_{h2}}-\varepsilon_{i_{h3}},\ldots,\varepsilon_{i_{h,t_h-1}}-\varepsilon_{i_{ht_h}}$.

So there are at most p components of R_λ, hence $\ell_1 = r\leq p$.

Choose $I\subset S$ such that R_I is of type $A_{\ell_1-1}\times A_{\ell_2-1}\times\ldots\times A_{\ell_s-1}$. Now $p\geq\ell_1$ implies $p>\langle\rho,\alpha^\vee\rangle$ for all $\alpha\in R_I$. So $(p,\dim H_I^0(w_\bullet\lambda)) = 1$ is equivalent to $w(R_\lambda)\cap R_I = \emptyset$ (for any $w\in W^I$).

Let W_λ be the group generated by all reflections s_α with $\alpha\in R_\lambda$. One can identify W with the symmetric group S_{n+1}. Then W_I and W_λ are identified with Young subgroups corresponding to dual partitions, so there is exactly one double coset $W_I w W_\lambda$ with $wW_\lambda w^{-1} \cap W_I = \{1\}$, cf.e.g.[15],1.35. This is then also the only double coset with $w(R_\lambda) \cap R_I = \emptyset$. By multiplying with an element from W_λ on the right (resp. from W_I on the left) one can assume that $w(\alpha) > 0$ for all $\alpha \in R_\lambda$, $\alpha > 0$ (resp. $w \in W^I$). One checks easily that the second multiplication does not destroy the first property as any $w_1 \in W_I$ permutes the positive roots not in R_I and as $w(R_\lambda) \cap R_I = \emptyset$.

We have thus found some $w \in W^I$ with $(p, \dim H_I^0(w \cdot \lambda)) = 1$. We claim that any $w' \in W^I$, $w' \neq w$ with $(p, \dim H_I^0(w' \cdot \lambda)) = 1$ satisfies $w' \cdot \lambda < w \cdot \lambda$ and $w \cdot \lambda - w' \cdot \lambda \notin \mathbb{Z}I$. Indeed, as $w' \in W_I w$ there has to be some $\alpha \in R_\lambda$, $\alpha > 0$ with $w'(\alpha) < 0$. Then also $w's_\alpha(R_\lambda) \cap R_I = \emptyset$ and $w's_\alpha \cdot \lambda = w' \cdot \lambda - \langle \lambda+\rho, \alpha^\vee \rangle w'(\alpha) > w' \cdot \lambda$ with $w's_\alpha \cdot \lambda - w' \cdot \lambda \notin \mathbb{Z} I$ as $w'(\alpha) \notin \mathbb{Z} I$. There is $w_1 \in W_I$ with $w_1 w' s_\alpha \in W^I$. Then $w_1 w' s_\alpha \cdot \lambda \geq w' s_\alpha \cdot \lambda$ and $w_1 w' s_\alpha \cdot \lambda - w' s_\alpha \cdot \lambda \in \mathbb{Z} I$. So $w_1 w' s_\alpha \cdot \lambda > w' \cdot \lambda$ and $w_1 w' s_\alpha \cdot \lambda - w' \cdot \lambda \notin \mathbb{Z} I$. Now the claim follows by induction.

Combining this with 3.6(3) we see that $r_I(\operatorname{ch} H^0(\lambda))$ cannot be divisible by p. So any nilpotent $x \in \underline{\underline{g}}_I$ belongs to $\underline{\underline{V}}_{\underline{\underline{g}}}(H^0(\lambda))$. (Observe that $x^{[p]} = 0$ because of $p \geq \ell_1$.) On the other hand $\underline{\underline{V}}_{\underline{\underline{g}}}(H^0(\lambda))$ is $\mathrm{Ad}(G)$-stable as $H^0(\lambda)$ is a G-module. For any partition π of $n+1$ let $\underline{\underline{O}}(\pi)$ be the $\mathrm{Ad}(G)$-orbit of all nilpotent elements having Jordan blocks with sizes given by the parts of π. Now $\underline{\underline{g}}_I$ contains some $x \in \underline{\underline{O}}({}^t\pi(\lambda))$. We get therefore:

(1) $\underline{\underline{O}}({}^t\pi(\lambda)) \subset \underline{\underline{V}}(H^0(\lambda))$.

It seems likely (cf.[13],4.16) that one has equality.

References

[1] H.H.Andersen,J.C.Jantzen: Cohomology of induced representations of algebraic groups, Math.Ann.269(1984), 487 - 525

[2] D.Benson: Modular Representation Theory: New Trends and Methods, Lecture Notes in Mathematics 1081, Berlin/Heidelberg/New York/ Tokyo 1984(Springer)

[3] N.Bourbaki: Groupes et algèbres de Lie, chap.4,5 et 6, Paris 1968 (Hermann)

[4] M.Demazure,P.Gabriel: Groupes Algébriques I, Paris/Amsterdam 1970 (Masson/North-Holland)

[5] E.Friedlander,B.Parshall: Cohomology of algebraic and related finite groups, Invent.math.74(1983), 85 - 117

[6] E.Friedlander,B.Parshall: Cohomology of Lie algebras and algebraic groups, Amer.J.Math.108(1986), 235 - 253

[7] E.Friedlander,B.Parshall: Cohomology of infinitesimal and discrete groups, Math.Ann.273(1986), 353 - 374

[8] E.Friedlander,B.Parshall: Geometry of p-unipotent Lie algebras, J.Algebra (to appear)

[9] E.Friedlander,B.Parshall: Support varieties for restricted Lie algebras, to appear

[10] W.Hesselink: Cohomology and the resolution of the nilpotent variety, Math.Ann.223(1976), 249 - 252

[11] G.Hochschild: Cohomology of restricted Lie algebras, Amer.J.Math. 76(1954), 555 - 580

[12] N.Jacobson: Lie Algebras, New York/London/Sydney 1962 (Interscience/ Wiley)

[13] J.C.Jantzen: Kohomologie von p-Lie-Algebren und nilpotente Elemente, Abh.Math.Sem.Univ.Hamburg 76(1986) (demnächst)

[14] J.C.Jantzen: Representations of algebraic groups (to appear)

[15] A.Kerber: Representations of Permutation Groups, Lecture Notes in Mathematics 240, Berlin/Heidelberg/New York 1971(Springer)

[16] B.Kostant: Lie group representations on polynomial rings, Amer.J. Math.85(1963), 327 - 404

[17] T.A.Springer: The unipotent variety of a semi-simple algebraic group, pp.373 - 391 in: Algebraic Geometry (Proc. Bombay 1968), London 1969 (Oxford Univ.Press)

On geometric invariant theory for infinite-dimensional groups.

Victor G. Kac[1] and Dale H. Peterson[2]

Dedicated to Tony Springer
on his 60th birthday.

Introduction.

Let G be a complex reductive algebraic group operating on a finite-dimensional vector space V. Given a 1-parameter subgroup $\alpha: \mathbb{C}^{\times} \longrightarrow G$, a point $v \in V$ is called α-unstable if $\lim_{t\to 0} \alpha(t)\cdot v = 0$. The set $\mathcal{N}_0$ of all points which are α-unstable for some α is contained in the null-cone $\mathcal{N} = \{v \in V | 0 \in \overline{G\cdot v}\}$. The Hilbert-Mumford theorem says that, in fact, $\mathcal{N} = \mathcal{N}_0$.

Kempf [10] elaborated on this. He constructed a G-equivariant map $\Psi: \mathbb{P}\mathcal{N} \longrightarrow \mathcal{P}$, where $\mathbb{P}\mathcal{N}$ denotes the projectivization of $\mathcal{N}$ and $\mathcal{P}$ denotes the set of all proper parabolic subgroups of G, on which G acts by conjugation. Given $\mathbb{C}v \in \mathbb{P}\mathcal{N}$, the set of maximal reductive subgroups of the associated parabolic subgroup $\Psi(\mathbb{C}v)$ consists of the centralizers of all "optimal" 1-parameter subgroups α for which v is α-unstable; $\Psi(\mathbb{C}v)$ is uniquely determined by this. The main purpose of his work was to show the existence of α over the field of definition. However, one obtains immediately another corollary, just from the existence of the map Ψ : the stabilizer of any $\mathbb{C}v \in \mathbb{P}\mathcal{N}$ is contained in a proper parabolic subgroup of G (namely, in $\Psi(\mathbb{C}v)$).

Now let G(A) be the complex Kac-Moody group associated to a

[1]Partially supported by the NSF grant DMS 8508953 and the Guggenheim foundation.

[2]Partially supported by the Sloan Foundation.

generalized Cartan matrix A. Recall the following two examples. Let $\underline{G}$ be a connected simply-connected (almost) simple algebraic group over $\mathbb{C}$, and let A be its Cartan matrix and $\tilde{A}$ its extended Cartan matrix. Given a commutative algebra R over $\mathbb{C}$, denote by $\underline{G}_R$ the group of R-points of G. Then (a) $G(A) \simeq \underline{G}_{\mathbb{C}}$ and (b) $G(\tilde{A})$ is a central extension by $\mathbb{C}^{\times}$ of the group $\underline{G}_{\mathbb{C}[t,t^{-1}]}$ (where $\mathbb{C}[t,t^{-1}]$ is the algebra of Laurent polynomials in t). The direct products of groups of example (a) are called finite type groups, and the groups of example (b) and their twisted analogues are called affine Kac-Moody groups.

We introduce a natural generalization, denoted by $\mathcal{K}$, of the category of direct sums of finite-dimensional modules over a reductive group to the case of an arbitrary Kac-Moody group G(A). If A is not of finite type then for any G(A)-module V from the category $\mathcal{K}$ the set of weights of V is contained in an open half-space, hence $V = \mathcal{N}_0$ and the Hilbert-Mumford theorem trivially holds. The question of picking out an optimal class of 1-parameter subgroups α of G(A) such that v is α-unstable remains non-trivial, however. This was the starting point of our work.

The main difficulty in the general setup of Kac-Moody groups is the absence of a simple function which would measure the "size" of α (the Killing form employed by Kempf does not always exist and if it does, need not be positive). We introduce a "distance function" instead (defined by properties (i) and (ii) of Proposition 1.1). Section 1 is devoted entirely to the study of properties of the distance function. A (probably sole) disadvantage of this function is that it is transcendental.

Using the distance function, we show in Section 2 that for any G(A)-module V from the category $\mathcal{K}$ the stabilizer of any point in $\mathbb{P}V$ is contained in a finite type parabolic subgroup (Proposition 2.2 (c)). This solves a problem posed by Slodowy [14]. The same techniques allow us to obtain several characterizations of the so-called bounded subgroups of G(A) (Theorem 1). A bounded subgroup, as defined by Bruhat and Tits [4] for an arbitrary Tits system, is a subgroup contained in a finite union of double cosets of the Borel subgroup. Our Theorem 1, in particular, claims that a bounded subgroup of G(A) is contained in a finite type parabolic. This was proved by Bruhat and Tits for affine Tits systems [4] (and seems to be an open problem for arbitrary Tits systems).

In Section 3, we use the theory of bounded subgroups developed in Section 2 to prove a variety of conjugacy theorems for a Kac-Moody group G(A) and its unitary form K(A). First, we show that every reductive subgroup of G(A) (resp. compact subgroup of K(A)) is conjugate into a "standard" reductive (resp. compact) subgroup (Propositions 3.1 and 3.5). In particular, we obtain the conjugacy of maximal complex tori (resp. compact tori) in G(A) (resp. K(A)). Second, we study Ad-locally finite subgroups of G(A) (Theorem 2) and, in particular, Ad-triangular subgroups (Proposition 3.3). In Theorem 3 we derive infinitesimal versions of the above results for Kac-Moody algebras. Some of these were obtained in [13]. Note that the proof of our "global" results do not use [13], whereas passing to the infinitesimal situation relies on [13] heavily. Finally, Proposition 3.6 deals with conjugacy classes of finite order automorphisms and conjugate-linear automorphisms of Kac-Moody algebras. Involutions (resp. finite order automorphisms of the first kind) of affine Kac-Moody algebras were classified by Levstein [11] (resp. Bausch [1]).

It is a well known open problem of algebraic geometry to show that every reductive subgroup of the group of biregular automorphisms of $\mathbb{C}^n$ can be conjugated into $GL_n(\mathbb{C})$. We are hopeful that techniques developed in this paper may help to make progress in the solution to this problem.

§1. A distance function on the Tits cone.

1.1. The proofs of all facts stated in this subsection may be found in [5, Chapters 3,4].

Let I be a set of n elements, and let $A = (a_{ij})_{i,j\in I}$ be a generalized Cartan matrix, i.e., $a_{ii} = 2$, a_{ij} are non-positive integers for $i \neq j$ and $a_{ij} = 0$ implies $a_{ji} = 0$. Then there exists a unique up to isomorphism triple $(\mathfrak{h}_{\mathbb{R}}, \Pi, \Pi^{\vee})$, called the <u>realization</u> of A, where $\mathfrak{h}_{\mathbb{R}}$ is a vector space over $\mathbb{R}$ of dimension 2n-rankA, and $\Pi = \{\alpha_i\}_{i\in I} \subset \mathfrak{h}_{\mathbb{R}}^*$, $\Pi^{\vee} = \{\alpha_i^{\vee}\}_{i\in I} \subset \mathfrak{h}_{\mathbb{R}}$ are linearly independent sets satisfying $\langle\alpha_j, \alpha_i^{\vee}\rangle = a_{ij}$. Put $Q = \sum_{i\in I} \mathbb{Z}\alpha_i$, $Q_+ = \sum_{i\in I} \mathbb{Z}_+\alpha_i$. Here and further on, $\mathbb{Z}_+ = \{0,1,2,\ldots\}$.

Let $r_i \in \mathrm{Aut}\, \mathfrak{h}_{\mathbb{R}}$, $i \in I$, be the <u>fundamental reflections</u>:

$$r_i \cdot h = h - \langle\alpha_i, h\rangle\alpha_i^{\vee}, \quad h \in \mathfrak{h}_{\mathbb{R}}.$$

The group $W \subset \mathrm{Aut}\, \mathfrak{h}_{\mathbb{R}}$ generated by all the r_i is called the <u>Weyl group</u>. Given a subset J of I, we denote by W_J the subgroup of W generated by $\{r_i\}_{i\in J}$.

For $J \subset I$ let $A_J = (a_{ij})_{i,j\in J}$. Then the group W_J is finite if and only if A_J is a matrix of finite type. In this case we say that J is a <u>subset of finite type</u> of I.

Let

$$C = \{h \in \mathfrak{h}_{\mathbb{R}} \mid \langle\alpha_i, h\rangle \geq 0,\ i \in I\}$$

be the <u>fundamental chamber</u>. For $h \in C$ put

$\mathbb{Z}_+(h) = \sum_{i\in I} \mathbb{Z}_+\langle\alpha_i, h\rangle$ and $Q_+^{\vee}(h) = \sum_{i\in I} \mathbb{Z}_+(h)\alpha_i^{\vee}$.

One has [5, Exercise 3.12]:

(1.1) $\quad h - w\cdot h \in Q_+^{\vee}(h)$ if $h \in C$, $w \in W$.

The set

$$X = \bigcup_{w\in W} w\cdot C$$

is a convex cone in $\mathfrak{h}_{\mathbb{R}}$ called the <u>Tits cone</u>. Every point of X is W-equivalent to a unique point of C. The stabilizer W_h of $h \in C$ is

W_J, where $J = \{i \in I | \langle\alpha_i, h\rangle = 0\}$. The stabilizer of any point of IntX is finite. Here and further IntY stands for the interior of a set Y.

The triple $(\mathfrak{h}^*_{\mathbb{R}}, \Pi^\vee, \Pi)$ is a realization of the matrix ${}^tA$, the automorphisms of $\mathfrak{h}^*_{\mathbb{R}}$ induced by the r_i being the corresponding fundamental reflections. The corresponding fundamental chamber and Tits cone (contained in $\mathfrak{h}^*_{\mathbb{R}}$) are denoted by $C^\vee$ and $X^\vee$.

1.2. A subset S of $\mathfrak{h}^*_{\mathbb{R}}$ is called <u>admissible</u> if it is W-invariant and is contained in the intersection of $\mathrm{Int}X^\vee$ with $M - Q_+$ for some finite subset M of $\mathfrak{h}^*_{\mathbb{R}}$.

In this subsection we shall prove the following

<u>Proposition 1.1</u>. There exists a real-analytic function $F\colon \mathrm{Int}\, X^\vee \longrightarrow (0,\infty)$ satisfying:

(i) F is W-invariant and strictly convex;

(ii) for every admissible $S \subset \mathfrak{h}^*_{\mathbb{R}}$, the series $\sum_{\substack{\lambda\in S\\ \mu\in S\cap C^\vee}} F(\frac{1}{2}(\lambda+\mu))$ converges.

A function F satisfying (i) and (ii) of Proposition 1.1 is called a <u>distance function</u>. Such a function can be constructed as follows. Let $\{x_j\}$ be a basis of $\mathfrak{h}_{\mathbb{R}}$ such that all x_j are in IntX. For $\lambda \in \mathfrak{h}^*_{\mathbb{R}}$ let

$$(1.2) \qquad F(\lambda) = \sum_j \sum_{w\in W} e^{\langle\lambda, w\cdot x_j\rangle}.$$

To prove that F is a distance function we need three lemmas.

<u>Lemma 1.1</u>. Let V be a finite-dimensional vector space over $\mathbb{R}$ and let ℓ_n, $n \in \mathbb{Z}$, span V. Let U be the subset of V^* consisting of all u such that the series $f(u) = \sum_{n\in\mathbb{Z}} e^{\langle u, \ell_n\rangle}$ converges. Then U is convex, the function f is strictly convex on U and real analytic on IntU.

<u>Proof</u>. The convexity assertions are obvious, since e^x is a

positive-valued and strictly convex function. We embed V^* in its complexification $V^*_{\mathbb{C}} = V^* + iV^*$ and extend the pairing $\langle , \rangle$ by linearity. Then the series $f(u)$ converges absolutely on $U + iV^*$. Moreover, using the convexity of e^x, this absolute convergence is uniform on compact polyhedra contained in $U + iV^*$. Hence $f(u)$ is a complex analytic function on $(\mathrm{Int}U) + iV^*$. □

Lemma 1.2. If $\lambda \in \mathrm{Int}\, X^\vee$ and $h \in X$, then the series $\sum_{h' \in W\cdot h} e^{\langle \lambda, h'\rangle}$ converges.

Proof. Since $\mathrm{Int}\, X^\vee$ is the convex hull of the union of the sets $w\cdot(\mathrm{Int}C^\vee)$, $w \in W$, and since the region of convergence of the series in question is W-invariant and convex (see Lemma 1.1), we may assume that $\lambda \in \mathrm{Int}C^\vee$. Since X is the union of the $w\cdot C$, we may assume that $h \in C$.

We have:

$$\sum_{h' \in W\cdot h} e^{\langle \lambda, h'\rangle} = e^{\langle \lambda, h\rangle} \sum_{h' \in W\cdot h} e^{\langle \lambda, h'-h\rangle} \leq e^{\langle \lambda, h\rangle} \sum_{h'' \in Q^\vee_+(h)} e^{-\langle \lambda, h''\rangle}$$

by (1.1). The last series converges, proving Lemma 1.2. □

Lemma 1.3. If $h \in \mathrm{Int}X$, then the series $\sum_{\substack{w\in W\\ \lambda\in S\\ \mu\in S\cap C^\vee}} e^{\langle \lambda+\mu, w\cdot h\rangle}$ converges, for any admissible $S \subset \mathfrak{h}^*_{\mathbb{R}}$.

Proof. The series in question may be rearranged into $\sum_{\substack{w\in W\\ \lambda\in S\\ \mu\in S\cap C^\vee}} e^{\langle \lambda+w\cdot\mu, h\rangle}$ and hence it is dominated by the series $N \sum_{\lambda,\mu\in S} e^{\langle \lambda+\mu, h\rangle}$, where $N = \max_{\mu\in(\mathrm{Int}X^\vee)\cap C^\vee} |W_\mu|$. As in the proof of Lemma 1.2, we may assume that $h \in \mathrm{Int}C$. By the definition of an admissible set, the last series is dominated by $NF_0(h) \sum_{\alpha,\beta\in Q_+} e^{-\langle \alpha+\beta, h\rangle}$, where F_0 is a finite exponential sum. This series converges for $h \in \mathrm{Int}C$. □

Proposition 1.1 now follows immediately from Lemmas 1.1, 1.2 and 1.3. □

1.3. For a subset Y of a vector space over $\mathbb{R}$ we shall denote by [Y] the convex hull of Y. Given a nonempty compact convex subset K of $\mathrm{Int}X^{\vee}$, by the strict convexity of the distance function F, there exists a unique point where F achieves its absolute minimum on K. We shall denote this point by μ_K. Now we can prove our first key proposition. We let $\mathbb{R}_+ = \{r \in \mathbb{R} | r \geq 0\}$.

Proposition 1.2. (a) Let K be a nonempty compact convex subset of $\mathrm{Int}X^{\vee}$ such that $\mu_K \in C^{\vee}$. Then μ_K is the unique point of absolute minimum of F on the set $\mathrm{Int}X^{\vee} \cap (K + \sum_{i\in I} \mathbb{R}_+\alpha_i)$.

(b) Let $S \subset \mathfrak{h}^*_{\mathbb{R}}$ be admissible. Then for any $\epsilon > 0$, the set of all nonempty finite subsets T of S such that $\mu_{[T]} \in C^{\vee}$ and $F(\mu_{[T]}) \geq \epsilon$ is finite.

Proof. To prove (a), we write $\lambda \in \mathrm{Int}X^{\vee} \cap (K + \sum_i \mathbb{R}_+\alpha_i)$ in the form: $\lambda = \mu_K + \alpha + \beta$, where $\mu_K + \alpha \in K$ and $\beta \in \sum_i \mathbb{R}_+\alpha_i$. Since $F(\mu_K + t\alpha) \geq F(\mu_K)$ for $0 \leq t \leq 1$, we have

$$(1.3) \qquad D_{\alpha} F(\mu_K) \geq 0.$$

Since F is strictly convex and assumes the same value at the endpoints of each interval $[\mu, r_i \cdot \mu]$, we have

$$D_{\alpha_i} F(\mu) > 0 \quad \text{if } \langle \mu, \alpha_i^{\vee} \rangle > 0.$$

Therefore, since $\mu_K \in C^{\vee}$, we have

$$(1.4) \qquad D_{\beta} F(\mu_K) \geq 0 \quad .$$

Combining (1.3) and (1.4), we get $D_{\alpha+\beta} F(\mu_K) \geq 0$. Since F is strictly convex, this forces $F(\mu_K + \alpha + \beta) \geq F(\mu_K)$, with equality iff $\alpha + \beta = 0$, proving (a).

To prove (b), we need to show that up to W-equivalence, there are only a finite number of finite subsets T of S with $F(\mu_{[T]}) \geq \epsilon$. Hence it suffices to show that the T with $T \cap C^{\vee} \neq \emptyset$ and $F(\mu_{[T]}) \geq \epsilon$ form a finite set. But this is clear since by (ii) of Proposition 1.1, there are only a finite number of possibilities for $\mu \in T \cap C^{\vee}$ and for any other $\lambda \in T$. □

Remarks. (a) If A is a matrix of finite type, then the function $F(\lambda) = (\lambda,\lambda)+1$ is a distance function. Here $(.,.)$ denotes a positive definite W-invariant symmetric bilinear form on $\mathfrak{h}^*_{\mathbb{R}}$.

(b) Let A be a matrix of affine type, let $(.,.)$ be a non-degenerate symmetric W-invariant bilinear form on $\mathfrak{h}^*_{\mathbb{R}}$ which is positive semidefinite on Q and let $\Lambda \in \mathrm{Int}X^{\vee}$. Then on the set $\{\lambda \in \mathfrak{h}^*_{\mathbb{R}} \mid (\lambda,\lambda) < 0 \text{ and } \Lambda - \lambda \in \Sigma\, \mathbb{R}\alpha_i\}$, the function $F(\lambda) = (-(\lambda,\lambda))^{-k}$ is a distance function for any $k > \frac{1}{2}(n+3)$.

§2. Bounded subgroups of a Kac-Moody group.

2.1. The proofs of all facts stated in this subsection may be found in [5, Chapters 1,3,9,10].

Let A be a generalized Cartan matrix and let $(\mathfrak{h}_{\mathbb{R}}, \Pi, \Pi^{\vee})$ be its realization. We put $\mathfrak{h} = \mathbb{C} \otimes_{\mathbb{R}} \mathfrak{h}_{\mathbb{R}}$ and identify $\mathfrak{h}^*_{\mathbb{R}}$ with $\{\lambda \in \mathfrak{h}^* \mid \langle\lambda,h\rangle \in \mathbb{R} \text{ for all } h \in \mathfrak{h}_{\mathbb{R}}\}$. The Kac-Moody algebra $\mathfrak{g}(A)$ is the Lie algebra over $\mathbb{C}$ generated by the vector space $\mathfrak{h}$ and symbols e_i and f_i, $i \in I$, with defining relations:

$[\mathfrak{h},\mathfrak{h}] = 0$; $[e_i,f_j] = \delta_{ij}\alpha_i^{\vee}$;

$[h,e_i] = \langle\alpha_i,h\rangle e_i$, $[h,f_i] = -\langle\alpha_i,h\rangle f_i$ for $h \in \mathfrak{h}$;

$(\mathrm{ad}\, e_i)^{1-a_{ij}} e_j = 0$, $(\mathrm{ad}\, f_i)^{1-a_{ij}} f_j = 0$ for $i \neq j$.

We have the canonical embedding $\mathfrak{h} \subset \mathfrak{g}(A)$. Let $\mathfrak{n}_+$ (resp. $\mathfrak{n}_-$) be the subalgebra of $\mathfrak{g}(A)$ generated by the e_i(resp. f_i), $i \in I$. We have the triangular decomposition: $\mathfrak{g}(A) = \mathfrak{n}_- \oplus \mathfrak{h} \oplus \mathfrak{n}_+$.

We have the <u>root space decomposition</u> $\mathfrak{g}(A) = \bigoplus_{\alpha \in \mathfrak{h}^*} \mathfrak{g}_\alpha$, where $\mathfrak{g}_\alpha = \{x \in \mathfrak{g} | [h,x] = \langle\alpha,h\rangle x$ for all $h \in \mathfrak{h}\}$, so that $\mathfrak{g}_{\alpha_i} = \mathbb{C}e_i$, $\mathfrak{g}_{-\alpha_i} = \mathbb{C}f_i$, $\mathfrak{g}_0 = \mathfrak{h}$. A <u>root</u> is an element of $\Delta := \{\alpha \in \mathfrak{h}^* | \alpha \neq 0, \mathfrak{g}_\alpha \neq 0\}$. Each root space $\mathfrak{g}_\alpha$, $\alpha \in \Delta$, is contained either in $\mathfrak{n}_+$ or in $\mathfrak{n}_-$; the root α is then called positive or negative respectively. Denote by Δ_+ and Δ_- the sets of positive and negative roots; then $\Delta_- = -\Delta_+$ and $\Delta_+ \subset Q_+$.

The set of roots Δ is W-invariant. A <u>real root</u> is an element of $\Delta^{re} = \{w\cdot\alpha | w \in W, \alpha \in \Pi\}$. If $\alpha \in \Delta^{re}$, then $\dim \mathfrak{g}_\alpha = 1$. Put $\Delta_+^{re} = \Delta^{re} \cap \Delta_+$.

A $\mathfrak{g}(A)$-module V is called <u>integrable</u> if the following two properties are satisfied:

(i) $V = \bigoplus_{\lambda \in \mathfrak{h}^*} V_\lambda$, where $V_\lambda = \{v \in V | \ h\cdot v = \langle\lambda,h\rangle v$ for all $h \in \mathfrak{h}\}$;

(ii) e_i and f_i are locally nilpotent on V for all $i \in I$.

The elements of $P(V) := \{\lambda \in \mathfrak{h}^* | V_\lambda \neq 0\}$ are called <u>weights</u> of the module V. The set P(V) is W-invariant. Also, the elements of $\mathfrak{g}_\alpha$ for $\alpha \in \Delta^{re}$ are locally nilpotent on V as well.

We shall work in the following category $\mathcal{K}$ of $\mathfrak{g}(A)$- modules. The objects of $\mathcal{K}$ are integrable $\mathfrak{g}(A)$-modules V which are $\mathfrak{n}_+$-locally finite (i.e. $U(\mathfrak{n}_+)v$ is finite-dimensional for every $v \in V$) and such that $P(V) \subset \operatorname{Int} X^v$; the morphisms are $\mathfrak{g}(A)$-homomorphisms. Note that $\mathcal{K}$ is closed under taking direct sums, tensor products, submodules and quotients. Note also that for any finitely-generated module V of $\mathcal{K}$, the set P(V) is admissible.

Note that the adjoint $\mathfrak{g}(A)$-module is integrable, but is not in $\mathcal{K}$ unless $\dim \mathfrak{g}(A) < \infty$ (or, equivalently, A is of finite type). The most important examples of modules from category $\mathcal{K}$ are (some of) the integrable highest weight modules $L(\Lambda)$ defined below.

Let $P_+ = \{\Lambda \in \mathfrak{h}^*_{\mathbb{R}} \mid \langle\Lambda,\alpha_i^\vee\rangle \in \mathbb{Z}_+, i \in I\}$. Given $\Lambda \in P_+$, there exists a unique up to isomorphism irreducible $\mathfrak{g}(A)$-module $L(\Lambda)$ which admits a non-zero vector v_Λ such that $\mathfrak{n}_+ \cdot v_\Lambda = 0$; $h \cdot v_\Lambda = \langle\Lambda,h\rangle v_\Lambda$ for all $h \in \mathfrak{h}$. This module is integrable; it is in the category $\mathcal{K}$ if and only if $\Lambda \in \operatorname{Int} X^\vee$ since

(2.1) $\quad P(L(\Lambda)) \subset [W\cdot\Lambda]$.

Note that for $\Lambda \in C^\vee$, the condition $\Lambda \in \operatorname{Int}X^\vee$ is equivalent to the condition : $J = \{i \in I \mid \langle\Lambda,\alpha_i^\vee\rangle = 0\}$ is of finite type.

2.2. We now turn to the construction of the group $G(A)$ associated to the Kac-Moody algebra $\mathfrak{g}(A)$. The proofs of the properties of $G(A)$ stated below may be found in [9].

Let G^* be the free product of the additive groups $\mathfrak{g}_\alpha$, $\alpha \in \Delta^{re}$. For an integrable $\mathfrak{g}(A)$-module $(V,d\pi)$ we define a G^*-module $(V,\tilde{\pi})$ by

$$\tilde{\pi}(x) = \exp d\pi(x) := \sum_{n\geq 0} (d\pi(x))^n/n! \ , \quad x \in \mathfrak{g}_\alpha, \alpha \in \Delta^{re}.$$

We put $G(A) = G^*/\cap \operatorname{Ker} \tilde{\pi}$, where the intersection is taken over all $\tilde{\pi}$ associated to integrable $\mathfrak{g}(A)$-modules. Thus, each module $(V,\tilde{\pi})$ is naturally a $G(A)$-module, which we denote by (V,π). We call $G(A)$ the <u>group associated to the Kac-Moody algebra</u> $\mathfrak{g}(A)$ and (V,π) the $G(A)$-<u>module associated to the integrable</u> $\mathfrak{g}(A)$-<u>module</u> $(V,d\pi)$. The $G(A)$-module associated to the adjoint $\mathfrak{g}(A)$-module $(\mathfrak{g}(A),\mathrm{ad})$ is denoted by $(\mathfrak{g}(A),\mathrm{Ad})$.

Given an element $x \in \mathfrak{g}_\alpha$, $\alpha \in \Delta^{re}$, we denote its image in $G(A)$ under the canonical homomorphism $G^* \longrightarrow G(A)$ by $\exp x$. We have by definition:

$$\pi(\exp x) = \exp d\pi(x), \quad x \in \mathfrak{g}_\alpha, \ \alpha \in \Delta^{re}.$$

Let $U_\alpha = \exp \mathfrak{g}_\alpha$ be the additive 1-parameter subgroup of $G(A)$ corresponding to the real root α. Then $G(A)$ is generated by the $U_{\pm\alpha_i}$, $i \in I$. Denote by U_+ (resp. U_-) the subgroup of $G(A)$ generated by all U_α (resp. $U_{-\alpha}$) with $\alpha \in \Delta_+^{re}$.

For each $i \in I$, we have a unique homomorphism φ_i: $SL_2(\mathbb{C}) \longrightarrow G(A)$ satisfying:

$$\varphi_i\begin{pmatrix}1 & t\\ 0 & 1\end{pmatrix} = \exp te_i,\ \varphi_i\begin{pmatrix}1 & 0\\ t & 1\end{pmatrix} = \exp tf_i \quad (t \in \mathbb{C}).$$

Let $G_i = \varphi_i(SL_2(\mathbb{C}))$, $H_i = \varphi_i(\{diag(t,t^{-1}) | t \in \mathbb{C}^\times\})$, and let N_i be the normalizer of H_i in G_i. Let H (resp. N) be the subgroup of G(A) generated by the H_i (resp. N_i); H is an abelian normal subgroup of N. The φ_i are monomorphisms and H is the direct product of the H_i. We have an isomorphism φ: $W \longrightarrow N/H$ such that $\varphi(r_i)$ is the coset $N_iH\backslash H$. We identify W and N/H using φ; this gives sense to expressions such as wH and wU_+w^{-1} occurring in the sequel. If $h \in \mathfrak{h}$, $w \in W$ and $n \in wH$, then $Ad(n)h = w\cdot h$. We put $B = HU_+$, $B_- = HU_-$. We have: $B \cap N = H$.

One of the basic facts about the group G(A) is that the quadruple $(G(A),B,N,\{r_i\}_{i\in I})$ is a Tits system. (The definition and basic properties of Tits systems may be found in [3].) In particular we have the Bruhat decomposition:

$$G(A) = \bigsqcup_{w\in W} BwB \quad \text{(disjoint union)}.$$

Another property of a Tits system is that given $J \subset I$, the set $P_J = \bigcup_{w\in W_J} BwB$ is a subgroup of G(A) called a <u>standard parabolic subgroup</u>. Conjugates of standard parabolic subgroups are called <u>parabolic subgroups</u>. A parabolic subgroup coincides with its normalizer. If J is of finite type, the group P_J and its conjugates are called <u>finite type</u> parabolic subgroups.

A more special property of G(A) is the Birkhoff decomposition:

$$G(A) = \bigsqcup_{w\in W} B_-wB \quad \text{(disjoint union)}.$$

There exists a finest topology on G(A) such that the homomorphisms φ_i are continuous and G(A) is a topological group. We fix this topology on G(A); then G(A) is a Hausdorff connected simply connected topological group (cf. [8]).

2.3 In this subsection V is a fixed $\mathfrak{g}(A)$-module from the category $\mathcal{K}$. Then V has the structure of the associated G(A)-module. Fix $v \in V$, $v \neq 0$. We decompose v relative to the weight space decomposition: $v = \sum_{\lambda \in P(V)} v_\lambda$, and put

$$\text{supp } v = \{\lambda \in P(V) \mid v_\lambda \neq 0\}.$$

Fix a distance function F on $X^\vee$. Given $g \in G(A)$, put

$$\gamma(g) = \mu_{[\text{supp } g\cdot v]} \text{ (see Section 1.3).}$$

<u>Proposition 2.1</u>. (a) If $n \in wH$ and $g \in G(A)$, then $\gamma(ng) = w\cdot\gamma(g)$.

(b) If $g \in G(A)$, $b \in B$ and $\gamma(g) \in C^\vee$, then $F(\gamma(bg)) \geq F(\gamma(g))$, with equality if and only if $\gamma(bg) = \gamma(g)$.

(c) There exists $g \in G(A)$ such that $\gamma(g) \in C^\vee$ and $F(\gamma(g')) \leq F(\gamma(g))$ for all $g' \in G(A)$.

(d) If $g \in G(A)$ is as in (c) and if $g' \in G(A)$, then g' is as in (c) if and only if $g'g^{-1} \in P_J$, where $J = \{i \in I \mid \langle \gamma(g), \alpha_i^\vee \rangle = 0\}$. Moreover, $\gamma(g') = \gamma(g)$ in this case.

<u>Proof.</u> (a) is clear since F is W-invariant. (b) follows from Proposition 1.2(a).

To prove (c), we may assume that $V = U(\mathfrak{g}(A))\cdot v$, so that P(V) is admissible. Using Proposition 1.2(b), there exists $g \in G(A)$ such that $F(\gamma(g')) \leq F(\gamma(g))$ for all $g' \in G(A)$. Using (a), we can take $\gamma(g) \in C^\vee$. This proves (c).

Now, let $g, g' \in G(A)$ be as in (c). Using the Bruhat decomposition, write $g'g^{-1} = bnb'$, where $b,b' \in B$ and $n \in N$, say $n \in wH$. Let $g'' = b'g$. By (b), $\gamma(g'') = \gamma(g)$. Similarly, $\gamma(b^{-1}g') = \gamma(g')$. Hence:

$$(2.2) \qquad \gamma(ng'') = \gamma(b^{-1}g') = \gamma(g').$$

But $w\cdot\gamma(g) = w\cdot\gamma(g'') = \gamma(ng'')$ by (a). Using (2.2) we get: $w\cdot\gamma(g) = \gamma(g')$, hence $w \in W_J$ amd $\gamma(g) = \gamma(g')$. This proves the "only if" part of (d). The "if" part follows immediately from (a) and (b). □

Let $\mathcal{P}$ denote the set of all finite type parabolic subgroups of $G(A)$ with the action of $G(A)$ by conjugation.

Proposition 2.2. (a) There exists a point v_0 on the orbit $G(A)\cdot v$ such that $\mu_{[\mathrm{supp}\ v_0]} \in C^{\vee}$, $F(\mu_{[\mathrm{supp}\ v_0]}) \geq F(\mu_{[\mathrm{supp}\ g\cdot v_0]})$ for all $g \in G(A)$, and the stabilizer of the line $\mathbb{C}v_0$ is contained in the finite type parabolic subgroup P_J, where
$J = \{i \in I | \langle \mu_{[\mathrm{supp}\ v_0]}, \alpha_i^{\vee}\rangle = 0\}$.

(b) There exists a $G(A)$-equivariant map

$$\varphi: \mathbb{P}V \longrightarrow \mathcal{P}.$$

(c) The stabilizer of $\mathbb{C}v$ is contained in a finite type parabolic subgroup of $G(A)$.

Proof. Let $v_0 = g\cdot v$, where g is as in Proposition 2.1 (c). If now $g'\cdot v_0 \in \mathbb{C}v_0$, then $g'g\cdot v \in \mathbb{C}g\cdot v$, hence $\gamma(g'g) = \gamma(g)$, and by Proposition 2.1(d) it follows that $g' = (g'g)g^{-1} \in P_J$, proving (a). (b) and (c) follow immediately from (a). □

Corollary of the proof (cf. [10]). If $\dim \mathfrak{g}(A) < \infty$ and $\dim V < \infty$, and if $v \in V$ is such that supp v is contained in an open half-space, then $G(A)_{\mathbb{C}v}$ is contained in a proper parabolic subgroup of $G(A)$. Therefore, by the Hilbert-Mumford theorem, if $0 \in \overline{G(A)\cdot v}$, then $G(A)_{\mathbb{C}v}$ is contained in a proper parabolic subgroup. □

2.4. Now we can prove our first main result on geometric invariant theory for Kac-Moody groups.

Theorem 1. The following conditions on a sub(semi)group P of $G(A)$ are equivalent:

(i) P is contained in a finite type parabolic subgroup;

(ii) P is contained in the union of a finite number of double cosets BwB, $w \in W$;

(iii) for every $G(A)$-module V from the category $\mathcal{K}$, every $v \in V$ is contained in a P-invariant finite-dimensional subspace;

(iv) P leaves invariant a non-zero finite-dimensional subspace of some G(A)-module V from the category $\mathcal{K}$;

(v) P leaves invariant a 1-dimensional subspace of some G(A)-module V from the category $\mathcal{K}$.

Proof. The implication (i) $\Rightarrow$ (ii) is clear (by properties of Tits systems). The implication (ii) $\Rightarrow$ (iii) is also clear since B acts locally-finitely on V. The implication (iii) $\Rightarrow$ (iv) is obvious. If U is a P-invariant m-dimensional subspace of V, then $\Lambda^m U$ is a P-invariant 1-dimensional subspace of $\Lambda^m V$; this proves the implication (iv) $\Rightarrow$ (v). Finally, the implication (v) $\Rightarrow$ (i) follows from Proposition 2.2. $\square$

Definition. A sub(semi)group P of G(A) satisfying one of the equivalent properties (i) - (v) of Theorem 1 is called a bounded sub(semi)group.

Remark. Bruhat and Tits [4] define bounded subgroups by property (i) and proved the equivalence of (i) and (ii) for any affine Tits system. Whether (i) is equivalent to (ii) for an arbitrary Tits system remains an open problem.

2.5. We now need a digression on $\mathfrak{g}'(A)$, the derived algebra of $\mathfrak{g}(A)$, also called a Kac-Moody algebra. It is generated by the e_i and f_i, $i \in I$; one has: $\mathfrak{g}(A) = \mathfrak{g}'(A) + \mathfrak{h}$. Let $\mathfrak{h}' = \mathfrak{g}(A) \cap \mathfrak{h}$ ($= \sum_{i \in I} \mathbb{C}\alpha_i^\vee$).

Given $J \subset I$, we denote by $\mathfrak{p}_J$ (resp. $\mathfrak{g}(A)_J$) the subalgebra of $\mathfrak{g}'(A)$ generated by $\mathfrak{h}' + \mathfrak{n}_+$ (resp. $\mathfrak{h}'$) and the e_i, f_i with $i \in J$.

A $\mathfrak{g}'(A)$-module V is called integrable if the e_i and f_i are locally nilpotent on V for all $i \in I$; then $\mathfrak{h}'$ is diagonalizable on V. Note that an integrable $\mathfrak{g}(A)$-module is an integrable $\mathfrak{g}'(A)$-module, and that the $\mathfrak{g}(A)$-modules $L(\Lambda)$ remain irreducible when restricted to

$\mathfrak{g}'(A)$. The construction of Section 2.2 applied to $\mathfrak{g}'(A)$ gives the same group $G(A)$ (cf. [9]).

Let AutA denote the group of automorphisms of the matrix A, i.e., $\mathrm{Aut}A = \{\sigma \in \mathrm{Aut}I \mid a_{\sigma(i)\sigma(j)} = a_{ij}, i,j \in I\}$. Then $\sigma \in \mathrm{Aut}A$ acts on Π by $\sigma\cdot\alpha_i = \alpha_{\sigma(i)}$. This action extends by linearity to Q and leaves invariant the sets Δ_+ and Δ_+^{re}. We have the corresponding action of AutA on $\mathfrak{g}'(A)$ defined by $\sigma\cdot e_i = e_{\sigma(i)}$, $\sigma\cdot f_i = f_{\sigma(i)}$, which lifts to $G(A)$ by $\sigma\cdot\exp x = \exp \sigma\cdot x$, $x \in \mathfrak{g}_\alpha \subset \mathfrak{g}'(A)$, $\alpha \in \Delta^{re}$.

Let $\tilde{H} = \mathrm{Hom}(Q,\mathbb{C}^\times)$. This group acts on $\mathfrak{g}'(A)$ and on $\mathfrak{g}(A)$ in an obvious way, and this action lifts to an action on $G(A)$. Let Aut$\mathbb{C}$ denote the group of continuous automorphisms of $\mathbb{C}$: $\mathrm{Aut}\mathbb{C} = \{1,\eta\}$, where η is the complex conjugation. Aut$\mathbb{C}$ acts on $\mathfrak{g}(A)$, $\mathfrak{g}'(A)$ and $G(A)$ in an obvious way.

The group $\mathrm{Aut}\mathbb{C} \times \mathrm{Aut}A$ acts in a natural way on $\tilde{H} \ltimes G(A)$, so that we can form the group

$$\tilde{G}(A) = (\mathrm{Aut}\,\mathbb{C} \times \mathrm{Aut}\,A) \ltimes (\tilde{H} \ltimes G(A)).$$

As has just been explained, $\tilde{G}(A)$ acts (canonically) on $\mathfrak{g}'(A)$. We extend this to a (non-canonical) action on $\mathfrak{g}(A)$ as follows. Let $\mathfrak{h}'_{\mathbb{R}} = \sum_{i\in I} \mathbb{R}\alpha_i^\vee$, $\mathfrak{h}^{\vee\prime}_{\mathbb{R}} = \sum_{i\in I} \mathbb{R}\alpha_i$, so that $\mathfrak{h}' = \mathbb{C} \otimes_{\mathbb{R}} \mathfrak{h}'_{\mathbb{R}}$. Define a linear map $\gamma\colon \mathfrak{h}'_{\mathbb{R}} \longrightarrow (\mathfrak{h}^{\vee\prime}_{\mathbb{R}})^*$ by $\langle \gamma(\alpha_i^\vee), \alpha_j\rangle = a_{ij}$. Clearly, γ is AutA-equivariant. Since AutA is a finite group, we may choose a subspace $\mathfrak{h}^o_{\mathbb{R}}$ of $(\mathfrak{h}^{\vee\prime}_{\mathbb{R}})^*$, complementary to $\gamma(\mathfrak{h}'_{\mathbb{R}})$ and AutA-invariant. Put $\mathfrak{h}^o = \mathbb{C} \otimes_{\mathbb{R}} \mathfrak{h}^o_{\mathbb{R}}$ and define a Lie algebra $\mathfrak{h}^o \ltimes \mathfrak{g}'(A)$ by $[h,x] = \langle\alpha,h\rangle x$, where $x \in \mathfrak{g}_\alpha \cap \mathfrak{g}'(A)$ and $h \in \mathfrak{h}^o$. By the uniqueness of the realization of A, we have a (non-canonical) identification of $\mathfrak{h}^o_{\mathbb{R}} \oplus \mathfrak{h}'_{\mathbb{R}}$ with $\mathfrak{h}_{\mathbb{R}}$, which extends to an isomorphism of $\mathfrak{h}^o \ltimes \mathfrak{g}'(A)$ with $\mathfrak{g}(A)$. This gives us an action of $\mathrm{Aut}\mathbb{C} \times \mathrm{Aut}A$ on $\mathfrak{g}(A)$, which extends to an action of $\tilde{G}(A)$ on $\mathfrak{g}(A)$.

Now, let $\tilde{\mathcal{X}}$ be the category of all $\mathfrak{g}(A)$- and $\tilde{G}(A)$-modules V satisfying (η acts conjugate-linearly):

(i) the actions of $\mathfrak{g}(A)$ and $\tilde{G}(A)$ are compatible;

(ii) V is in $\mathcal{X}$ and the action of G(A) is induced by that of $\mathfrak{g}'(A)$;

(iii) $\tilde{H}$ acts locally-finitely on V.

The morphisms of $\tilde{\mathcal{X}}$ are the obvious ones. Then $\tilde{\mathcal{X}}$ is closed under direct sums, tensor products, submodules and quotient modules.

Using the identification $\mathfrak{h} = \mathfrak{h}^o \oplus \mathfrak{h}'$, define fundamental weights Λ_i, $i \in I$, by:

$$\Lambda_i|_{\mathfrak{h}^o} = 0, \ \langle \Lambda_i, \alpha_j^\vee \rangle = \delta_{ij}.$$

Then $\sigma \cdot \Lambda_i = \Lambda_{\sigma(i)}$. Hence, if $\Lambda = \Sigma k_i \Lambda_i \in P_+$ is such that $k_{\sigma(i)} = k_i$ for all $\sigma \in \mathrm{Aut}\, A$, then $L(\Lambda)$ is from the category $\tilde{\mathcal{X}}$ if $\Lambda \in \mathrm{Int} X^\vee$. (Here we let $g(v) = g(\lambda - \Lambda)v$ if $g \in \tilde{H}$ and $v \in L(\Lambda)_\lambda$.) For example $L(\rho)$, where $\rho = \sum_{i \in I} \Lambda_i$, is in $\tilde{\mathcal{X}}$.

Note that the action of $\mathrm{Aut}\mathbb{C} \times \mathrm{Aut}A$ on $\mathfrak{h}$ normalizes W, so that we can form a group

$$\tilde{W} = \mathrm{Aut}A \ltimes W.$$

The group $\tilde{W}$ leaves $X^\vee$ invariant; putting (cf. (1.2)):

$$\tilde{F}(\lambda) = \sum_{\sigma \in \mathrm{Aut}A} F \circ \sigma,$$

we get a $\tilde{W}$-invariant distance function.

Let $\tilde{B} = (\mathrm{Aut}\,\mathbb{C})\tilde{H}B$, $\tilde{B}_- = (\mathrm{Aut}\,\mathbb{C})\tilde{H}B_-$, $\tilde{P}_J = (\mathrm{Aut}\,\mathbb{C})\tilde{H}P_J$, etc. Then we have the following variants of the Bruhat and Birkhoff decompositions:

$$\tilde{G}(A) = \coprod_{w \in \tilde{W}} \tilde{B}w\tilde{B} \quad \text{(disjoint union)},$$

$$\tilde{G}(A) = \coprod_{w \in \tilde{W}} \tilde{B}_- w\tilde{B} \quad \text{(disjoint union)}.$$

The following variant of Theorem 1 is useful for applications.

<u>Theorem $\tilde{1}$.</u> The following conditions on a sub(semi)group P of $\tilde{G}(A)$ are equivalent:

(i) P normalizes a finite type parabolic subgroup of G(A);

(ii) P is contained in the union of a finite number of double cosets $\tilde{B}w\tilde{B}$, $w \in \tilde{W}$;

(iii) for any $\tilde{G}(A)$-module V from the category $\tilde{\mathcal{X}}$, any $v \in V$ is contained in a P-invariant finite-dimensional subspace;

(iv) P leaves invariant a non-zero finite-dimensional subspace of some $\tilde{G}(A)$-module V from the category $\tilde{\mathcal{X}}$;

(v) P leaves invariant a 1-dimensional subspace of some $\tilde{G}(A)$-module V from the category $\tilde{\mathcal{X}}$.

Proof is essentially the same as that of Theorem 1. □

A subgroup P of $\tilde{G}(A)$ satisfying one of (i)-(v) of Theorem $\tilde{1}$ is called bounded.

Remarks. (a) Using the properties of Tits systems, it is clear that (ii) of Theorem $\tilde{1}$ is equivalent to: $\bigcup_{g \in P} gBg^{-1}$ is contained in a finite union of the BwB.

(b) With essentially the same formulation and proof, Theorem $\tilde{1}$ holds in a version applying to all automorphisms of $\mathbb{C}$.

(c) A subset of G(A) is called bounded if it is contained in a finite union of the BwB. A set S of automorphisms of G(A) is called uniformly bounded if $\bigcup_{s \in S} s \cdot P$ is bounded for every bounded subset P. We conjecture that a subsemigroup of automorphisms of G(A) is uniformly bounded if and only if it normalizes a finite type parabolic subgroup (cf. [4]).

§3. Applications to conjugacy theorems.

3.1. In order to prove our next theorem, we need the following two lemmas.

Lemma 3.1. Let P be a finite type parabolic subgroup of G(A) and let V be an integrable $\mathfrak{g}'(A)$-module. Then for every $v \in V$, there exists a finite-dimensional subspace V' of V such that

(3.1) $$P \cdot v \subset U(\mathfrak{n}_+)V'.$$

Proof. Let $P = gP_Jg^{-1}$, where $g \in G(A)$ and P_J is a standard finite-type parabolic. Write $g = b_1 n b_2$, where $b_1, b_2 \in B$ and $n \in N$. Write $v' = b_1^{-1}\cdot v$. Then (3.1) is equivalent to :

(3.2) $$nP_J n^{-1}\cdot v' \subset U(\mathfrak{n}_+)V' .$$

Now, it is easy to check that

$$P_J\cdot v_o \subset U(\mathfrak{p}_J)v_o$$

for any $v_o \in V$. We have the vector space decomposition

$$\mathfrak{p}_J = \mathfrak{h}'\oplus \mathfrak{a}_1 \oplus \mathfrak{a}_2 \oplus \mathfrak{a}_3, \text{ where}$$

$$\mathfrak{a}_1 = \mathfrak{n}_+ \cap \mathrm{Ad}(n)^{-1}\mathfrak{n}_+,\ \mathfrak{a}_2 = \mathfrak{n}_+ \cap \mathrm{Ad}(n)^{-1}\mathfrak{n}_-, \mathfrak{a}_3 = \mathfrak{n}_- \cap \mathfrak{p}_J.$$

Recall that $\mathfrak{h}'$ acts locally finitely on V. Clearly, $\mathfrak{a}_2$ and $\mathfrak{a}_3$ are finite-dimensional and are spanned by real root vectors, and hence act locally-finitely on V. Finally, $\mathrm{Ad}(n)\ \mathfrak{a}_1 \subset \mathfrak{n}_+$. Putting $V' = nU(\mathfrak{h}')U(\mathfrak{a}_2)U(\mathfrak{a}_3)n^{-1}\cdot v'$, we obtain (3.2). □

<u>Lemma 3.2.</u> Let $a_k = b_k n_k b_k'$, $k = 1,2,\dots,$ be an infinite sequence of elements of G(A) such that b_k, $b_k' \in B$, and $n_k \in N$ are such that their images w_k in W are all distinct. Then the linear span of all $(\mathrm{Ad}\ a_k)e_i$ and $(\mathrm{Ad}\ a_k)f_i$, $i \in I$, $k=1,2,\dots,$ is infinite-dimensional.

<u>Proof.</u> Suppose the contrary; then for all $\alpha \in \Delta$:

(3.3) $\bigcup_k \mathrm{supp}(\mathrm{Ad}\ a_k)\mathfrak{g}_\alpha$ is a finite set.

Let β_k be of minimal height in $\mathrm{supp}(\mathrm{Ad}\ a_k)\mathfrak{g}_\alpha$. It is clear that we have:

(3.4) $\mathrm{height}(w_k\cdot\alpha) \geq \mathrm{height}(\beta_k)$.

From (3.3), (3.4) and (3.4) for $-\alpha$ we deduce

(3.5) $|\mathrm{height}(w_k\cdot\alpha)| \leq c(\alpha)$,

where $c(\alpha)$ is a constant depending on α but not on k. Since $w\cdot\Pi$ determines $w \in W$, (3.5) contradicts the distinctness of the w_k. □

Given $J \subset I$, the set $P_J^- = \bigcup_{w\in W_J} B_-wB_-$ is a subgroup of G(A) called a standard <u>opposite</u> parabolic subgroup; its conjugates are called opposite parabolic subgroups. We also may introduce the category $\mathcal{K}_-$ of $\mathfrak{g}(A)$-modules, whose objects are $\mathfrak{n}_-$-finite integrable modules V such that $P(V) \subset -\mathrm{Int}\, X^{v}$. Then Theorems 1 and $\tilde{1}$ hold if we replace parabolics by opposite parabolics, B by B_{-}, $\mathcal{X}$ by $\mathcal{X}_{-}$, etc; these will be called Theorems 1^{-} and $\tilde{1}^{-}$. A subgroup satisfying the equivalent conditions of Theorem 1^{-} or $\tilde{1}^{-}$ is called <u>antibounded.</u>

<u>Remark</u>. A more adequate terminology would be bounded below (resp. above) in place of bounded (resp. antibounded).

<u>Theorem 2.</u> The following conditions on a subgroup S of G(A) (resp. $\tilde{G}(A)$) are equivalent:

(i) S is bounded and antibounded;

(ii) there exist finite type subsets J and J' of I and $w \in W$ such that S can be conjugated into $P_J \cap wP^{-}_{J'}w^{-1}$(resp. a subgroup that normalizes P_J and $wP^{-}_{J'}w^{-1}$) by an element of G(A);

(iii) the adjoint action of S on $\mathfrak{g}'(A)$ is locally finite.

<u>Proof</u>. We first give the proof for G(A) and then explain how to adapt it to $\tilde{G}(A)$. Let S be a bounded and antibounded subgroup of G(A). Then, by Theorems 1 and 1^{-} there exist g_1 and g_2 in G(A) such that $g_1Sg_1^{-1} \subset P_J$ and $g_2Sg_2^{-1} \subset P^{-}_{J'}$, where J and J' are finite type subsets of I. According to the Birkhoff decomposition, we can write $g_2g_1^{-1} = b_{-}n^{-1}b$, where $b_{-} \in B_{-}$, $b \in B$ and $n \in N$. Putting

$$g = bg_1 \ (= nb_{-}^{-1}g_2),$$

we get $gSg^{-1} \subset P_J \cap nP^{-}_{J'}n^{-1}$. This proves the implication (i) $\Longrightarrow$ (ii). The implication (ii) $\Longrightarrow$ (iii) is clear by Lemmas 3.1 and 3.1^{-}. The implication (iii) $\Longrightarrow$ (i) follows from Lemmas 3.2 and 3.2^{-}.

Now we explain how to modify arguments so that they will apply to $\tilde{G}(A)$. To prove (i) $\Longrightarrow$ (ii) we use Theorems $\tilde{1}$ and $\tilde{1}^{-}$. The implication (ii) $\Longrightarrow$ (iii) is proved as above for G(A) by replacing P_J and $P^{-}_{J'}$ by their normalizers in $\tilde{G}(A)$. The implication (iii) $\Longrightarrow$ (i) follows from the obvious analogue of Lemma 3.2. □

Remarks. (a) Using Lemma 3.3 below, it is easy to show that any subgroup S of G(A) satisfying the conditions of Theorem 2 acts locally finitely on every integrable $\mathfrak{g}'(A)$- module.

(b) We conjecture that if an element $g \in G(A)$ is divisible in G(A) then the subgroup generated by g is bounded.

3.2. Let $\mathfrak{c} \subset \mathfrak{h}'$ be the center of $\mathfrak{g}'(A)$. A subgroup S of G(A) (resp. $\tilde{G}(A)$) is called reductive if with respect to Ad S, $\mathfrak{g}'(A)/\mathfrak{c}$ decomposes into a direct sum of irreducible finite-dimensional representations. Similarly, we define a reductive subalgebra of $\mathfrak{g}'(A)$.

Let $J \subset I$ be a subset of finite type. Then the subgroup $G(A)_J$ of G(A) generated by H and all U_{α_i}, $U_{-\alpha_i}$ with $i \in J$ is a reductive subgroup, called a standard reductive subgroup. Similarly, define a subalgebra $\mathfrak{g}(A)_J$ of $\mathfrak{g}'(A)$ generated by $\mathfrak{h}'$ and all e_i, f_i, $i \in J$.

To prove our next result we need one more lemma. Given $J \subset I$, put $\Delta_J = \Delta_+ \cup ((\sum_{j\in J} \mathbb{Z}\alpha_j) \cap \Delta)$ and $\Delta_+^J = (\sum_{j\in J} \mathbb{Z}\alpha_j) \cap \Delta_+$.

Lemma 3.3. Let J and J' be subsets of I of finite type and let $w \in W$. Then the group $P = P_J \cap wP_{J'}^- w^{-1}$ is generated by H and the U_α with $\alpha \in \Delta_J \cap (-w\cdot\Delta_{J'})$.

The proof of Lemma 3.3 is based on three sublemmas.

Sublemma 3.3.1. Fix $w_1 \in W$ and $J' \subset I$, and put $P = w_1 P_{J'} w_1^{-1}$. Then:

(a) If $u_- \in U_-$, $n \in N$ and $u_- n \in P$, then u_-, $n \in P$.

(b) If $J \subset I$ is of finite type, then:

(i) there exists $w \in W_J$ such that $P_J^- \cap U_+ \subset wPw^{-1}$,

(ii) for all $w' \in W$, there exists $w \in W_J$ such that $P_J \cap ww'U_-w'^{-1}w^{-1} \subset U_+$.

Proof. To prove (a), choose $n_1 \in w_1 H$. Since [9] $w_1^{-1}U_-w_1 \subset U_+U_-$, we may write $n_1^{-1}u_-n_1 = u_+u_-'$, where $u_+ \in U_+$, $u_-' \in U_-$. So: $u_+u_-'n_1^{-1}nn_1 \in P_{J'}$. Since $u_+ \in U_+ \subset P_{J'}$, we get

$$u_-'n_1^{-1}nn_1 \in P_{J'}. \tag{3.6}$$

Using the fact that $P_{J'}$ inherits the structure of a refined Tits system from that of G(A) [9], we have $P_{J'} = \coprod_{n' \in N \cap P_{J'}} (U_- \cap P_{J'})n'U_+$. We also have [9], $G(A) = \coprod_{n' \in N} U_- n' U_+$. We deduce from these and (3.6) that $n_1^{-1} n n_1 \in P_{J'}$, so that $n \in P$ and hence also $u_- \in P$. This proves (a).

To prove (b)(i), choose $w \in W_J$ so as to minimize $\ell(ww_1)$. Then for $j \in J$, $\ell(r_j w w_1) > \ell(ww_1)$ and so $\alpha_j \in ww_1(\Delta_+)$. Since $P_J^- \cap U_+$ is generated by the U_α, $\alpha \in \Delta_+^J$, we deduce that $P_J^- \cap U_+ \subset ww_1 U_+ w_1^{-1} w^{-1} \subset wPw^{-1}$, proving (b)(i).

To prove (b)(ii), choose $w \in W_J$ to maximize $\ell(ww')$ and put $U' = ww'U_- w'^{-1} w^{-1}$. Then [9] $U' = (U' \cap U_-)(U' \cap U_+)$. So, we must show that $P_J \cap U_- \cap U' = \{1\}$. This is clear since, by the choice of w, $P_J \cap U_- \subset ww'U_+ w'^{-1} w^{-1}$. □

__Sublemma 3.3.2__. If $w \in W$ and if U is a subgroup of $U_w := U_+ \cap wU_- w^{-1}$ which is normalized by $\tilde{H}$, then U is closed and is generated by the subgroups U_α, $\alpha \in \Delta_+^{re}$, which it contains.

__Proof__. We proceed by induction on $\ell(w)$. The cases $\ell(w) = 0$ or 1 are trivial. Let $\ell(w) > 0$; choose $i \in I$ such that $\ell(r_i w) < \ell(w)$. Using [9], we have a homeomorphism

$$U_w = U_1 \ltimes U_2, \text{ where } U_1 = U_{r_i} \text{ and } U_2 = r_i U_{r_i w} r_i^{-1}.$$

Define $h \in \tilde{H} = \mathrm{Hom}(Q, \mathbb{C}^\times)$ by $\langle \alpha_j, h \rangle = \exp(1 - \delta_{ij})$. Fix $u \in U$ and write: $u = u_1 u_2$, where $u_k \in U_k$. Then for $n = 1, 2, \ldots$ we have $u_2^{-1} h^{-n} u_2 h^n = u^{-1} h^{-n} u h^n \in U \cap U_2$ and $\lim_{n \to \infty} h^{-n} u_2 h^n = 1$. Since $U \cap U_2$ is closed by the inductive assumption, we deduce that $u_2 \in U \cap U_2$. It follows that $U = (U \cap U_1) \ltimes (U \cap U_2)$, and so the sublemma follows by induction. □

__Sublemma 3.3.3.__ If W_1 and W_2 are parabolic subgroups of W (i.e. W-conjugates of subgroups of the form W_J, $J \subset I$), then $W_1 \cap W_2$ is a parabolic subgroup of W.

__Proof__. Choose $h_1, h_2 \in X$ with stabilizers W_1 and W_2 in W. Choose

$t > 0$ such that $w\cdot(h_1 + th_2) = h_1 + th_2$ implies $w \in W_1 \cap W_2$, which is possible since card $\mathbb{R}$ > card W, and put $h = h_1 + th_2$. Then $W_1 \cap W_2 = W_h$, proving the sublemma. □

<u>Proof of Lemma 3.3.</u> Let Sublemma $3.3.1^-$ be Sublemma 3.3.1 with + and − interchanged.

Put $P' = wP^-_{J'}w^{-1}$. By Sublemma $3.3.1^-$ (b)(i), we may assume that $P_J \cap U_- \subset P^-$. For $g \in P_J \cap P'$, write $g = u_-nu_+$, where $u_- \in P_J \cap U_-$, $nH \in W_J$ and $u_+ \in U_+$. Since $u_- \in P'$, we have $nu_+ \in P'$, and so $n, u_+ \in P'$ by Sublemma $3.3.1^-$(a). So:

$$P_J \cap P' = (P_J \cap U_-)(P_J \cap P' \cap N)(P' \cap U_+).$$

But $P_J \cap U_-$ is generated by the U_α's it contains, and by using Sublemmas $3.3.1^-$ (b)(ii) and 3.3.2, we see that $P' \cap U_+$ is generated by the U_α's it contains. Finally, using Sublemma 3.3.3, $(P_J \cap P' \cap N)/H$ is generated by some reflections r_α, $\alpha \in \Delta^{re}$, and so $P_J \cap P' \cap N$ is contained in the subgroup of $G(A)$ generated by H and the U_α's contained in $P_J \cap P'$. Combining these statements, the lemma is proved, since U_α is contained in $P_J \cap P'$ iff $\alpha \in \Delta_J \cap (-w\cdot\Delta_{J'})$. □

<u>Proposition 3.1</u>. Every reductive subgroup S of $G(A)$ can be conjugated into some standard reductive subgroup $G(A)_J$.

<u>Proof</u>. We may assume by Theorem 2 that S is a subgroup of $P: = P_J \cap wP^-_{J'}w^{-1}$ as in Lemma 3.3. Denoting by U^J the smallest normal subgroup of P_J containing all U_α, $\alpha \in \Delta_+^{re}$, that are not contained in $G(A)_J$, we have [9, Proposition 4.6]:

$$P_J = G(A)_J \ltimes U^J, \tag{3.7}$$

so that we get a homomorphism φ: $P \longrightarrow G(A)_J$. Using Lemma 3.3, we see that $\varphi(P) \subset P$. Therefore, $P = P' \ltimes U'$, where $P' = P \cap G(A)_J$ and $U' = P \cap U^J$.

Let $Z \subset H$ be the center of $G(A)$, so that $G(A)/Z$ acts faithfully on $\mathfrak{g}'(A)/c$ [9]. Let V' be the linear span of the $\mathrm{AdP}\cdot e_i$, $\mathrm{AdP}\cdot f_i$, $i \in I$, and let $V = (V'+c)/c$. By Lemmas 3.1 and 3.1^-, V is finite-dimensional, so that V is a finite-dimensional subspace of $\mathfrak{g}'(A)/c$ on which Ad P = P/Z acts faithfully. Since P is generated by H and some of the U_α (Lemma 3.3), we may regard P/Z as a connected algebraic subgroup of GL(V). Similarly, P'/Z is a connected algebraic subgroup of P/Z. Let U"/Z be the unipotent radical of P'/Z and let G/Z be a maximal reductive subgroup of P'/Z, so that

$$P'/Z = G/Z \ltimes U''/Z.$$

Using this and (3.7), we deduce that

$$P/Z = G/Z \ltimes (U''/Z \ltimes U'/Z).$$

But the first factor is reductive and the second one is unipotent, so that G/Z is a maximal reductive subgroup of P/Z.

Let $\overline{S}$ be the Zariski closure of SZ/Z in GL(V). Since V is a completely reducible $\overline{S}$-module, $\overline{S}$ is a reductive subgroup of GL(V) and so of P/Z. But any reductive subgroup of P/Z is contained in a maximal reductive subgroup, and any two maximal reductive subgroups of P/Z are P/Z-conjugate. Hence, S is P-conjugate to a subgroup of $G \subset P' \subset G(A)_J$. □

We have the following corollary of Proposition 3.1:

<u>Proposition 3.2</u>. Every complex torus (i.e. Ad-diagonalizable connected subgroup) of G(A) is G(A)-conjugate into H. The subgroup H is a maximal complex torus of G(A). Every Ad-diagonalizable subgroup of G(A) which is generated by two elements is G(A)-conjugate into H.

□

3.3. Given a representation π of G(A) on V, a subgroup R of G(A) is called π-<u>triangular</u> if every $v \in V$ is contained in an R-invariant finite-dimensional subspace on which $\pi(R)$ is triangular in some basis. For example, given $w \in W$, the subgroup $B_w := B \cap wB\_w^{-1}$ of G(A) is triangular in any integrable module. This is because $B_w = H \ltimes U_w$, where $U_w = U_+ \cap wU_-w^{-1}$ acts locally unipotently on any integrable $\mathfrak{g}'(A)$-module.

Another corollary of Theorems 1 and 2 is

Proposition 3.3. Every $\mathrm{Ad}_{\mathfrak{g}'(A)}$-triangular subgroup R of G(A) can be conjugated into one of the subgroups B_w.

Proof. Since R is bounded (Theorem 2), it can be conjugated into a parabolic of finite type P_J. We may assume that $R \subset P_J$. Since R is $\mathrm{Ad}_{\mathfrak{g}'(A)}$-triangular, its image $\bar{R}$ in $G(A)_J$ defined by using (3.7) is $\mathrm{Ad}_{\mathfrak{g}(A)_J}$-triangular, hence can be conjugated by an element of $G(A)_J$ into the Borel subgroup $G(A)_J \cap B$ of $G(A)_J$. It follows that R can be conjugated into B. Similarly, R can be conjugated into B_-. The same argument as in the proof of Theorem 2 now gives that R can be conjugated into some B_w. □

Remark. Infinitesimal versions of Propositions 3.2 and 3.3 have been proved in [13].

3.4. Now we shall prove an infinitesimal version of Theorem 2 and Proposition 3.1.

Theorem 3. Let A be a symmetrizable generalized Cartan matrix. Then:

(a) If $\mathfrak{g}$ is an ad-locally finite subalgebra of $\mathfrak{g}'(A)$, then $\mathfrak{g}$ can be conjugated by G(A) into a subalgebra $\mathfrak{p}_J \cap (\mathrm{Ad}\, n)\mathfrak{p}^-_{J'}$, where J,J' are finite type subsets of I and $n \in N$.

(b) If $\mathfrak{g}$ is a reductive subalgebra of $\mathfrak{g}'(A)$, then $\mathfrak{g}$ can be conjugated by G(A) into some $\mathfrak{g}(A)_J$, where J is a finite type subset of I.

Proof. If $\mathfrak{g}$ is an ad-locally finite subalgebra of $\mathfrak{g}'(A)$, then, by [13], $\exp \mathrm{ad}\mathfrak{g} \subset \mathrm{Ad}\, G(A)$. Hence, by applying Theorem 2, we may assume that the group G generated by $\exp \mathrm{ad}\mathfrak{g}$ is contained in $P_J \cap nP^-_{J'}n^{-1}$, where J and J' are of finite type and $n \in N$. But then G, and hence $\mathfrak{g}$

normalizes $\mathfrak{p}_J$ and $(\mathrm{Ad}\ n)\mathfrak{p}_J^-$, so that $\mathfrak{g} \subset \mathfrak{p}_J \cap (\mathrm{Ad}\ n)\mathfrak{p}_J^-$, proving (a). If, in addition, $\mathfrak{g}$ is reductive, then G is reductive and so by Proposition 3.1 we may assume that $G \subset G(A)_J$. But then G, and hence $\mathfrak{g}$, normalizes $\mathfrak{p}_J$ and $\mathfrak{p}_J^-$, so that $\mathfrak{g} \subset \mathfrak{p}_J \cap \mathfrak{p}_J^- = \mathfrak{g}(A)_J$. □

3.5 One can easily construct examples of semisimple finite-dimensional subalgebras of a Kac-Moody algebra which are not reductive. The situation is better however in the affine case:

Proposition 3.4. Let $\mathfrak{g}$ be a semisimple finite-dimensional subalgebra of an affine Kac-Moody algebra $\mathfrak{g}'(A)$. Then $\mathfrak{g}$ is a reductive subalgebra and hence can be conjugated into a subalgebra of $\mathfrak{g}(A)_J$, for some finite type subset J of I.

Proof. Let x be a nilpotent element of the Lie algebra $\mathfrak{g}$. The inclusion homomorphism $\varphi: \mathfrak{g} \longrightarrow \mathfrak{g}'(A)$ induces a homomorphism $\bar{\varphi}: \mathfrak{g} \longrightarrow \mathfrak{g}'(A)/c$. But $\mathfrak{g}'(A)/c$ is a subalgebra of $\hat{L} \otimes_{\mathbb{C}} \dot{\mathfrak{g}}$, where $\dot{\mathfrak{g}}$ is a simple finite dimensional Lie algebra and $\hat{L}$ is the field of formal Laurent series over $\mathbb{C}$. Thus, $\bar{\varphi}$ induces a homomorphism $\hat{\varphi}: \hat{L} \otimes_{\mathbb{C}} \mathfrak{g} \longrightarrow \hat{L} \otimes_{\mathbb{C}} \dot{\mathfrak{g}}$ over $\hat{L}$ of finite- dimensional (over $\hat{L}$) semi-simple Lie algebras (this type of argument may be found in [12]). Hence $\bar{\varphi}(x)$ is nilpotent on $\hat{L} \otimes_{\mathbb{C}} \dot{\mathfrak{g}}$, so that $\bar{\varphi}(x)$ is nilpotent on $\mathfrak{g}'(A)/c$ and $\varphi(x)$ is nilpotent on $\mathfrak{g}'(A)$.

But $\mathfrak{g}$ is generated by its nilpotent elements, hence $\varphi(\mathfrak{g})$ is generated by elements ad-nilpotent on $\mathfrak{g}(A)$. By [6, Lemma on p. 170] it follows that $\mathfrak{g}$ is ad-locally finite, proving the proposition. □

Remark. In the case when $\mathfrak{g}$ is a subalgebra of type X_ℓ of $X_\ell^{(1)}$, the conjugacy stated by Proposition 3.4 is claimed in [12]; the proof there is not quite correct however.

3.6. In this subsection we derive conjugacy theorems for the unitary form K(A) of the group G(A).

Recall that the Lie algebra $\mathfrak{g}(A)$ carries a unique conjugate-linear involution ω, called the <u>compact involution</u>, such that $\omega|_{\mathfrak{h}_{\mathbb{R}}} = -1$, $\omega(e_i) = -f_i$, $\omega(f_i) = -e_i$. One has the corresponding involution of G(A), also denoted by ω, such that $\omega(\exp x) = \exp \omega(x)$ for $x \in \mathfrak{g}_\alpha$, $\alpha \in \Delta^{re}$. The fixed point set of this involution is denoted by K(A) and is called the <u>unitary form</u> of G(A).

Recall some of the properties of K(A), proofs of which may be found in [9]. Put $K_i = G_i \cap K(A)$, $T = H \cap K(A)$, $K(A)_J = G(A)_J \cap K(A)$, $K_w = BwB \cap K(A)$. Then $K_i = \varphi_i(SU_2)$; $K(A)_{\{i\}} = TK_i$; $K(A)_J$ is generated by T and the $K(A)_i$ with $i \in J$; and K(A) is the disjoint union of the K_w, $w \in W$. Furthermore, we have:

(3.8) $$K(A)_J = K(A) \cap P_J.$$

We also have the Iwasawa decomposition, which often allows one to prove results about K(A) by using related results about G(A):

(3.9) $$G(A) = K(A)B.$$

The topology of G(A) induces a topology on K(A) which has the following properties. The subgroup T is a torus (i.e. a compact connected abelian subgroup); the K_w are locally closed; a closed subset of K(A) is compact if and only if it intersects only a finite number of the K_w; and the subgroups $K(A)_J$ with J of finite type are compact.

<u>Proposition 3.5</u>. (a) Every compact subgroup K of K(A) can be conjugated into one of the $K(A)_J$ where J is a finite type subset of I.
(b) Every torus of K(A) can be conjugated into T, which is a maximal torus of K(A). Any element g of K(A) such that the closure of $\{g^n | n \geq 1\}$ is compact can be conjugated into T.
(c) The stabilizer in K(A) of any finite-dimensional subspace of a G(A)-module from the category $\mathcal{X}$ is compact.
(d) Let Ω be a G(A)-orbit in a G(A)-module V from the category $\mathcal{X}$. Let $\Omega_0 = \{v \in \Omega | F(\mu_{[\mathrm{supp}\ v]})$ is maximal and $\mu_{[\mathrm{supp}\ v]} \in C^\vee\}$. Then μ:

$= \mu_{[\mathrm{supp}\ v]}$ is independent of the choice of $v \in \Omega_0$, Ω_0 is a $BW_\mu B$-orbit and $\Omega = K\cdot\Omega_0$.

Proof. (d) follows from Proposition 2.1 and the Iwasawa decomposition. If $v \in V$ and $\Omega = G(A)\cdot v$, then, by (d), there exists $k \in K(A)$ such that $k\cdot v \in \Omega_0$. Hence, by Proposition 2.2(a), the stabilizer of $\mathbb{C}k\cdot v$ in $G(A)$ lies in P_J, where J is of finite type, hence the stablizer of $\mathbb{C}k\cdot v$ in $K(A)$ lies in $P_J \cap K(A) = K(A)_J$ (see (3.8)). (c) follows by using the argument proving the implication (iv) $\Longrightarrow$ (v) of Theorem 1. (a) also follows easily. Indeed, K from (a) is a bounded subgroup of $G(A)$ since it is covered by a finite number of the K_w. Hence there exists a $G(A)$-module from the category $\mathcal{X}$ and $v \in V$ such that $\mathbb{C}v$ is K-invariant. But, by the above argument, the stabilizer of $\mathbb{C}k\cdot v$ is contained in $K(A)_J$ for some $k \in K(A)$. Thus, $kKk^{-1} \subset K(A)_J$. Finally (b) follows, using (a), from the corresponding facts about compact connected Lie groups. □

Remarks. (a) Let S denote the set of all conjugates of the subgroups $K(A)_J$, with J of finite type, of the group $K(A)$. Then, as in the case of $G(A)$, for any module V from the category $\mathcal{X}$, there exists a $K(A)$-equivariant map $\mathbb{P}V \longrightarrow S$.

(b) Every compact subgroup of $G(A)$ is reductive.

(c) Every compact subgroup of $K(A)$ (resp. $G(A)$) is contained in a maximal compact subgroup of $K(A)$ (resp. $G(A)$). Every maximal compact subgroup of $K(A)$ (resp. $G(A)$) can be conjugated onto $K(A)_J$ for a unique finite-type subset J of I. $K(A)_J$ is a maximal compact subgroup of $K(A)$ (resp. $G(A)$) if and only if J is maximal among all finite-type subsets of I. The proof of these facts is based on Proposition 3.5 and the following (presumably well-known) lemma:

Lemma. (a) Every finite subgroup of W is W-conjugate into W_J for some $J \subset I$ of finite type.

(b) If J is maximal among all finite-type subsets of I, then W_J is a maximal finite subgroup of W.

(c) If J and J' are maximal among all finite-type subsets of I and if $J \neq J'$, then W_J and $W_{J'}$ are not W-conjugate.

<u>Proof</u>. To prove (a), let W_0 be a finite subgroup of W, and let $h \in \operatorname{Int} X$. Then the stabilizer in W of $\sum_{w \in W_0} w \cdot h$ is a finite parabolic subgroup of W containing W_0, proving (a).

To prove (b), let J be maximal among all finite-type subsets of I, and suppose that W_0 is a finite subgroup of W strictly containing W_J. Choose $h \in C$ with stabilizer W_J, and $h' \in \operatorname{Int} X$ with stabilizer containing W_0. By the maximality of J, $h' \notin C$. Among the $i \in I$ with $\langle \alpha_i, h' \rangle < 0$, choose one with minimal $-\langle \alpha_i, h' \rangle^{-1} \langle \alpha_i, h \rangle$, and put $h'' = \langle \alpha_i, h \rangle h' - \langle \alpha_i, h' \rangle h$. Then $h'' \in C \cap \operatorname{Int} X$ and $W_{h''} \supset W_J$, so that $W_{h''} = W_J$ by the maximality of J. This contradicts $r_i \in W_{h''}$, proving (b).

To prove (c), let J and J' be maximal among all subsets of I of finite type and satisfy $J \neq J'$. Then the fixed-point sets of W_J and $W_{J'}$ on $\operatorname{Int} X$ are nonempty disjoint subsets of C, so that W_J and $W_{J'}$ cannot be W-conjugate, proving (c). □

(d) The analogue of Remark (c) for reductive subgroups of $G(A)$ is also true.

(e) Every continuous automorphism of $G(A)$ is in the group of automorphisms generated by the actions of $G(A)$, $\operatorname{Aut} A$, $\tilde{H}$, η and ω.

(f) Every continuous automorphism of $K(A)$ is in the group of automorphisms generated by the actions of $K(A)$, $\operatorname{Aut} A$, η and $\{g \in \tilde{H} | g(Q) \subset S^1 = \{t \in \mathbb{C} | \ |t| = 1\}\}$.

(g) We conjecture that the group of <u>all</u> automorphisms of $G(A)$ is generated by the action of $\tilde{G}(A)$, ω, and the group of <u>all</u> automorphisms of $\mathbb{C}$. In particular, we believe that every automorphism of $G(A)$

leaves $\{gBg^{-1} | g \in G(A)\} \cup \{gB_-g^{-1} | g \in G(A)\}$ invariant.

(h) The closure of a subgroup K of K(A) is compact iff K is bounded.

(i) If $g \in \mathrm{Aut}A \ltimes (\tilde{H} \ltimes G(A))$ and if the subgroup generated by g is bounded then g normalizes a G(A)-conjugate of B.

(j) Proposition 3.5(c) fails for arbitrary integrable G(A)-modules. For example if $\mathfrak{g}(A)$ is of affine type then the centralizer in K(A) of the subspace $\mathfrak{h}'/\mathbb{C}$ of $\mathfrak{g}'(A)/\mathbb{C}$ is not compact.

<u>Example</u>. Let K be a connected simply connected simple compact Lie group and let $\tilde{K}$ be the group of maps of S^1 into K with finite Fourier series, with the box topology (induced by an inclusion $\tilde{K} \subset \mathrm{Mat}_N(\mathbb{C}[t,t^{-1}])$. K is identified with the subgroup of constant loops of $\tilde{K}$. Let $\tilde{A} = (a_{ij})_{i,j=0}^{\ell}$ be the extended Cartan matrix of K; then the Kac-Moody group $K(\tilde{A})$ is a central extension by S^1 of the group $\tilde{K}$ [7]. Let T be a maximal torus of K; then it is a maximal torus of $\tilde{K}$ and, by using Proposition 3.5(b), we see that any torus of $\tilde{K}$ is conjugate into T. Let $\varphi_{\alpha_i} : SU_2 \longrightarrow K$, $i = 1,\dots,\ell$, and φ_θ: $SU_2 \longrightarrow K$ be homomorphisms associated to the simple roots and to the highest root θ. Define φ_0: $SU_2 \longrightarrow \tilde{K}$ by:

$$\varphi_0\begin{pmatrix} \alpha & \beta \\ -\bar{\beta} & \bar{\alpha} \end{pmatrix}(t) = \varphi_\theta\begin{pmatrix} \alpha & \beta t \\ -\bar{\beta}\bar{t} & \bar{\alpha} \end{pmatrix}, \quad |t| = 1,$$

and let $K_i = \varphi_i(SU_2)$, $i = 0,\dots,\ell$. Denote by K^j, $j = 0,\dots,\ell$, the subgroup of $\tilde{K}$ generated by all K_i with $i \neq j$ (so that $K^0 = K$). K^j is a compact connected subgroup of $\tilde{K}$ whose Cartan matrix is obtained from $\tilde{A}$ by deleting the j-th row and column. Then by Proposition 3.5(a), any compact subgroup of $\tilde{K}$ can be conjugated into one of the compact subgroups K^j. Moreover, by Remark (c), the K^j are mutually nonconjugate maximal compact subgroups of $\tilde{K}$. Note that evaluation at t=1 defines injective homomorphisms ψ_j: $K^j \longrightarrow K$ and that $\pi_1(K^j)$ is cyclic of order $a_j^\vee$. Note that the quotient of the root lattice of K by that of $\psi_j(K^j)$ is cyclic of order a_j. It is easy to show that K^j

can be conjugated into $K^0 = K$ by a continuous automorphism if and only if $a_j = 1$. (The values of the a_j and $a_j^\vee$ may be found in [5, Chapter 4].)

3.7. In this subsection, we derive from Theorems $\tilde{1}$ and 2 a conjugacy theorem for finite order automorphisms and conjugate-linear automorphisms of a Kac-Moody algebra $\mathfrak{g}'(A)$ with symmetrizable A. Let ω be the compact involution of $\mathfrak{g}'(A)$. It follows from [13, Theorem 2(c)] that the group $\{1, \omega\} \ltimes \mathrm{Ad}\,\tilde{G}(A)$ contains all automorphisms and conjugate-linear automorphisms of $\mathfrak{g}'(A)$. Those from $\mathrm{Ad}\,\tilde{G}(A)$ (resp. from $\omega\,\mathrm{Ad}\,\tilde{G}(A)$) are called of the <u>1'st kind</u> (resp. <u>2'nd kind</u>). Note that σ is of the 1'st (resp. 2'nd) kind if and only if $\sigma \cdot \mathfrak{n}_+$ is commensurable with $\mathfrak{n}_+$ (resp. $\mathfrak{n}_-$).

Recall that N is the normalizer of H and hence of $\mathfrak{h}'$ in G(A) [9]. It follows that the normalizer of $\mathfrak{h}'$ in $\mathrm{Ad}\,\tilde{G}(A)$ is $\tilde{N} := \mathrm{Ad}((\mathrm{Aut}\mathbb{C} \times \mathrm{Aut}A) \ltimes (N \ltimes \tilde{H}))$. Put $\tilde{N}_o = \mathrm{Ad}(\mathrm{Aut}A \ltimes (N \ltimes \tilde{H}))$.

<u>Proposition 3.6</u>. Let A be a symmetrizable generalized Cartan matrix.

(a) Every finite order automorphism of the 1'st kind of $\mathfrak{g}'(A)$ can be conjugated into $\mathrm{Aut}A \ltimes \tilde{H}$.

(b) Every finite order automorphism of the 2'nd kind of $\mathfrak{g}'(A)$ can be conjugated into $\omega\,\tilde{N}_o$.

(c) Every finite order conjugate-linear automorphism of the 1'st kind of $\mathfrak{g}'(A)$ can be conjugated into $\eta\,\tilde{N}_o$.

(d) Every finite order conjugate-linear automorphism of the 2'nd kind of $\mathfrak{g}'(A)$ can be conjugated into $\omega\,\tilde{N}_o$.

<u>Proof</u>. Let σ be a finite order automorphism or conjugate- linear automorphism of the 1'st kind of $\mathfrak{g}'(A)$. Then $\sigma \in \mathrm{Ad}\,\tilde{G}(A)$ and by Theorem 2 we may assume, replacing σ by its conjugate, that σ normalizes $\mathfrak{p}_J$ and its finite-dimensional subalgebra $\mathfrak{a} = \mathfrak{p}_J \cap (\mathrm{Ad}\,w)\mathfrak{p}_{J'}^-$, where $w \in W$ and $J, J' \subset I$ are finite type subsets. By the

finite-dimensional theory (see e.g. [2, Theorem 4.5]), σ normalizes a Cartan subalgebra of $\mathfrak{a}$. Since $\mathfrak{h}'$ is the reductive part of a Cartan subalgebra of $\mathfrak{a}$, replacing σ by its conjugate, we may assume that σ normalizes $\mathfrak{h}'$. Hence $\sigma \in \tilde{N}_o$ if σ is an automorphism and $\sigma \in \eta\, \tilde{N}_o$ if σ is a conjugate-linear automorphism. This proves (c). To complete the proof of (a), note that σ acts on $\mathfrak{p}_J/\mathfrak{n}^J \cong \mathfrak{g}(A)_J$. By [2, Theorem 4.5], we may assume that σ fixes an element of $\mathfrak{h}'$ with positive eigenvalues on $\mathfrak{n}_+/\mathfrak{n}^J$. It follows that σ normalizes $\mathfrak{h}'$ and $\mathfrak{n}_+$. This completes the proof of (a).

Now let σ be an automorphism or conjugate-linear automorphism of the 2'nd kind and of finite order. By (a) we may assume that $\sigma^2(B) = B$. It follows from [13, Theorem 3], that $\sigma \cdot B = gB_-g^{-1}$ for some $g \in G(A)$. Writing $g = b_+nb_-$ according to the Birkhoff decomposition, we have: $B \cap \sigma \cdot B \supset b_+Hb_+^{-1}$. Replacing σ by its conjugate, we obtain that σ leaves stable an Ad-triangular subgroup $R: = B \cap \sigma \cdot B$ containing H. Applying [2, Theorem 7] (the proof of which works for conjugate linear automorphisms as well) we deduce that σ can be replaced by a conjugate by an element of R so that H becomes σ-invariant. This proves (b) and (d). □.

<u>Proposition 3.7</u>. Let σ be a conjugate-linear involution of the 2'nd kind of $\mathfrak{g}'(A)$. Then σ can be conjugated to a conjugate-linear involution of the following form for some involution $\nu \in \text{Aut}\, A$: $\sigma \cdot e_i = f_{\nu \cdot i}$, $\sigma \cdot f_i = e_{\nu \cdot i}$ if $\nu \cdot i \neq i$; $\sigma \cdot e_i = \pm f_i$, $\sigma \cdot f_i = \pm e_i$ if $\nu \cdot i = i$.

<u>Proof</u>. By Proposition 3.6(d) we may assume that $\sigma \in \omega\tilde{N}_o$. Then for each $\alpha \in \Delta_+$, the subspace $V_\alpha: = (\mathfrak{g}_\alpha + \mathfrak{g}_{-\alpha}) + \sigma \cdot (\mathfrak{g}_\alpha + \mathfrak{g}_{-\alpha})$ is $\sigma\omega$-invariant, the Hermitian form $H(x,y) = -(x|\omega \cdot y)$ is positive definite on V_α [7] (where $(\cdot|\cdot)$ is a standard invarinat bilinear form on $\mathfrak{g}'(A)$), and $\sigma\omega$ is self-adjoint with respect to this form. Since also $\sigma\omega|_{\mathfrak{h}'} = 1$, it follows that $\sigma\omega$ is diagonalizable on $\mathfrak{g}'(A)$ with real eigenvalues. Put $\mathfrak{r} = ((\sigma\omega)^2)^{1/4}$; this is an automorphism of

$\mathfrak{g}'(A)$ and one checks that $\varphi^{-1}\sigma\varphi$ commutes with ω (this argument goes back to E. Cartan). Thus, we may assume that σ commutes with ω. (So far we have not used that σ is of the 2'nd kind.) Then $\sigma\omega$ is an involution of the 1'st kind of $\mathfrak{g}'(A)$ which normalizes K(A). Using Theorem 2, we obtain that a K(A)-conjugate of $\sigma\omega$ normalizes a parabolic subalgebra $\mathfrak{p}_J$ where J is of finite type, and hence normalizes $\mathfrak{h}'$ and $\mathfrak{n}_+$, by the argument proving Proposition 3.6(a). This completes the proof. □

References.

1. Bausch J., Automorphismes des algèbres de Kac-Moody affines,C.R. Acad. Sci. Paris (1986).

2. Borel A., Mostow G.D., On semisimple automorphisms of Lie algebras, Ann. Math. (2) 61(1955), 389-405.

3. Bourbaki N., Groupes et algèbres de Lie, Ch. 4-5, Hermann, Paris, 1968.

4. Bruhat F., Tits J., Groupes réductifs sur un corps local, Publ. Math. IHES, 41(1972), 5-251.

5. Kac V.G. Infinite-dimensional Lie algebras, Progress in Math. 44, Birkhauser, Boston, 1983, Second edition: Cambridge University Press, 1985.

6. Kac V.G., Constructing groups associated to infinite-dimensioanl Lie algebras, Proceedings of the conference on Infinite-diminsional Lie groups, Berkeley 1984, MSRI Publ. #4, 1985, 167-216.

7.Kac V.G., Peterson D.H., Unitary structure in representations of infinite-dimensional groups and a convexity theorem, Invent. math. 76(1984), 1-14.

8. Kac V.G., Peterson D.H. Regular functions on certain infinite-dimensional groups, Arithmetic and Geometry (ed. M. Artin and J. Tate), Progress in Math. 36, Birhauser, Boston, 141-166, 1983.

9. Kac V.G., Peterson D.H., Defining relations of certain infinite-dimensional groups, Proceedings of the Cartan conference, Lyon 1984, Asterisque, Numero hors serie, 1985, 165-208.

10. Kempf G., Instability in invariant theory, Ann. Math. 108 (1978), 299-316.

11. Levstein F, A classification of involutive automorphisms of affine Kac-Moody Lie algebras, Thesis MIT, 1983.

12. Morita J., Conjugacy classes of the subalgebras X_ℓ in the affine Lie algebra $X_\ell^{(1)}$, preprint, 1986.

13. Peterson D.H., Kac V.G., Infinite flag varieties and conjugacy theorems, Proc. Natl. Acad. Sci. USA 80(1983), 1778-1782.

14. Slodowy P., An adjoint quotient for certain groups attached to Kac-Moody algebras, Proceedings of the conference on Infinite-dimensional groups, Berkeley 1984, MSRI Publ. #4, 1985, 307-333.

Victor G. Kac
Department of Mathematics
M.I.T.
Cambridge MA 02139
U.S.A.

Dale H. Peterson
Department of Mathematics
University of British Columbia
Vancouver
Canada

ETALE LOCAL STRUCTURE OF MATRIX INVARIANTS AND CONCOMITANTS

Lieven Le Bruyn (*)

University of Antwerp,Belgium

Claudio Procesi

University of Rome,Italy

(*) : work supported by an NFWO-grant

0 : INTRODUCTION

One of the basic problems in linear algebra is to study the equivalence classes of m-tuples of n by n matrices under simultaneous conjugation. This problem is readily seen to be equivalent to that of studying n-dimensional representations of the free algebra $\mathbb{C} < X_1, ..., X_m >$ upto equivalence.

Geometrically, one has to sudy the orbit structure of mn^2-dimensional affine space

$$X_{m,n} = M_n(\mathbb{C}) \bigoplus ... \bigoplus M_n(\mathbb{C})$$

under action by componentswise conjugation of the general linear group $GL_n(\mathbb{C})$. The space of orbits of $X_{m,n}$ is not Hausdorff due to the existence of non closed orbits. The classical approach is to approximate this space using $GL_n(\mathbb{C})$ invariants as parameters of a variety $V_{m,n}$ which will also be denoted $X_{m,n}/GL_n(\mathbb{C})$. The resulting map

$$\pi : X_{m,n} \to V_{m,n}$$

is surjective and the fiber $\pi^{-1}(p)$ of each point $p \in V_{m,n}$ contains a unique closed orbit. Thus $V_{m,n}$ parametrizes naturally the closed orbits of the action which , by Artin's fundamental paper [Ar] , are in one-to-one correspondence with the isomorphism classes of semi-simple n-dimensional representations of the free algebra $\mathbb{C} < X_1, ..., X_m >$.

From [Pr2] we get that the coordinate ring of $V_{m,n}$ is the center of the so called trace ring of m generic n by n matrices $\mathbb{T}_{m,n}$. By this we mean the following : consider the coordinate ring

$$\mathbb{C}[X_{m,n}] = \mathbb{C}[x_{ij}(l) : 1 \leq i,j \leq n; 1 \leq l \leq m]$$

then we can consider in $M_n(\mathbb{C}[X_{m,n}])$ the so called generic matrices

$$X_l = (x_{ij}(l))_{i,j} \in M_n(\mathbb{C}[X_{m,n}])$$

The $\mathbb{C}$-algebra generated by these elements is called the ring of m generic n by n matrices , $\mathbb{G}_{m,n}$. Then $\mathbb{T}_{m,n}$ is the subalgebra of $M_n(\mathbb{C}[X_{m,n}])$ generated by $\mathbb{G}_{m,n}$ and $Tr(\mathbb{G}_{m,n})$.

It is known that $V_{m,n}$ is a unirational variety and that the Formanek center of $\mathbb{T}_{m,n}$ (see [Pr3]) determines an open smooth subvariety of $V_{m,n}$.

In this paper we aim to initiate the geometric study of the varieties $V_{m,n}$ both globally and locally based on some powerful results of D. Luna [Lu]. We will now briefly describe the main results :

We say that a point $\xi \in V_{m,n}$ is of representation type $\tau = (e_1, k_1; ...; e_r, k_r)$ if the corresponding isomorphism class of semi-simple representations is built from r distinct simple components of dimensions k_i occuring with multiplicities e_i.If τ is such a representation type we call $V_{m,n}(\tau)$ the subset of $V_{m,n}$ consisting of all points of representation type τ. We will prove that the sets $V_{m,n}(\tau)$ form a finite stratification into locally closed smooth subvarieties. Furthermore, we prove that $V_{m,n}(\tau)$ lies in the closure of $V_{m,n}(\tau')$ if and only if τ is a degeneration (or refinement) of τ' (theorem II.1.1).

Next, we determine explicitly the étale local structure in a point $\xi \in V_{m,n}$ of representation type $(1, k_1; ...; 1, k_r)$ (theorem II.2.1). A generalization of this result to arbitrary representation type can be phrased in the setting of representations of quivers and their invariants in the following way. Let $\Delta = (\Delta_0, \Delta_1)$ be a finite quiver where Δ_0 is the set of vertices and Δ_1 the set of directed arrows between these vertices (we allow loops and multiple edges). A representation of Δ is a couple $V = (V_x, V_\alpha)$ where each V_x for $x \in \Delta_0$ is a finite dimensional vector space and the V_α for $\alpha \in \Delta_1$ are linear maps between the corresponding vectorspaces. $d = (dim(V_x))_x$ is called the dimension vector of the representation. Conversely, for any dimension vector d we can look at the space of all representations of Δ with this dimension vector. This is an affine space admitting a natural action of the reductive group

$$GL(\Delta, d) = \prod GL_{d_x}(\mathbb{C})$$

and we denote the corresponding quotient-variety by $V(\Delta, d)$. Now, let $\xi \in V_{m,n}$ be of representation type $(e_1, k_1; ...; e_r, k_r)$ then we can form a quiver Δ_ξ consisting of r vertices $\{x_1, ..., x_r\}$ and $(m-1)k_i^2 + 1$ loops in vertex x_i and $(m-1)k_i k_j$ directed edges from vertex x_i to x_j. Let d_ξ be the dimension vector $(e_1, ..., e_r)$. Then, one can show as in II.2 that a neighborhood of the origin in $V(\Delta_\xi, d_\xi)$ is analytically isomorphic to a neighborhood of ξ in $V_{m,n}$. Furthermore, we claim that the coordinate ring of these quotient varieties $V(\Delta, d)$ are generated by traces of oriented cycles in the quiver. In this paper we prove this claim for the special case Δ_ξ, d_ξ if ξ is of representation type $(1, k_1; ...; 1, k_r)$ using the results on tori-invariants.

In section II.3 we combine these results to prove that the singular locus of $V_{m,n}$ is determined by the Formanek center of $\mathbb{T}_{m,n}$.

In the final chapter we give an explicit description of the étale local structure of the trace ring of m generic n by n matrices $\mathbb{T}_{m,n}$ in a point $\xi \in V_{m,n}$ of representation type $(1, k_1; ...; 1, k_r)$. We prove that this 'noncommutative slice' is a Cohen-Macaulay module and that its Poincaré series satisfies a specific functional equation. In the final section, these results are applied in order to solve the regularity problem for trace rings of generic matrices : $gldim(\mathbb{T}_{m,n}) < \infty$ if and only if m or n is equal to one or $(m,n) = (2,2),(2,3)$ or $(3,2)$.

Acknowledgement

We thank D. Luna (Grenoble) for advising us to use his étale slice method to tackle regularity questions for trace rings of generic matrices and their centra.

The results of this paper were obtained while both authors visited the University of California at San Diego. We like to thank the department of mathematics and especially L. Small for the hospitality.

I : PRELIMINARIES

For the reader's convenience, we have collected in this first chapter the basic results which will be used throughout the paper. Further,we fix notation and terminology which we will use freely in the next two chapters.

I.1 : THE ETALE SLICE THEOREM

Throughout this paper, the basefield will be $\mathbb{C}$. Let G be a linear algebraic group. A representation of G is a finite dimensional complex vectorspace V (the representation space) together with a homomorphism of algebraic groups

$$\rho : G \rightarrow GL(V)$$

We denote this representation by ρ or (V,G). We will always assume that G is reductive, that is, every representation of G is completely reducible.

An action of G on a complex algebraic variety X is said to be rational if the canonical map

$$G \times X \rightarrow X$$

is a morphism of varieties. If the action is rational, G acts on the coordinate ring $\mathbb{C}[X]$ of X and one can define an algebraic variety X/G by $\mathbb{C}[X/G] = \mathbb{C}[X]^G$, the invariant ring under this action which happens to be affine. The natural inclusion $\mathbb{C}[X]^G \subset \mathbb{C}[X]$ gives rise to a morphism of varieties

$$\pi_X : X \to X/G$$

which is onto and separates disjoint G-stable closed subsets of X, [Mu,Ch.1,2].

Let $\xi \in X/G$, then it follows from these facts that the fiber $\pi_X^{-1}(\xi)$ contains exactly one closed orbit (that of minimal dimension) which we will denote by $T(\xi)$. Therefore, the points of the quotient variety X/G parametrize the closed G-orbits in X.

If x is a point of X , we will denote by $G(x)$ the orbit of x under G and by G_x the isotropy group of x, that is, the stabilizer of the point x. The theorem of Matsuchima [Ma] asserts that whenever the orbit $G(x)$ is closed, the isotropy group G_x is itself a reductive group.

Before stating the étale slice theorem, we need to recall some facts about fibers. Let H be a reductive group acting rationally on an affine variety Y and let G be a reductive group containing H as a subgroup. The group H acts on G by translations on the right which makes G into the total space of a principal fibration over H (locally trivial in the étale topology). That is, there exists an affine variety Z, an étale and onto morphism $Z \to G/H$ and an H-isomorphism

$$H \times Z \simeq G \times_{G/H} Z$$

Recall that a map between smooth complex algebraic varieties is étale if its differential is everywhere an isomorphism.

Now, H acts on $G \times Y$ by $h.(g, y) = (gh^{-1}, hy)$ and we denote the quotient under this action by $G \times^H Y$. The action of G on itself by translations on the left passes through this quotient in such a way that the projection

$$G \times^H Y \to G/H$$

commutes with the action of G. The action of G on $G \times^H Y$ is totally determined by the action of H on Y and $(G \times^H Y)/G$ is identical to Y/H.

Again, let X be a rational affine G-variety and let X' be a G-invariant subset of X. We say that X' is a G-saturated subset of X if $\pi_X^{-1}(\pi_X(X')) = X'$. If X' is open (resp. closed) we also have that the natural map $X'/G \to X/G$ is an open (resp. closed) embedding.

We can now state a version of Luna's slice theorem [Lu,p.97] or [Sc,Th.5.3,p.55]

Theorem I.1.1 : Let X be a representation space of G. Let $x \in X$ be such that the orbit $G(x)$ is closed. Choose a G_x-splitting of $X \simeq T_x(X)$ as $T_x(G(x)) \oplus N_x$ and let ϕ denote the canonical equivariant map

$$\phi : G \times^{G_x} N_x \to X$$

$$[g, n] \to g.(x + n)$$

Then there exists an affine open G-saturated subset Y of X and an affine open G_x-saturated neighborhood B_x of 0 in N_x such that the maps

$$\phi : G \times^{G_x} B_x \to Y$$

$$\overline{\phi} : (G \times^{G_x} B_x)/G \to Y/G$$

are both étale, where $\overline{\phi}$ denotes the map induced by ϕ. Also, ϕ and the natural map $G \times^{G_x} B_x \to B_x/G_x$ give rise to a fiber product diagram

$$\begin{array}{ccc} G \times^{G_x} B_x & \longrightarrow & B_x/G_x \simeq (G \times^{G_x} B_x)/G \\ \Big\downarrow \phi & & \Big\downarrow \overline{\phi} \\ Y & \longrightarrow & Y/G \end{array}$$

with a G-isomorphism

$$G \times^{G_x} B_x \cong Y \times_{Y/G} B_x/G_x$$

Whereas ϕ is a morphism between smooth varieties, $\overline{\phi}$ usually is not. We recall that a morphism between two affine complex algebraic varieties is étale in a point iff it is a local isomorphism

We also wish to interpret the module $\mathbb{C}E_{\Phi,\alpha}$ in terms of invariant theory. Suppose that the equation $\Phi.\beta = \alpha$ has at least one integral solution $\beta \in \mathbb{Z}^s$, then the map

$$\chi_\alpha : T_r \rightarrow \mathbb{C}^*$$

defined by $\chi_\alpha(\tau_u) = u^\alpha$ is a one- dimensional representation (or character) of T_r. Then, one can verify that $\mathbb{C}E_{\Phi,\alpha}$ is the module of semi-invariants or relative invariants of T_r with respect to the character χ_α, i.e.

$$CE_{\Phi,\alpha} =$$

$$\{f(x_1, ..., x_s) \in \mathbb{C}[x_1, ..., x_s] : \forall \tau_u \in T_r : \tau_u.f(x_1, ..., x_s) = \chi_\alpha(\tau_u) f(x_1, ..., x_s)\}$$

From this one can deduce that $\mathbb{C}E_{\Phi,\alpha}$ is a finitely generated $\mathbb{C}E_\Phi$-module. However, it is not true in general that $\mathbb{C}E_{\Phi,\alpha}$ is a Cohen-Macaulay module. Stanley [St,Th.3.2] has proved the following sufficient condition :

Proposition I.2.1 : Suppose there exists a rational solution $\beta = (\beta_1, ..., \beta_s)^\tau \in \mathbb{Q}^s$ to $\Phi.\beta = \alpha$ satisfying $-1 < \beta_i \leq 0$ for all i , then $\mathbb{C}E_{\Phi,\alpha}$ is a Cohen-Macaulay module.

In [St] and [Ho] it was shown that ring- and moduletheoretic properties of $\mathbb{C}E_\Phi$ and $\mathbb{C}E_{\Phi,\alpha}$ are closely linked to topological properties of the $(s-r)$-dimensional convex polyhedral cone C_Φ in $\mathbb{R}^s$ consisting of the set of all solutions $\beta \in \mathbb{R}^s$ to $\Phi.\beta = 0$.

Let P_Φ be the non-degenerate cross-section $C_\Phi \cap \{(\beta_1, ..., \beta_s) \in \mathbb{R}^s : \sum \beta_i = 1\}$ of C_Φ , then P_Φ is an $(s-r-1)$-dimensional convex polytope.

If $\beta \in \mathbb{R}^s_+$, we define its support $supp(\beta) = \{i : \beta_i > 0\}$ and if $\beta \in \mathbb{R}^s$ we define its negative support $supp_-(\beta) = \{i : \beta_i < 0\}$. If $\mathcal{F}$ is a face of P_Φ , then all elements of the relative interior of $\mathcal{F}$ have the same support, $supp(\mathcal{F})$. It follows that the faces of P_Φ are in one-to-one correspondence with the supports of elements $\beta \in E_\Phi$ and that two faces $\mathcal{F}, \mathcal{G}$ satisfy $\mathcal{F} \subset \mathcal{G}$ iff $supp(\mathcal{F}) \subset supp(\mathcal{G})$. If v is a vertex of P_Φ , then those elements $\beta \in E_\Phi$ satisfying $supp(\beta) = supp(v)$ are $\mathbb{N}$-multiples of a unique element $\beta_v \in E_\Phi$, the so called completely fundamental solution corresponding to the vertex v. Recall that $\beta \in E_\Phi$ is said to be completely fundamental if whenever $m\beta = \gamma + \delta$ where $m \geq 1$ and $\gamma, \delta \in E_\Phi$, then $\gamma = i\beta$ for some $0 \leq i \leq m$. With $CF(E_\Phi)$ one denotes the set of completely fundamental solutions. The Krull dimension of $\mathbb{C}E_\Phi$ is equal to the dimension of the $\mathbb{Q}$- vectorspace spanned by $CF(E_\Phi)$, i.e. is equal to $s - r$.

Let $< E_\Phi >$ denote the group generated by E_Φ in $\mathbb{Z}^s$ and let $< E_{\Phi,\alpha} > =$ $< E_\Phi > + E_{\Phi,\alpha}$, the coset of $< E_\Phi >$ in $\mathbb{Z}^s$ containing $E_{\Phi,\alpha}$. Now, let $\mathcal{P}_\Phi^*$ be the dual polytope of $\mathcal{P}_\Phi$ and define

$$\Gamma_\beta = \bigcup \{\mathcal{F}^* : \mathcal{F} \text{ face of } \mathcal{P}_\Phi \text{ s.t. } supp_-(\beta) \subset supp(\mathcal{F})\}$$

From [St2,p.47] we recall that if $\mathbb{C}E_{\Phi,\alpha}$ is a Cohen-Macaulay module of dimension d , its d-th local cohomology module with respect to the irrelevant ideal of $\mathbb{C}E_\Phi$ is given by

$$H^d(\mathbb{C}E_{\Phi,\alpha}) = \mathbb{C}\{x^\beta : \beta \in < E_{\Phi,\alpha} >, \Gamma_\beta = \emptyset\}$$

The $\mathbb{C}E_\Phi$-module structure on it is given by

$$x^\gamma . x^\beta = x^{\beta+\gamma} \text{ if } \Gamma_{\gamma+\beta} = \emptyset$$

$$= 0 \text{ if } \Gamma_{\gamma+\beta} \neq \emptyset$$

for all $\gamma \in E_\Phi$ and $x^\beta \in H^d(\mathbb{C}E_{\Phi,\alpha})$.

Suppose there exists $\gamma = (\gamma_1, ..., \gamma_s) \in \mathbb{Q}^s$ with $-1 < \gamma_i \leq 0$ such that $\Phi.\beta = \alpha$, then by Prop.I.2.1, $\mathbb{C}E_{\Phi,\alpha}$ is a Cohen-Macaulay module. In this case, $H^d(\mathbb{C}E_{\Phi,\alpha})$ can be described as follows : choose $m \in \mathbb{N}$ such that $m\gamma$ is integral and let $\beta \in < E_{\Phi,\alpha} >$, then $\Phi.m(\beta - \gamma) = 0$ and $supp_-(\beta) = supp_-(\beta - \gamma) = supp_- m(\beta - \gamma)$. Since Γ_β depends only upon $supp_-(\beta)$ we have $\Gamma_\beta = \Gamma_{m(\beta-\gamma)}$. If we assume,further,that there exists a solution $\delta = (\delta_1, ..., \delta_s) \in E_\Phi$ such that $\delta_i > 0$ for all i , then since $m(\beta - \gamma) \in < E_\Phi >$ we have [St2,p.47] : $\Gamma_{m(\beta-\gamma)} = \emptyset$ iff $m(\beta - \gamma) < 0$. Therefore, we have

$$H^d(\mathbb{C}E_{\Phi,\alpha}) = \mathbb{C}\{x^\beta : \beta \in < E_{\Phi,\alpha} >: \beta < 0\}$$

I.3 : FINITE DIMENSIONAL REPRESENTATIONS

An n-dimensional representation of the free algebra in m variables is an algebra morphism

$$\phi : \mathbb{C} < X_1, ..., X_m > \rightarrow M_n(\mathbb{C})$$

Note that this is equivalent to giving m elements $\phi(X_1), ..., \phi(X_m)$ in $M_n(\mathbb{C})$, i.e. n-dimensional representations of $\mathbb{C} < X_1, ..., X_m >$ can be parametrized by an affine variety $X_{m,n}$ which is just mn^2-dimensional affine space.

Two n-dimensional representations are said to be equivalent if they differ upto a $\mathbb{C}$-automorphism of $M_n(\mathbb{C})$. So, the projective linear group $PGL_n(\mathbb{C})$ acts on $X_{m,n}$ and the orbits under this action are the equivalence classes of representations.

If ϕ is an n-dimensional representation, then $\mathbb{C}^{(n)}$ becomes a $\mathbb{C} < X_1, ..., X_m >$-module via ϕ. If $\mathbb{C}^{(n)}$ is completely reducible as such, then ϕ is said to be semi-simple. In general one has a composition series

$$0 = V_t \subset V_{t-1} \subset ... \subset V_1 \subset V_0 = \mathbb{C}^{(n)}$$

of $\mathbb{C} < X_1, ..., X_m >$-modules. Then, $W = \oplus(V_i/V_{i+1})$ is completely reducible and n-dimensional. For a suitable choice of basis of $\mathbb{C}^{(n)}$,ϕ can be expressed in the matrix-form

$$\phi = \begin{pmatrix} \phi_1 & & N \\ & \ddots & \\ 0 & & \phi_t \end{pmatrix}$$

where the $\phi_j : \mathbb{C} < X_1, ..., X_m > \rightarrow M_{k_j}(\mathbb{C})$ for $k_j = dim_{\mathbb{C}}(V_{j-1}/V_j)$ are the irreducible components. With ϕ we can associate the semi-simple representation

$$\phi^{ss} = \phi_1 \oplus ... \oplus \phi_t$$

isomorphic to W as module.

Artin [Ar] proved that ϕ^{ss} lies in the closure of the orbit $GL_n(\mathbb{C})(\phi)$.Moreover he proved that the closed orbits in $X_{m,n}$ under action of $GL_n(\mathbb{C})$ correspond precisely to the semi-simple representations. Therefore, the quotient variety

$$V_{m,n} = X_{m,n}/GL_n(\mathbb{C})$$

parametrizes the equivalence classes of semi-simple n- dimensional representations of $\mathbb{C} < X_1, ..., X_m >$.

A concrete description of the coordinate ring $\mathbb{C}[V_{m,n}]$ can be given as follows. Consider the polynomial ring

$$\mathcal{P}_{m,n} = \mathbb{C}[x_{ij}(l) : 1 \leq i, j \leq n; 1 \leq l \leq m]$$

and let $\mathbb{G}_{m,n}$ be the ring of m generic n by n matrices, that is the subalgebra of $M_n(\mathcal{P}_{m,n})$ generated by the m elements

$$X_l = (x_{ij}(l))_{i,j} \in M_n(\mathcal{P}_{m,n})$$

The ring of matrixinvariants $\mathcal{R}_{m,n}$ is the subalgebra of $\mathcal{P}_{m,n}$ generated by the traces of elements of $\mathbb{G}_{m,n}$. See for example [Pr] for a proof that $\mathcal{R}_{m,n}$ is an affine algebra and $\mathbb{C}[V_{m,n}] = \mathcal{R}_{m,n}$.

Not much is known about the geometry of $V_{m,n}$. In [Pr] it was proved that the points corresponding to the equivalence classes of irreducible n-dimensional representations form an open smooth subvariety $V_{m,n}^{irr}$ of dimension $(m-1)n^2 + 1$.

If one is not only interested in the semi-simple representations but also in their irreducible components one has to study a certain 'noncommutative algebraic variety' $U_{m,n}$. The trace ring of m generic n by n matrices is the subalgebra of $M_n(\mathcal{P}_{m,n})$ generated by $\mathcal{R}_{m,n}$ and $\mathbb{G}_{m,n}$ and will be denoted by $\mathbb{T}_{m,n}$.

The points of $U_{m,n}$ are the maximal twosided ideals of the noncommutative but affine p.i.-algebra $\mathbb{T}_{m,n}$. We can equip $U_{m,n}$ with the usual Zariski topology [Pr3], that is a typical open set consists of those maximal ideals not containing a given twosided ideal of $\mathbb{T}_{m,n}$. Since $\mathcal{R}_{m,n} \subset \mathbb{T}_{m,n}$ is a central extension, there is a canonical continuous map

$$i : U_{m,n} \rightarrow V_{m,n}$$

In [AS] the fibers of i were described in the following way. Let $\xi \in V_{m,n}$, then ξ corresponds to the equivalence class of a semi-simple representation

$$\phi = \phi_1 \oplus ... \oplus \phi_t$$

Let $\chi_1, ..., \chi_r$ be the distinct irreducible components, where χ_i is a k_i-dimensional representation occuring with multiplicity e_i in ϕ. That is, $\sum_{i=1}^{r} e_i = t$ and $\sum_{i=1}^{r} e_i.k_i = n$. We can always assume that $k_1 \geq k_2 \geq ... \geq k_r$ and then we say that ξ or ϕ is of representation-type $(e_1, k_1; e_2, k_2; ...; e_r, k_r)$.

The fiber $i^{-1}(\xi)$ consists of r points (ϕ, ξ_i) each corresponding to one of the distinct irreducible components. The morphism i is then given by sending a point $\xi = (\phi, \xi_i)$ to ξ.

II : THE VARIETIES $V_{m,n}$

In this chapter we aim to initiate the geometrical study of the varieties $V_{m,n}$. In the first section we will show that the different representation types give a finite stratification of $V_{m,n}$ into locally closed smooth subvarieties. In the second section we describe the étale local structure of $V_{m,n}$ in points corresponding to semi-simple representations with distinct irreducible components. In the last section we show that, except when $(m, n) = (2, 2)$, $V_{m,n}$ is always singular and the singular locus is precisely the difference $V_{m,n} - V_{m,n}^{irr}$.

II.1 : STRATIFICATION OF $V_{m,n}$

Recall that a point $\xi \in V_{m,n}$ is said to be of representation-type $\tau = (e_1, k_1; ...; e_r, k_r)$ if the corresponding semi-simple n-dimensional representation has r distict irreducible components χ_i of dimension k_i and multiplicity e_i.

Another representation-type $\tau' = (e'_1, k'_1; ...; e'_r, k'_r)$ is said to be a refinement of τ if there is a permutation σ on $\{1, ..., r'\}$ such that there exist natural numbers

$$j_0 = 1 \leq j_1 < j_2 < ... < j_r = r'$$

such that for every $1 \leq i \leq r$ we have

$$\begin{cases} e_i k_i = \displaystyle\sum_{j=j_{i-1}+1}^{j_i} e'_{\sigma(j)} k'_{\sigma(j)} \\ e_i \mid e'_{\sigma(j)} \text{ for all } j_{i-1} < j \leq j_i \end{cases}$$

This defines a partial ordering on the set of all representation- types for n-dimensional representations : RT_n. For example, RT_4 has the following Hasse-diagram :

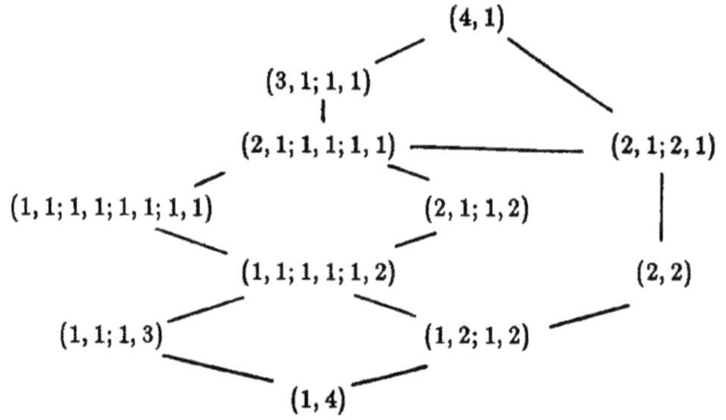

For a representation-type $\tau \in T_n$, we denote by $V_{m,n}(\tau)$ the set of all points $\xi \in V_{m,n}$ of type τ. The main result of this section can now be stated as :

Theorem II.1.1 : With notations as above we have :

(1) : $\{V_{m,n}(\tau) : \tau \in RT_n\}$ is a finite stratification of $V_{m,n}$ into locally closed irreducible smooth algebraic subvarieties.

(2) : $V_{m,n}(\tau')$ lies in the closure of $V_{m,n}(\tau)$ if and only if τ' is a refinement of τ.

Proof :

Let $\tau = (e_1, k_1; ...; e_r, k_r)$ and $\xi \in V_{m,n}(\tau)$, then the fiber of ξ under the morphism $\pi_{m,n}$: $X_{m,n} \to V_{m,n}$ contains one closed orbit $T(\xi)$. In this orbit, one can find a point $x = (x_1, ..., x_m)$ where each n by n matrix x_i is of the form

$$\begin{pmatrix} m_1 \otimes 1_{e_1} & 0 & \\ 0 & m_2 \otimes 1_{e_2} & 0 \\ 0 & & m_r \otimes 1_{e_r} \end{pmatrix} = x_i$$

where each $m_i \in M_{k_i}(\mathbb{C})$.

We will now compute the isotropy group in this point : $GL_n(\mathbb{C})_x$. An element $\alpha \in GL_n(\mathbb{C})$ leaves x fixed if and only if it commutes with each of the x_i. Therefore, $GL_n(\mathbb{C})_x$ is the multiplicative group of units of the centralizer of

$$\begin{pmatrix} M_{k_1}(\mathbb{C}) \otimes 1_{e_1} & & 0 \\ & \ddots & \\ 0 & & M_{k_r}(\mathbb{C}) \otimes 1_{e_r} \end{pmatrix}$$

which is the algebra generated by the x_i's by assumption. It is easy to verify that this group is equal to

$$GL_n(\mathbb{C})_x \simeq GL_{e_1}(\mathbb{C}) \times ... \times GL_{e_r}(\mathbb{C})$$

where the embedding in $GL_n(\mathbb{C})$ is given by

$$GL_{e_1}(\mathbb{C}).1_{k_1} \times ... \times GL_{e_r}(\mathbb{C}).1_{k_r}$$

Of course, a different choice of the element x in $T(\xi)$ gives a group conjugated to $GL_n(\mathbb{C})_x$ in $GL_n(\mathbb{C})$. Further, if $x' \in T(\xi')$ is chosen such that

$$GL_n(\mathbb{C})_{x'} = GL_{e'_1}(\mathbb{C}).1_{k'_1} \times ... \times GL_{e'_r}(\mathbb{C}).1_{k'_r}$$

and if $GL_n(\mathbb{C})_{x'}$ is conjugated to $GL_n(\mathbb{C})_x$ in $GL_n(\mathbb{C})$, then it is clear that $r = r'$ and there exists a permutation σ on $\{1, ..., r\}$ such that $(e_i, k_i) = (e'_{\sigma(i)}, k'_{\sigma(i)})$, i.e. ξ and ξ' belong to the same set $V_{m,n}(\tau)$.

The statement now follows immediatly from Theorem I.1.2.(1).

(2) : In the first part we have shown that the isotropy corresponding to $V_{m,n}(\tau)$, where $\tau = (e_1, k_1; ..., e_r, k_r)$, is tho conjugacy class of

$$GL_{e_1}(\mathbb{C}).1_{k_1} \times ... \times GL_{e_r}.1_{k_r} \stackrel{\text{def}}{=} GL_n(\mathbb{C})_\tau$$

in $GL_n(\mathbb{C})$. From theorem I.1.2.(2) we know that $V_{m,n}(\tau')$ lies in the closure of $V_{m,n}(\tau)$ if and only if the group $GL_n(\mathbb{C})_\tau$ is conjugated to a subgroup of $GL_n(\mathbb{C})_{\tau'}$.It is easy to verify that this happens precisely when τ' is a refinement of τ.

For example, the closed subvariety of $V_{m,n}$ determined by the Formanek center of the trace ring of m generic n by n matrices (in out terminology $V_{m,n} - V_{m,n}(1, n)$) is in general reducible.

Each of its $\lceil \frac{n}{2} \rceil$ irreducible components contains an open set induced by $V_{m,n}(1,i;1,n-i)$.

Finally, we note that the dimension of the subvariety $V_{m,n}(\tau)$ where $\tau = (e_1,k_1;...;e_r,k_r)$ is equal to $(m-1)(k_1^2+...+k_r^2)+r$.

II.2 : LOCAL STRUCTURE OF $V_{m,n}$

According to the étale slice theorem,the local structure of the variety $V_{m,n} = X_{m,n}/GL_n(\mathbb{C})$ near a point ξ is isomorphic to that of the quotient of the slice representation near the origin , i.e. with $N_x/GL_n(\mathbb{C})_x$ where $x \in T(\xi)$ and N_x is the normal space in $X_{m,n}$ to the orbit $GL_n(\mathbb{C})(x)$.

Suppose ξ is a point of type $(e_1,k_1;...;e_r,k_r)$, then we can take for $x = (x_1,...,x_m) \in X_{m,n}$ such that each of the x_i has the form

$$x_i = \begin{pmatrix} m_1 \otimes 1_{e_1} & & \\ 0 & \ddots & 0 \\ & & m_r \otimes 1_{e_r} \end{pmatrix}$$

where $m_i \in M_{k_i}(\mathbb{C})$.

In the foregoing section we have calculated the isotropy group in such a point

$$GL_n(\mathbb{C})_x = GL_{e_1}(\mathbb{C}).1_{k_1} \times ... \times GL_{e_r}(\mathbb{C}).1_{k_r}$$

The tangent space $T_x(GL_n(\mathbb{C})(x))$ in $X_{m,n}$ to the orbit $GL_n(\mathbb{C})(x)$ is equal to the image of the linear map

$$M_n(\mathbb{C}) \to M_n(\mathbb{C}) \bigoplus ... \bigoplus M_n(\mathbb{C})$$

$$y \to [y,x_1] \bigoplus ... \bigoplus [y,x_m]$$

see for example [Mo].

The kernel of this map is clearly the centralizer of the subalgebra of $M_n(\mathbb{C})$ generated by $\{x_1,...,x_m\}$.

So, we obtain an exact sequence of $GL_n(\mathbb{C})_x$-modules

$$0 \to C_x \to M_n(\mathbb{C}) \to T_x(GL_n(\mathbb{C})(x)) \to 0$$

where

$$C_x = \begin{pmatrix} M_{e_1}(\mathbb{C}) \otimes 1_{k_1} & & \\ 0 & \ddots & 0 \\ & & M_{e_r}(\mathbb{C}) \otimes 1_{k_r} \end{pmatrix}$$

But then, since $GL_n(\mathbb{C})_x$ is a reductive group (so every $GL_n(\mathbb{C})_x$- module is determined upto isomorphism by its irreducible components), the normal space N_x to $T_x(GL_n(\mathbb{C})(x))$ is isomorphic to the $GL_n(\mathbb{C})_x$-module

$$N_x = M_n(\mathbb{C}) \bigoplus ... \bigoplus M_n(\mathbb{C}) \bigoplus C_x$$

where we have $m-1$ copies of $M_n(\mathbb{C})$ and the action of the isotropy group $GL_n(\mathbb{C})_x$ is,of course, given by componentswise conjugation.

The étale slice is then the variety corresponding to the ring of invariant polynomial mappings from N_x to $\mathbb{C}$ under this action of $GL_n(\mathbb{C})_x$.

We will now describe this ring in the special case that all the irreducible components χ_i of the to ξ associated semi-simple representation are distinct, that is, ξ is of type $(1, k_1; ...; 1, k_r)$ where $\sum k_i = n$.

In this case, the isotropy group of x is the r-dimensional torus T_r which is embedded in $GL_n(\mathbb{C})$ as

$$T_r = \mathbb{C}^*.1_{k_1} \times ... \times \mathbb{C}^*.1_{k_r}$$

Clearly, T_r acts trivially on the following subspace N_1 of N_x

$$N_1 = \bigoplus_{i=1}^{m-1} \begin{pmatrix} M_{k_1}(\mathbb{C}) & & \\ 0 & \ddots & 0 \\ & & M_{k_r}(\mathbb{C}) \end{pmatrix} \bigoplus \begin{pmatrix} \mathbb{C}.1_{k_1} & & \\ 0 & \ddots & 0 \\ & & \mathbb{C}.1_{k_r} \end{pmatrix}$$

so, the étale slice is

$$N_x/T_r = \mathbb{A}^d \times N_2/T_r$$

where $d = (m-1)(k_1^2 + ... + k_r^2) + r$ and

$$N_2 = \bigoplus_{i \neq j}^{r} V_{ij}$$

where V_{ij} is an $(m-1)k_ik_j$-dimensional vectorspace on which an element $t = (\alpha_1, ..., \alpha_r) \in T_r$ acts by sending an element $v \in V_{ij}$ to $\alpha_i\alpha_j^{-1}v$. Let

$$S = \mathbb{C}[N_2] = \mathbb{C}[v_{ij}(\alpha) : 1 \leq i \neq j \leq r; 1 \leq \alpha \leq (m-1)k_ik_j]$$

then the action of T_r on S is diagonal and is therefore determined by an r by s matrix with integer coefficients where $s = Kdim(S) = 2(m-1)\sum_{i\neq j}^{r} k_i k_j$.

The column corresponding to the variable $v_{ij}(\alpha)$ consists of zeroes except at the i-th row $+1$ and at the j-th row -1.

One easily verifies that the last row is a linear combination of the others so we can restrict attention to the $r-1$ by s matrix Φ which is obtained by erasing this last row. One verifies that $rk(\Phi) = r-1$.

The ring of invariants, $\mathbb{C}[N_2/T_r]$, is obtained from the set of integer solutions $\beta \in \mathbb{N}^s$ to $\Phi.\beta = 0$. As we have seen before, it suffices to consider the fundamental solutions. In this case, they are also completely fundamental. The corresponding monomials in $S = \mathbb{C}[N_2]$ are obtained by the following procedure :

Let $2 \leq k \leq r$ and let $(i_1, ..., i_k)$ be a cycle of k distinct elements from $\{1, ..., r\}$ s.t. its minimal element is i_1. Then,we get the invariants

$$V_{i_1 i_2} \times V_{i_2 i_3} \times ... \times V_{i_{k-1} i_k} \times V_{i_k i_1}$$

which are generated by the elements

$$v_{i_1 i_2}(\alpha_1)...v_{i_{k-1} i_k}(\alpha_{k-1}) v_{i_k i_1}(\alpha_k)$$

where the α_j run over all admissible values.

Finally, we note that

$$dim(N_2/T_r) = s - (r-1) = 2(m-1)\sum k_i k_j - r + 1$$

which is compatible with the fact that

$$(m-1)n^2 + 1 = dim(V_{m,n} = d + dim(N_2/T_r)$$

In the next section we will give a more precise description in the special case that $r = 2$. Let us summarize things in

Theorem II.2.1 :

If ξ is a point in $V_{m,n}$ of type $(1,k_1;...;1,k_r)$, then a neighbourhood of ξ is isomorphic to a neighborhood of the origin in $\mathbb{A}^d \times N_2/T_r$ where $d = (m-1)(k_1^2 + ... + k_r^2) + r$, $N_2 = \bigoplus_{i \neq j}^r V_{ij}$ where V_{ij} is $(m-1)k_ik_j$-dimensional and T_r acts on it by $(\alpha_1,...,\alpha_r).v = \alpha_i\alpha_j^{-1}v$.

Further, the coordinate ring $\mathbb{C}[N_2/T_r]$ is the subring of $\mathbb{C}[N_2] = \mathbb{C}[v_{ij}(\alpha) : 1 \leq i \neq j \leq r, 1 \leq \alpha \leq (m-1)k_ik_j]$ generated by all monomials of the form $v_{i_1i_2}(\alpha_1)...v_{i_{k-1}i_k}(\alpha_{k-1})v_{i_ki_1}(\alpha_k)$ where $(i_1,i_2,...,i_k)$ is a cycle of length $2 \leq k \leq r$ of distinct elements from $\{1,...,r\}$. Its Krull dimension is $2(m-1)\sum k_ik_j - r + 1$.

II.3 : SINGULAR LOCUS OF $V_{m,n}$

The main result of this section states that the closed subvariety of $V_{m,n}$ determined by the Formanek center of the trace ring of m generic n by n matrices (or,equivalently, the set of reducible semi-simple representations) is precisely the singular locus of $V_{m,n}$.

If m or n is equal to 1, $V_{m,n}$ is clearly nonsingular, so we may assume that m and $n \geq 2$.

Proposition II.3.1 : The variety $V_{m,n}$ is singular except when $(m,n) = (2,2)$.

Proof :

Assume that $V_{m,n}$ is nonsingular. Since $\mathcal{R}_{m,n}$ is a positively graded affine algebra, it has to be a polynomial ring in $(m-1)n^2+1$ variables over $\mathbb{C}$. So, the Brauer group $Br(V_{m,n})$ is just $Br(\mathbb{C}) = 1$.

R. Hoobler proved in [Hb] the Auslander-Goldman conjecture stating that the Brauer group of a smooth affine variety is determined by the codimension one irreducible subvarieties. Therefore, we have

$$Br(V_{m,n}) = \bigcap_{p \in X^{(1)}} Br((\mathcal{R}_{m,n})_p)$$

where $X^{(1)}$ is the set of all height one prime ideals of $\mathcal{R}_{m,n}$ and the intersection is taken in the

Brauer group of the field of fractions $\mathcal{K}_{m,n}$.

Now, we know that the localization $(\mathbb{T}_{m,n})_p$ of $\mathbb{T}_{m,n}$ at any height one prime ideal p of $\mathcal{R}_{m,n}$ is Azumaya except for $(m,n) = (2,2)$ and $p = (X_1X_2 - X_2X_1)^2$.

For, the dimension of the closed subvariety determined by p is equal to $(m-1)n^2$. Suppose that the corresponding localization of $\mathbb{T}_{m,n}$ is not Azumaya, then the points lying on this closed subvariety correspond to reducible semi-simple representations.

It follows from our stratification result (Th.II.1.1) that such a variety of maximal dimension has an open subset consisting of semi-simple representations with two irreducible components of dimensions r and $n-r$. Therefore, the dimension of such a variety is at most $(m-1)[r^2+(n-r)^2]+2$. Clearly, the equation

$$(m-1)[r^2 + (n-r)^2] + 2 = (m-1)n^2$$

has only an integer solution if $(m,n) = (2,2)$ and $r = 1$. It is well known from 19-th century algebra that $\mathcal{R}_{2,2} = \mathbb{C}[Tr(X_1), Tr(X_2), D(X_1), D(X_2), Tr(X_1X_2)]$. Therefore, the class of the generic division algebra $\Delta_{m,n}$ in $Br(\mathcal{K}_{m,n})$ belongs to $\bigcap_{p\in X^{(1)}} Br((\mathcal{R})_p)$ provided $(m,n) \neq (2,2)$, and so we obtain a contradiction.

We will now investigate when $V_{m,n}$ is smooth in a point ξ corresponding to an equivalence class of a semi-simple representation having two distinct irreducible components.

Proposition II.3.2 : The étale slice of $V_{m,n}$ in a point ξ of type $(1,r;1,n-r)$ is $\mathbb{A}^d \times W$ where $d = (m-1)[r^2 + (n-r)^2] + 2$ and

$$\mathbb{C}[W] = \mathbb{C}[t_{ij} : 1 \leq i,j \leq (m-1)r(n-r)]/I_2$$

where I_2 is the ideal generated by all 2 by 2 minors of the generic matrix $(t_{ij})_{i,j}$.

Proof :

By the calculations of the foregoing section we know that the slice in ξ is equal to

$$\mathbb{A}^d \times N_2/T_2$$

where $N_2 = V_{12} \bigoplus V_{21}$ and both components are $(m-1)r(n-r)$-dimensional and an element $(\alpha, \beta) \in T_2$ acts on a generator $x_i \in S(V_{12})$ (resp. $y_j \in S(V_{21})$) by sending it to $\alpha\beta^{-1}x_i$ (resp. $\alpha^{-1}\beta y_j$). Therefore, the invariant ring $\mathbb{C}[N_2/T_2]$ is generated by the monomials $x_i y_j$ for all admissible values for i and j. The relations among these invariants are easily seen to be generated by the 2 by 2 minors of the matrix $(x_i y_j)_{i,j}$. Sending the indeterminate t_{ij} to $x_i y_j$ we get the required statement.

Proposition II.3.3 : If ξ is a point of $V_{m,n}$ of type $(1, r; 1, n-r)$, then $V_{m,n}$ is singular in ξ except when $(m, n) = (2, 2)$.

Proof :

By étale descent, it suffices to show that the étale slice is singular in the origin. This follows from the fact that I_2 is a nontrivial ideal (if $(m, n) \neq (2, 2)$) not generated by degree one elements (in the obvious gradation on $\mathbb{C}[t_{ij}]$.

Using this fact and the stratification result of II.1, we can now prove the main result of this section :

Theorem II.3.4 : The singular locus of the variety $V_{m,n}$ coincides with the complement $V_{m,n} - V_{m,n}^{irr}$ except when $(m, n) = (2, 2)$.

Proof :

Let $F\mathbb{T}_{m,n}$ be the Formanek center of the trace ring of m generic n by n matrices, i.e. the ideal defining the open set $V_{m,n}^{irr}$. Let $\overline{V_{m,n}}$ be the variety defined by $\mathbb{C}[\overline{V_{m,n}}] = \mathcal{R}_{m,n}/F\mathbb{T}_{m,n}$, then by theorem II.1.1 we know that each of the irreducible components of $\overline{V_{m,n}}$ has an open set determined by semi-simple representations having two distinct irreducible components.

Suppose that $V_{m,n}^{reg}$, the open set of all regular points in $V_{m,n}$, is strictly larger than $V_{m,n}^{irr}$, then $V^{reg}_{m,n}$ induces a proper open subvariety in at least one of the irreducible components of $\overline{V_{m,n}}$. This entails that $V^{reg}_{m,n}$ contains points corresponding to semi-simple representations having two distinct irreducible components, but this is impossible by proposition II.3.3.

III : TRACE RINGS OF GENERIC MATRICES.

In this chapter we will investigate the étale local structure of the trace ring of m generic n by n matrices, $\mathbb{T}_{m,n}$. If $\xi \in V^{irr}_{m,n}$, it is well known that this étale local structure is just n by n matrices over a commutative (regular) domain. We will describe explicitly the structure when ξ is a point corresponding to a semi-simple representation with distinct irreducible components. It will turn out that the 'noncommutative slice' in such a point is Cohen-Macaulay and its Poincaré series satisfies a certain functional equation. In the final section, these results are applied to solve the regularity problem for trace rings, i.e. $gldim(\mathbb{T}_{m,n}) < \infty$ iff m or n is 1 or $(m,n) = (2,2), (2,3)$ or $(3,2)$.

III.1 : LOCAL STRUCTURE OF $\mathbb{T}_{m,n}$

Recall from [Pr2] that the trace ring of m generic n by n matrices is the ring of equivariant maps :

$$\phi : X_{m,n} = M_n(\mathbb{C}) \bigoplus \ldots \bigoplus M_n(\mathbb{C}) \rightarrow M_n(\mathbb{C})$$

i.e. polynomial maps such that for every $\alpha \in GL_n(\mathbb{C})$ the following diagram is commutative :

$$\begin{array}{ccc} X_{m,n} & \xrightarrow{\phi} & M_n(\mathbb{C}) \\ \alpha \downarrow & & \downarrow \alpha \\ X_{m,n} & \xrightarrow{\phi} & M_n(\mathbb{C}) \end{array}$$

where the action of $GL_n(\mathbb{C})$ on $M_n(\mathbb{C})$ is given by conjugation. Now, let ξ be any point in $V_{m,n}$ and $x \in T(\xi)$, then we know that the diagram below is defined and commutative in a neighborhood of x

$$\begin{array}{ccc} GL_n(\mathbb{C}) \times^{GL_n(\mathbb{C})_x} N_x & \longrightarrow & N_x/GL_n(\mathbb{C})_x \cong (GL_n(\mathbb{C}) \times^{GL_n(\mathbb{C})_x} N_x)/GL_n(\mathbb{C}) \\ \downarrow & & \downarrow \\ X_{m,n} & \longrightarrow & V_{m,n} \end{array}$$

where the morphism $N_x \to X_{m,n}$ is defined by sending a point n to $x + n$. There exists an open affine $GL_n(\mathbb{C})_x$ stable neighborhood N_x^o of the origin of N_x and an open affine $GL_n(\mathbb{C})$ stable neighborhood $X_{m,n}^o$ of x in $X_{m,n}$ so that

$$GL_n(\mathbb{C}) \times^{GL_n(\mathbb{C})_x} N_x^o \cong X_{m,n}^o \times_{U_{m,n}} (N_x^o/GL_n(\mathbb{C})_x)$$

From this the following can be easily proved : let B be the coordinate ring of $N_x^o/GL_n(\mathbb{C})_x$ which is an algebra over $\mathbb{C}[V_{m,n}]$ then $\mathbb{T}_{m,n} \otimes_{\mathbb{C}[V_{m,n}]} B$ (the noncommutative ring $\mathbb{T}_{m,n}$ localized in the given étale neighborhood of ξ) is isomorphic to the ring of equivariant maps from $GL_n(\mathbb{C}) \times^{GL_n(\mathbb{C})_x} N_x^o$ to $M_n(\mathbb{C})$. Furthermore, we can assume that N_x^o is the set of elements of N_x where an invariant polynomial f (under $GL_n(\mathbb{C})_x$) on N_x is not zero.

Then, if R is the ring of equivariant maps from

$$GL_n(\mathbb{C}) \times^{GL_n(\mathbb{C})_x} N_x$$

to $M_n(\mathbb{C})$ we have that

$$\mathbb{T}_{m,n} \bigotimes_{\mathbb{C}[V_{m,n}]} B \cong R[\frac{1}{f}]$$

The ring R can be called the noncommutative slice (in the point ξ).

We will now restrict attention to the case that ξ is of representation type $(1, k_1; \ldots; 1, k_r)$, that is when $GL_n(\mathbb{C})_x = T_r$. Then we have to describe the ring of equivariant maps

$$f : M_n(\mathbb{C}) \bigoplus \ldots \bigoplus M_n(\mathbb{C}) \bigoplus C_x \to M_n(\mathbb{C})$$

where T_r acts on every component by conjugation. This study is essentially the study of all polynomial maps

$$g : M_n(\mathbb{C}) \bigoplus \ldots \bigoplus M_n(\mathbb{C}) \bigoplus C_x \bigoplus M_n(\mathbb{C})^* \to \mathbb{C}$$

which are invariant under T_r and homogeneous of degree one in the indeterminates corresponding to the component $M_n(\mathbb{C})^*$. As a T_r-module, $M_n(\mathbb{C})^*$ decomposes into a direct sum of one-dimensional vectorspaces $\bigoplus \mathbb{C}e_{ij}$. If

$$k_1 + \ldots + k_s < i \leq k_1 + \ldots + k_{s+1}$$

$$k_1 + \ldots + k_t < j \leq k_1 + \ldots + k_{t+1}$$

then $\tau = (u_1, \ldots, u_r)$ acts on $\mathbb{C}e_{ij}$ by sending e_{ij} to $u_s^{-1}u_t e_{ij}$.

This allows us to determine the part of the invariant ring which is homogeneous of degree one in the variable corresponding to e_{ij} : z_{ij}. A typical element is of the form $h.z_{ij}$ where h is a semiinvariant on N_x with respect to the character

$$\chi_{st} : T_r \to \mathbb{C}^*$$

determined by sending $\tau = (u_1, \ldots, u_r)$ to $u_s.u_t^{-1}$. With notations as in II.2 this module of semi-invariants is

$$\mathbb{C}[E_{\Phi,\alpha}][y_1, \ldots, y_d]$$

where $d = (m-1)(k_1^2 + \ldots + k_r^2) + r$ and α is the $(r-1)$-dimensional column vector with a -1 on the i-th row, $+1$ at the j-th row and zeroes elsewhere. Therefore, $\mathbb{C}[E_{\Phi,\alpha}]$ is the subvectorspace of $\mathbb{C}[N_2]$ consisting of all polynomials h such that $h.v_{ts}(a) \in \mathbb{C}[N_2/T_r]$ for any $1 \leq a \leq k_s k_t(m-1)$. These observations prove

Theorem III.1.1 : The noncommutative slice of the trace ring of m generic n by n matrices, $\mathbb{T}_{m,n}$, in a point $\xi \in V_{m,n}$ corresponding to a semi-simple representation of type $(1, k_1; \ldots; 1, k_r)$ is isomorphic to a polynomial ring in $(m-1)(k_1^2 + \ldots + k_r^2) + r$ indeterminates over the $\mathbb{C}$- algebra

$$\Gamma_{m,n}(\xi) = \begin{pmatrix} M_{k_1}(\mathbb{C}[N_2/T_r]) & W_{12} & \ldots & W_{1r} \\ W_{21} & M_{k_2}(\mathbb{C}[N_2/T_r]) & \ldots & W_{2r} \\ \\ W_{r1} & W_{r2} & \ldots & M_{k_r}(\mathbb{C}[N_2/T_r]) \end{pmatrix}$$

where W_{ij} is a k_i by k_j block of $\mathbb{C}[N_2/T_r; \chi_{ij}]$ which is the module of semi-invariants on N_2 with respect to the character $\chi_{ij} : T_r \to \mathbb{C}^*$ or, equivalently, the subvectorspace of $\mathbb{C}[N_2]$ consisting of polynomials h such that $h.v_{ji}(a) \in \mathbb{C}[N_2/T_r]$ for all $1 \leq a \leq (m-1)k_i k_j$.

III.2 : THE FUNCTIONAL EQUATION

In [Le] it was shown that the trace ring of m generic 2 by 2 matrices is always a Cohen-Macaulay module over its center , i.e. a (graded) free module of finite rank over a polynomial subring of the center.

Unfortunately, the étale-slice machinary cannot be used to prove Cohen-Macaulayness for arbitrary p.i.-degree. The reason is that the study of the noncommutative slice in a point of type $(n,1)$ is as hard as the study of the trace ring in the origin. Nevertheless, we can give some weight to the conjecture that trace rings are Cohen-Macaulay by proving that the Cohen-Macaulay locus is large.

Theorem III.2.1 : The trace ring of m generic n by n matrices is a Cohen-Macaulay module in a point ξ of $V_{m,n}$ of type $(1,k_1;...;1,k_r)$.

Proof :

By étale descent, it is sufficient to prove Cohen-Macaulayness of the noncommutative slice in ξ. By the Hochster-Roberts theorem and theorem III.1.1 this amounts to showing that $\mathbb{C}[N_2/T_r;\chi_{ij}]$ is a Cohen-Macaulay module. In the foregoing section we have seen that $\mathbb{C}[N_2/T_r\} \chi_{ij}] = \mathbb{C}[E_{\Phi,\alpha}]$ where $\alpha(i) = -1$,$\alpha(j) = 1$ and $\alpha(k) = 0$ elsewhere.By proposition I.2.1 we know that $\mathbb{C}[E_{\Phi,\alpha}]$ is Cohen- Macaulay if there exists a rational solution $\beta = (\beta_1,...,\beta_s)^\tau$ to $\Phi.\beta = \alpha$ satisfying $-1 < \beta_i \leq 0$. It is easy to see that such a solution always exists (with entries $\frac{1}{2}$ or 0) if $n > 2$ or $m > 2$.

The remaining case, $\mathbb{T}_{2,2}$ is easily seen to be a free module of rank four generated by $\{1, x_1, x_2, x_1x_2\}$ over its center which is a polynomial ring.

We will now study the Poincaré series of the noncommutative slice, or equivalently that of $\Gamma_{mn,n}(\xi)$.Usually, it is rather hard to determine the power series $P(\Gamma_{m,n}(\xi);t)$. There is, however, one important exception :

Proposition III.2.2 : Let $\xi \in V_{m,n}$ be a point of type $(1,k;1,l)$, then the noncommutative slice in ξ has the following Poincaré series

$$\frac{1}{(1-d)^d}.\{2kl.\sum_{s>0}^{\infty}\binom{(m-1)kl+s-1}{s}\binom{(m-1)kl+s-1}{s-1}t^{2s-1}$$
$$+(k^2+l^2)\sum_{s=0}^{\infty}\binom{(m-1)kl+s-1}{s}^2 t^{2s}\}$$

Proof :

We will first determine the Poincaré series of the ring of all polynomial maps

$$M_n(\mathbb{C})\bigoplus \ldots \bigoplus M_n(\mathbb{C})\bigoplus \begin{pmatrix} \mathbb{C}\bigotimes 1_k & 0 \\ 0 & \mathbb{C}\bigotimes 1_l \end{pmatrix} \bigoplus M_n(\mathbb{C})^* \rightarrow \mathbb{C}$$

which are invariant under T_2. Here we give the indeterminates corresponding to the first m factors degree t and those to the last factor degree x. This invariant ring is a polynomial ring in $(m-1)(k^2+l^2)+2$ indeterminates of degree t and k^2+l^2 indeterminates of degree x over the ring of invariants of

$$(V_{12}\bigoplus W_{21})\bigoplus(V_{21}\bigoplus W_{12})$$

where $dim(V_{12}) = dim(V_{21}) = (m-1)kl$ and $dim(W_{12}) = dim(W_{21}) = kl$ and $(\alpha,\beta) \in T_2$ acts on the first component by multiplication with $\alpha\beta^{-1}$ and with $\alpha^{-1}\beta$ on the second. The ring of invariants is then the subring of

$$\mathbb{C}[x_i, y_i : 1 \leq i \leq kl(m-1); u_j, v_j : 1 \leq j \leq kl]$$

generated by all products $x_iy_i, x_iv_j, y_iu_j, y_iv_j$. The ideal of relations between them is generated by the 2 by 2 minors of the matrix

$$\begin{pmatrix} x_1y_1 & \ldots & x_1y_{kl(m-1)} & x_1v_1 & \ldots & x_1v_{kl} \\ \\ x_{kl(m-1)}y_1 & \ldots & x_{kl(m-1)}y_{kl(m-1)} & x_{kl(m-1)}v_1 & \ldots & x_{kl(m-1)}v_{kl} \\ y_1u_1 & \ldots & y_{kl(m-1)}u_1 & u_1v_1 & \ldots & u_1v_{kl} \\ \\ y_1u_{kl} & \ldots & y_{kl(m-1)}u_{kl} & u_{kl}v_1 & \ldots & u_{kl}v_{kl} \end{pmatrix}$$

So, all relations are homogeneous in the (t,x)-gradation. It is then fairly easy to see that the Poincaré series of this invariant ring is (use Plethysm formula)

$$\sum_{s=0}^{\infty} S^s(\mathbb{C}^{klm}) \bigotimes S^s(\mathbb{C}^{klm})$$

where $\mathbb{C}^{klm} = U \bigoplus V$ with $dim(U) = (m-1)kl$, $dim(V) = kl$ and the coordinates of U (resp. V) have degree t (resp. x). So,

$$S^s(U \bigoplus V) = \sum_{i=0}^{s} S^i(U) \bigotimes S^{s-i}(V)$$
$$= \sum_{i=0}^{s} \binom{(m-1)kl+i-1}{i} \binom{kl+s-i-1}{s-i} t^i x^{s-i}$$

Therefore, the Poincaré series in the (t,x)-gradation of the invariant ring is equal to

$$\sum_{s=0}^{\infty} \left(\sum_{i=0}^{s} \binom{(m-1)kl+i-1}{i} \binom{kl+s-i-1}{s-i} t^i x^{s-i} \right)^2$$

and therefore, this expression multiplied with

$$(1-t)^{-((m-1)(k^2+l^2)+2)}(1-x)^{-(k^2+l^2)}$$

is the Poincaré series of the total ring of invariants in the (t,x)-gradation. The Poincaré series of the noncommutative slice is then the partial derivative of this expression with respect to x and evaluated at $x = 0$, which gives us the claimed expression.

We have seen in the foregoing section that $\mathbb{C}[N_2/T_r; \chi_{ij}] = \mathbb{C}E_{\Phi,\alpha}$, for certain α, is a Cohen-Macaulay module which is clearly of dimension $2(m-1)\sum k_i k_j - r + 1 = h$.

Assume that $\theta_1, ..., \theta_h$ is a homogeneous system of parameters for $\mathbb{C}[N_2/T_r; \chi_{ij}]$ and let $S = \mathbb{C}[\theta_1, ..., \theta_h]$, then the canonical module of $\mathbb{C}[N_2/T_r; \chi_{ij}]$ is defined to be

$$\Omega(\mathbb{C}[N_2/T_r; \chi_{ij}] = HOM_S(\mathbb{C}[N_2/T_r; \chi_{ij}], S)$$

which is $\mathbb{Z}$-graded in the obvious way.

Form [St,th 4.4] we retain that $\Omega(\mathbb{C}[N_2/T_r; \chi_{ij}])$ is the subvectorspace of $\mathbb{C}[N_2]$ spanned by all monomials $\prod v_{ij}(k)^{b_{ij}(k)}$ s.t. $(\beta_{ij}(k)) \in \overline{E_{\Phi,\alpha}}$ (for notation see I.2) and $\Gamma_{-\beta_{ij}(k)} = \emptyset$.

But, if $(\beta_{ij}(k)) \in \overline{E_{\Phi,\alpha}}$ then $(-\beta_{ij}(k)) \in \overline{E_{\Phi,-\alpha}}$ and since $\mathbb{C}[E_{\Phi,-\alpha}] = \mathbb{C}[N_2/T_r; \chi_{ji}]$ we have again a solution to $\Phi.\beta = -\alpha$ satisfying $-1 < \beta_i \leq 0$ for all i, whence we can apply the argument at the end of I.2 to ensure that $\Gamma_{(-\beta_{ij}(k))} = \emptyset$ iff $-\beta_{ij}(k) < 0$ for all i, j, k.

So, we have shown that

$$\Omega(\mathbb{C}[N_2/T_r; \chi_{ij}]) = \bigoplus\{\mathbb{C}\prod v_{ij}(k)^{\beta_{ij}(k)} : \beta_{ij}(k) > 0\}$$

Luckily, there is a unique strictly positive solution to $\Phi.\beta = 0$ namely $\beta_i = 1$ for all $1 \leq i \leq 2(m-1)\sum k_i k_j$. Therefore,

$$\Omega(\mathbb{C}[N_2/T_r; \chi_{ij}]) = \prod v_{ij}(k).\mathbb{C}[N_2/T_r; \chi_{ij}]$$

Translating this equality to Poincaré series gives us

$$\mathcal{P}(\Omega(\mathbb{C}[N_2/T_r; \chi_{ij}]); t) = t^s.\mathcal{P}(\mathbb{C}[N_2/T_r; \chi_{ij}]; t)$$

where $s = 2(m-1)\sum k_i k_j$.

Applying [St2,p.58] we get the functional equation

$$\mathcal{P}(\mathbb{C}[N_2/T_r; \chi_{ij}]; \frac{1}{t}) = (-1)^h.t^s.\mathcal{P}(\mathbb{C}[N_2/T_r; \chi_{ij}]; t)$$

where $h = 2(m-1)\sum k_i k_j - r + 1$.

This concludes the proof of the following

Theorem III.2.3 : Let $\xi \in V_{m,n}$ be of representation type $(1, k_1; \ldots; 1, k_r)$, then the Poincaré series of the noncommutative slice R at ξ satisfies the following functional equation

$$\mathcal{P}(R; \frac{1}{t}) = (-1)^d.t^{(m-1)n^2+r}.\mathcal{P}(R; t)$$

where $d = Kdim(\mathbb{T}_{m,n}) = (m-1)n^2 + 1$.

III.3 : REGULARITY OF $\mathbb{T}_{m,n}$

When m or n is equal to one, $\mathbb{T}_{m,n}$ is a commutative polynomial ring and hence has finite global dimension. In [SS] L.Small and J.T. Stafford proved that $gldim(\mathbb{T}_{2,2}) = 5$. In [LV] it was shown that for $n \leq 4$, $gldim(\mathbb{T}_{m,n}) < \infty$ if and only if $(m,n) = (2,2); (2,3)$ or $(3,2)$. In this section we aim to show that these are the only noncommutative trace rings of generic matrices having finite global dimension.

The strategy of the proof is the following : choose a suitable point ξ in $V_{m,n}$. If $gldim(\mathbb{T}_{m,n}) < \infty$, then the global dimension of the noncommutative slice in ξ has to be finite,too. Suppose that ξ is chosen in such a way that all indecomposable graded projective modules of the noncommutative slice have the same Poincaré series, then finite global dimension can be tested by the fact that the Poincaré series of such a projective module needs to have the rational form $\frac{k}{f(t)}$ for some $f(t) \in \mathbb{Z}[t]$ and $k \in \mathbb{N}$, by the standard argument.

Theorem III.3.1 : The trace ring of m generic n by n matrices, $\mathbb{T}_{m,n}$, has finite global dimension if and only if

(1) : m or n is equal to one

(2) : $(m,n) = (2,2); (2,3)$ or $(3,2)$

Proof :

Consider a point $\xi \in V_{m,n}$ corresponding to a sum of n distinct one-dimensional irreducible representations. By theorem II.2.1 , the central slice is a polynomial ring in $m.n$ variables over the invariant ring $\mathbb{C}[N_2/T_n]$ described in II.2.1.

By theorem III.1.1 , the noncommutative slice is also a polynomial ring in $m.n$ variables over the ring $\Gamma_{m,n}(\xi)$ described in III.1.1 . From this description it is clear that every indecomposable graded projective left module over the noncommutative slice is of the form

$$P_i = \bigoplus_{j=1}^{n} \mathbb{C}[N_2/T_r; \chi_{ji}] \underset{\mathbb{C}[N_2/T_r]}{\otimes} \mathbb{C}[N_x/T_r]$$

Further, by symmetry it is clear that the Poincaré series of all $\mathbb{C}[N_2/T_r; \chi_{ij}]$, $i \neq j$, are equal. So, all indecomposable graded projectives have the same Poincaré series and we have to verify when

$\wp(P_i; t)$

$$= \frac{1}{(1-t)^{mn}} \cdot \wp(\mathbb{C}[N_2/T_n]; t) + (n-1)\wp(\mathbb{C}[N_2/T_n; \chi_{12}]; t))$$

has a rational expression of the form $(f(t))^{-1}$ where $f(t) \in \mathbb{Z}[t]$.

It follows from the description of $\mathbb{C}[N_2/T_n]$ and from the fact that $\mathbb{C}[N_2/T_n; \chi_{12}]$ is a finitely generated module over $\mathbb{C}[N_2/T_n]$ that their Poincaré series have the rational form

$$\wp(\mathbb{C}[N_2/T_n]; t) = \frac{p(t)}{(1-t)^{\alpha_1} \ldots (1-t^n)^{\alpha_n}}$$

$$\wp(\mathbb{C}[N_2/T_n; \chi_{ij}]; t) = \frac{q(t)}{(1-t)^{\alpha_1} \ldots (1-t^n)^{\alpha_n}}$$

for some $p(t), q(t) \in \mathbb{Z}[t]$ and $\alpha_i \in \mathbb{N}$. Therefore, $f(t)$ has to be a product of irreducible factors of $1 - t^i$ where $1 \leq i \leq n$.

Further , the residue of the pole in $t = 1$ must be equal to $Kdim(P_i) = Kdim(\mathbb{C}[N_x/T_n]) = (m-1)n^2 + 1 \doteq d$. Therefore,

$$f(t) = (1-t)^d \prod (\text{ irreducible factors } \neq 1 - t)$$

In any case, $f(t)^{-1}$ satisfies the functional equation

$$(1) : f(\frac{1}{t}) = (-1)^d t^{def(f)} f(t)$$

On the other hand, we have seen in the proof of theorem III.2.3 that the Poincaré series of $\mathbb{C}[N_x/T_n]$ and $\mathbb{C}[N_2/T_n; \chi_{12}] \otimes \mathbb{C}[N_x/T_n]$ satisfy the functional equation

$$(2) : \wp(-; \frac{1}{t}) = (-1)^d t^{(m-1)n^2+n} \wp(-; t)$$

Therefore, if the noncommutative slice has finite global dimension, or if $\mathbb{C}$ has finite projective dimension, we get from a combination of (1) and (2) that

$$deg(f) = (m-1)n^2 + n$$

whence $f(t) = (1-t)^d(1 + a_1t + a_2t^2 + ... + a_{n-1}t^{n-1})$. This entails that the first terms in the power series expansion of $f(t)^{-1}$ are

$$(3) : 1 + (d - a_1)t + (\frac{d(d+1)}{2} - da_1 + a_1^2 - a_2)t^2 + ...$$

On the other hand, it follows from the description of $\mathbb{C}[N_2/T_n]$ and $\mathbb{C}[N_2/T_n; \chi_{12}]$ that

$$\wp(\mathbb{C}[N_2/T_n]; t) = 1 + (m-1)\frac{n(n-1)}{2}t^2 + ...$$

$$\wp(\mathbb{C}[N_2/T_n; \chi_{12}]; t) = (m-1)t + (n-2)(m-1)^2t^2 + ...$$

Therefore,$\wp(P_i; t)$ is equal to

$$(4) : \frac{1}{(1-t)^n} 1 + (n-1)(m-1)t + (n-1)(m-1)(\frac{n}{2} + (m-1)(n-2))t^2 + ...$$

Comparing the coefficient of t in (3) and (4) gives the equation

$$a_1 = (n-1)[(n-1)(m-1) - 1]$$

But we know that a_1 has to be smaller or equal to the number of irreducible factors in the remaining part of $f(t)$.Therefore, $a_1 \leq n-1$ giving the inequality

$$(n-1)(m-1) - 1 \leq 1$$

It is clear that the only integer solutions with m or $n > 1$ are

$$(m, n) = (2,2), (2,3) \text{ or } (3,2)$$

So, in all other cases $\mathbb{C}$ is of infinite projective dimension over the noncommutative slice R, and so also R/R_+. In view of the slice theorem, we have to verify that

$$\mathbb{T}_{m,n} \bigotimes_{\mathbb{C}[V_{m,n}]} B \simeq R[\frac{1}{f}]$$

has infinite global dimension. Now, assume $R[\frac{1}{f}]$ has global dimension d . If $k > d$ and M is an R-module we have

$$0 = Ext^k_{R[\frac{1}{f}]}((R/R_+)[\frac{1}{f}], M[\frac{1}{f}])$$
$$= Ext^k_R(R/R_+, M)[\frac{1}{f}] = Ext^k_R(R/R_+, M)$$

since $f(0) \neq o$ and R/R_+ is annihilated by R_+.So, we obtain a contradiction.

In view of [LV,Prop 1,Prop 2,Prop 7] this finishes the proof.

We will conclude this paper with an example. Consider the trace ring of 3 generic 2 by 2 matrices and let $\xi \in V_{3,2}$ of representation type $(1,1;1,1)$. Then, $\mathbb{C}[N_2] = \mathbb{C}[v_1, v_2, v_3, v_4]$ and $\mathbb{C}[N_2/T_2]$ is the subring generated by the monomials $v_i v_j$, $i \neq j$. Hence, $\mathbb{C}[N_2/T_2] = \mathbb{C}[x,y,z,t]/(xy - zt)$ and therefore

$$\wp(\mathbb{C}[N_2/T_2]; t) = \frac{1+t^2}{(1-t^2)^3}$$

Let P be the height one prime ideal of $\mathbb{C}[N_2/T_2]$ generated by (x, z), then

$$\Gamma_{3,2}(\xi) \cong \begin{pmatrix} \mathbb{C}[N_2/T_2] & P \\ P^{-1} & \mathbb{C}[N_2/T_2] \end{pmatrix}$$

Since $v_1.\mathbb{C}[N_2/T_2; \chi_{12}] = P$ and the quotient $\mathbb{C}[N_2/T_2]/P$ is the polynomial ring in z and t we obtain

$$t\wp(\mathbb{C}[N_2/T_2; \chi_{12}]; t) = \frac{1+t^2}{(1-t^2)^3} - \frac{1}{(1-t^2)^2} = \frac{2t^2}{(1-t^2)^3}$$

and

$$\wp(P_i; t) = \frac{1}{(1-t)^6}\wp(\mathbb{C}[N_2/T_2]; t) + \wp(\mathbb{C}[N_2/T_2; \chi_{12}]; t)$$
$$= \frac{1+2t+t^2}{(1-t)^6(1-t^2)^3} = \frac{1}{(1-t)^8(1-t^2)}$$

Since $\mathbb{T}_{3,2}$ has finite global dimension, $\Gamma_{3,2}(\xi)$ is regular too, giving an example of a reflexive Azumaya algebra of global dimension three having an height one prime ideal $(\Gamma_{3,2}(\xi)(x, z))^{**}$ which is not generated by a normalizing element since (x, z) is the generator of $Cl(\mathbb{C}[N_2/T_2]) \cong \mathbb{Z}$

REFERENCES :

[Ar] : Artin M. ; On Azumaya algebras and finite dimensional representations of rings; J.Alg 11 (1969),pp 532-563

[AS] : Artin M.,Schelter W.; Integral ring homomorphisms; Adv. Math 39 (1981),pp 289-329

[Hb] : Hoobler R. ; When is $Br(X) = Br'(X)$? ; Springer LNM 917 (1982),pp 231-244

[Ho] : Hochster M. ; Rings of invariants of tori,Cohen-Macaulay rings generated by monomials and polytopes; Ann. Math 96 (1972),pp 318-337

[HR] : Hochster M.,Roberts J.; Rings of invariants of reductive groups acting on regular rings are Cohen-Macaulay; Adv. Math 13 (1974),pp 115-175

[Le] : Le Bruyn L. ; Trace rings of generic 2 by 2 matrices; Memoirs AMS (to appear)

[Lu] : Luna D.; Slices étales; Bull Soc Math France Mém 33 (1973),pp 81-105

[LV] : Le Bruyn L.,Van den Bergh M. ; Regularity of trace rings of generic matrices; J.Alg (to appear)

[Mo] : Morrison K.; The scheme of finite dimensional representations of an algebra; Pac. J Math 91 (1980),pp 199-218

[Mu] : Mumford D.; Geometric invariant theory; Springer (1965)

[Ma] : Matsuchima Y.; Espaces homogènes de Stein des groupes de Lie complexes; Nagoya Math J 16 (1960),pp 205-218

[Pr] : Procesi C.; Finite dimensional representations of algebras; Israel J Math 19 (1974),pp 169-182

[Pr2] : Procesi C.; Invariant theory of n by n matrices; Adv. Math 19 (1976),pp 306-381

[Pr3] : Procesi C.; Rings with polynomial identities; Marcel Dekker (1973)

[Sc] : Schwartz G. ; Lifting smooth homotopies of orbit spaces; Publ IHES

[SS] : Small L.,Stafford J.T. ; Homological properties of generic matrices; Israel J Math (1985)

[St] : Stanley R.; Linear diophantine equations and local cohomology; Inv. Math 68 (1982),pp 175-193

[St2] : Stanley R.; Combinatorics and commutative algebra; Birkhäuser PM 41 (1983)

Fourier transforms on a semisimple Lie algebra over F_q

George Lusztig
M.I.T., Cambridge MA 02139 USA
and II Università degli studi di Roma

To Tonny Springer on his 60th birthday

1. Let $\mathfrak{g}$ be the Lie algebra of a reductive connected algebraic group G over k, an algebraic closure of the finite prime field F_p. In this paper we shall assume that p is large. We assume chosen a non-singular G-invariant symmetric bilinear form $\langle , \rangle : \mathfrak{g} \times \mathfrak{g} \to k$. If we are given

(a) an F_q-rational structure on G (hence on $\mathfrak{g}$) with Frobenius map F, such that $\langle , \rangle$ is defined over F_q,

we can define the Fourier transform of a function $f : \mathfrak{g}^F \to \bar{Q}_\ell$ to be the function

(b) $\hat{f} : \mathfrak{g}^F \to \bar{Q}_\ell$, $\hat{f}(\xi) = \sum_{\xi' \in \mathfrak{g}^F} \Psi\langle\xi,\xi'\rangle f(\xi')$, where $\Psi: F_q \to \bar{Q}_\ell^*$ is a fixed non-trivial character.

(c) Let N be the variety of nilpotent elements in $\mathfrak{g}$.
The purpose of this paper is to describe those G^F- invariant functions $f:\mathfrak{g}^F \to \bar{Q}_\ell$ such that both f and $\hat{f}$ vanish on $\mathfrak{g}^F - N^F$.

It turns out that there are very few such functions, other than 0. They are very closely related to the cuspidal character sheaves of $[L_2]$.
For example, if $G = Sp_{2n}(k)$ (resp. $SO_n(k)$), there is (up to a scalar) at most one function $f \neq 0$ as above; it exists if and only if $n = \frac{i(i+1)}{2}$ (resp. $n = i^2$) for some integer $i \geq 0$, and in that case it is supported by the nilpotent elements with Jordan blocks of sizes $2,4,6,\ldots,2i$ (resp. $1,3,5,\ldots,2i-1$) in the standard representation of G.

The study of the Fourier transform of G^F-invariant functions on $\mathfrak{g}^F$ has been initiated in Springer's work [S] in connection with the geometry of nilpotent orbits. He obtained very interesting applications to the theory of Green functions of reductive groups over F_q and the representation theory of Weyl groups. (Earlier, Harish-Chandra has discovered the connection of Fourier transforms on real and p-adic Lie algebras with the character theory of Lie groups.) This has been further pursued by Kazhdan [Kz].

The theory of $\mathcal{D}$-modules and perverse sheaves [BBD] has provided some new tools for the study of Fourier transform, see [B], [HK], [KLa]; in this paper we shall make use of this theory as well as of the results in $[L_2]$ on character sheaves.

Here are some of the notations used in this paper. We shall denote by $\mathcal{M}(X)$ the abelian category of ℓ-adic perverse sheaves on an algebraic variety X over k; we assume that ℓ is a fixed prime $\neq$ p. If G acts algebraically on X we have the concept of G-equivariant perverse sheaves on X see $[L_2, 1.9]$; these form a full subcategory $\mathcal{M}_G(X)$ of $\mathcal{M}(X)$. In particular, $\mathcal{M}_G(\mathfrak{g})$ is defined in terms of the adjoint action $(g,\xi) \to Ad(g)\xi$ of G on $\mathfrak{g}$.

2. In $[L_1]$, $[L_2]$ we have studied a class of irreducible perverse sheaves on G called admissible complexes. We wish to define an analogous concept for $\mathfrak{g}$ instead of G. We first define the process of induction. Let:

(a) P be a parabolic subgroup of G with Levi subgroup L and unipotent radical U; let $\mathfrak{p}$, $\mathfrak{l}$, $\mathfrak{u}$ be the corresponding Lie algebras; let $\rho: \mathfrak{p} \to \mathfrak{l}$ be the canonical projection.

Consider the diagram

$$\mathfrak{l} \xleftarrow{\pi} V_1 \xrightarrow{\pi'} V_2 \xrightarrow{\pi''} \mathfrak{g}$$

where

$$V_1 = \{(\xi,h) \in \mathfrak{g} \times G \mid Ad(h^{-1})\xi \in \mathfrak{p}\}$$
$$V_2 = \{(\xi,hP) \in \mathfrak{g} \times G/P \mid Ad(h^{-1})\xi \in \mathfrak{p}\}$$
$$\pi''(\xi,hP) = \xi, \quad \pi'(\xi,h) = (\xi,hP), \quad \pi(\xi,h) = \rho(Ad(h^{-1})\xi).$$

Let A be an object of $\mathcal{M}_L(\mathfrak{l})$. There is a well defined perverse sheaf A_1 on V_2 such that $\tilde{\pi} A \cong \tilde{\pi}' A_1$. (Here $\tilde{\pi}$, $\tilde{\pi}'$ denote inverse images with a shift, as in $[L_2$, (1.7.4)]. We define

$$i_L^G A = \pi''_! A_1.$$

This is a complex of sheaves on $\mathfrak{g}$; it is said to be obtained from A by induction. Let $K \in \mathcal{M}_G(\mathfrak{g})$ be irreducible.

If G is semisimple, we say that K is cuspidal if its support is a closure of a single nilpotent orbit in $\mathfrak{g}$ and if for any $P \subsetneqq G$ as in (a) we have $\rho_!(K|\mathfrak{p}) = 0$ as a complex of sheaves on $\mathfrak{l}$ (notation of (a)).

We now drop the assumption that $\mathfrak{g}$ is semisimple and write

(b) $\mathfrak{g} = \mathfrak{z} \oplus \mathfrak{g}'$ where $\mathfrak{z}$ is the Lie algebra of Z_G^0 (= connected centre of G) and $\mathfrak{g}'$ is the Lie algebra of G/Z_G^0; we say that K is cuspidal if it is of form $K_1 \boxtimes K_2$ where $K_2 \in \mathcal{M}_{G/Z_G^0}(\mathfrak{g}')$ is cuspidal in the sense of the previous definition and $K_1 \in \mathcal{M}(\mathfrak{z})$ is (up to shift) a local system of the form $h^* \mathcal{E}_{\Psi_0}$ where $h: \mathfrak{z} \to k$ is a k-linear form and $\mathcal{E}_{\Psi_0}$ is the local system on k defined by the F_p-covering $x^p - x = y$ of k and by Ψ_0, a fixed imbedding $F_p \hookrightarrow \bar{Q}_\ell^*$.

An object $A \in \mathcal{M}_G(\mathfrak{g})$ is said to be admissible if it is irreducible and if there exists P, L, $\mathfrak{p}$, $\mathfrak{l}$ as in (a) and a cuspidal $K \in \mathcal{M}_L(\mathfrak{l})$ such that A is a direct summand of $i_L^G K$. In particular, a cuspidal perverse sheaf on $\mathfrak{g}$ is admissible.

3. Here are some properties of admissible objects in $M_G(\mathfrak{g})$.

(a) If P, L, $\mathfrak{p}$, $\mathfrak{l}$ are as in 2(a) and $A_0 \in M_L(\mathfrak{l})$ is admissible then $i_L^G A_0$ is a direct sum of finitely many admissible objects in $M_G(\mathfrak{g})$; if in addition, we have $P \neq G$, then any direct summand K of $i_L^G A_0$ satisfies supp $K \not\subset \mathfrak{z} + N$ (see 2(b)) and hence is not cuspidal.

(b) If $A \in M_G(\mathfrak{g})$ is admissible then $A|N$ extended by zero on $\mathfrak{g} - N$ (shifted by codim N) is a semisimple object of $M_G(\mathfrak{g})$.

(c) Induction of admissible perverse sheaves is transitive.

(d) Let $A \in M_G(\mathfrak{g})$ be irreducible. Write $\mathfrak{g} = \mathfrak{z} \oplus \mathfrak{g}'$ as in 2(b). Then A is admissible if and only if it is of form $h^* E_{\Psi_0} \boxtimes A'$ where $h^* E_{\Psi_0}$ is as in no. 2 and $A' \in M_{G/Z_G^0}(\mathfrak{g}')$ is admissible.

(e) If G is semisimple, there is at most one cuspidal object in $M_G(\mathfrak{g})$ on which the centre of G acts by a prescribed character.

(f) Assume that G is semisimple and that $K \in M_G(\mathfrak{g})$ is irreducible with support the closure of a single nilpotent orbit C. Then there exist P, L, $\mathfrak{p}$, $\mathfrak{l}$ as in 2(a) and a cuspidal object $A_0 \in M_L(\mathfrak{l})$ such that K is a direct summand of $i_L^G(A_0)|N$ extended by zero on $\mathfrak{g} - N$ (shifted by codim N).

(g) Let G, K; C be as in (f); assume that K is cuspidal. Then the restriction of K to $\overline{C} - C$ is zero.

(h) Let G, K, C be as in (g) and let $A \in M_G(\mathfrak{g})$ be admissible, non-cuspidal. Let L be the irreducible local system on C such that $K|C$ is L (up to shift). Then no homology sheaf of A restricted to C contains L as a direct summand.

We now make some comments on the proofs of (a) - (h). Let $\log: G \to \mathfrak{g}$ be a logarithm map as in [BR]. From the definitions, it follows that for G semisimple, $\log^*$ defines a bijection between the set of cuspidal objects in $M_G(\mathfrak{g})$ and the set of "strongly cuspidal" perverse sheaves [L_2, II(7.1.5)] on G whose support is the closure of a single unipotent class of G. By [L_2, I 6.9(b), V(23.1(b))] the condition "strongly cuspidal" above is equivalent to "cuspidal" [L_2, II(7.1.1)] and to "cuspidal character sheaf" [L_2, I 2.10, I 3.10]. Hence the classification of cuspidal objects in $M_G(\mathfrak{g})$ (for G semisimple) is the same as the classification of cuspidal character sheaves on G with support in the unipotent variety of G. Hence (e) follows from [L_1], [L_2]. Similarly (g) follows from [L_2, V 23.1(a)]. Similarly, using the definitions, we see that the restrictions of character sheaves of G to the unipotent variety of G correspond under $\log^*$ to the restrictions to N of admissible objects in $M_G(\mathfrak{g})$. Therefore (b), (f), (h) follow from analogous properties of character sheaves on G, see [L_1, (6.6.1)], [L_1, 6.5], [L_2, III(14.3)]. Properties (a), (c) are proved in an entirely similar way as the corresponding properties of character sheaves on G, see [L_2, I 4.4(b), (4.3.2), 4.2]. Property (d) follows from definitions.

4. Let V be a finite dimensional k-vector space with a given non-singular bilinear form $V \times V \to k$. Deligne has defined the Fourier transform $\mathcal{F}K$ of a perverse sheaf $K \in \mathcal{M}(V)$; then $\mathcal{F}K \in \mathcal{M}(V)$. The definition is in terms of a fixed embedding $\Psi_0 : F_p \to \overline{Q}_\ell^*$. We refer to [B] and [KLa] for the precise definition and properties of $\mathcal{F}$.

We shall use this construction for $\mathfrak{g}$ and $\langle , \rangle$. It is known that $\mathcal{F}$ is additive and

(a) $\mathcal{F}\mathcal{F}K \cong j^*K$, where $j: \mathfrak{g} \to \mathfrak{g}$ is defined by $j\,\xi = -\xi$. It follows that $\mathcal{F}$ takes irreducible (resp. semisimple) objects in $\mathcal{M}(\mathfrak{g})$ to irreducible (resp. semisimple) objects in $\mathcal{M}(\mathfrak{g})$. When considering the transformation $\mathcal{F}$ on a subalgebra ℓ of $\mathfrak{g}$ as in 2(a), we shall take it with respect to the restriction of $\langle , \rangle$ to ℓ.

Note also that $\mathcal{F}$ takes an object of $\mathcal{M}_G(\mathfrak{g})$ to an object of $\mathcal{M}_G(\mathfrak{g})$. An irreducible object of $\mathcal{M}_G(\mathfrak{g})$ is said to be <u>orbital</u> if its support is the closure of a single G-orbit in $\mathfrak{g}$. An irreducible object of $\mathcal{M}_G(\mathfrak{g})$ is said to be <u>anti-orbital</u> if it is of the form $\mathcal{F}K$ where $K \in \mathcal{M}_G(\mathfrak{g})$ is orbital.

5. <u>Theorem. Let</u> $A \in \mathcal{M}_G(\mathfrak{g})$ <u>be irreducible.</u>

(a) A <u>is admissible if and only if it is anti-orbital.</u>

(b) <u>If</u> G <u>is semisimple and</u> A <u>is cuspidal, then</u> $\mathcal{F}A \cong A$. <u>Thus</u> A <u>is both orbital and anti-orbital.</u>

The proof will be given in no. 9.

6. Assume given an F_q-structure on G, $\mathfrak{g}$ as in 1(a). If K is a perverse sheaf on $\mathfrak{g}$ such that $F^*K \cong K$, we choose an isomorphism $\varphi: F^*K \xrightarrow{\sim} K$ and we define the <u>characteristic function</u> $\chi_{K,\varphi}: \mathfrak{g}^F \to \overline{Q}_\ell$ by

$$\chi_{K,\varphi}(\xi) = \sum_i (-1)^i \mathrm{Tr}(\varphi, \mathcal{H}^i_\xi K)$$

where $\mathcal{H}^i K$ are the cohomology sheaves of K and the subscript ξ denotes the stalk at ξ. We shall use several times the following principle.

If K, K' are two semisimple perverse sheaves on $\mathfrak{g}$, in order to prove that $K \cong K'$, it is enough to check that one can choose an F_q-structure as above and $\varphi: F^*K \xrightarrow{\sim} K$, $\varphi': F^*K' \to K'$ such that $\chi_{K,\varphi^i} = \chi_{K',\varphi'^i} : \mathfrak{g}^{F^i} \to \overline{Q}_\ell$ for $i = 1,2,3,\ldots$; here $\varphi^i : (F^i)^*K \xrightarrow{\sim} K$ is defined by iterating φ and φ'^i is defined similarly. The Fourier transform $\hat{f}$ of a function $f: \mathfrak{g}^F \to \overline{Q}_\ell$ is defined by 1(b) where $\Psi: F_q \to \overline{Q}_\ell^*$ is $\Psi_0 \circ \mathrm{Tr}_{F_q/F_p}$. If $f = \chi_{K,\varphi}$ (as above) then $\hat{f} = \chi_{\mathcal{F}K,\varphi'}$ for a suitable $\varphi': F^*\mathcal{F}K \xrightarrow{\sim} \mathcal{F}K$.

7. Let K be an orbital object in $\mathcal{M}_G(\mathfrak{g})$ with support $\overline{C}$ where C is the G-orbit of $\sigma + \nu \in \mathfrak{g}$ (σ semisimple, ν nilpotent, $[\sigma,\nu] = 0$). Let L be the centralizer of σ in G, P a parabolic subgroup of G with Levi subgroup L and let $U, \mathfrak{l}, \mathfrak{p}, \mathfrak{u}$ be as in 2(a). Let K_0 be the orbital object in $\mathcal{M}_L(\mathfrak{l})$ whose support is the closure of the L-orbit C_0 of $\sigma + \nu$ and is such that $K_0|\overline{C}_0$ is (up to shift) the same as $K|\overline{C}_0$. Assuming that 5(a) holds for L, we shall prove that

(a) $\mathcal{F}K \cong i_L^G(\mathcal{F}K_0)$.

Note that $\mathcal{F}K_\circ$ is anti-orbital hence by our assumption, it is admissible in $M_L(\mathfrak{l})$. Hence $i_L^G(\mathcal{F}K_\circ)$ is a semisimple perverse sheaf on $\mathfrak{g}$, see 3(a). Since $\mathcal{F}K$ is a semisimple (in fact irreducible) perverse sheaf on $\mathfrak{g}$, to prove (a) it is enough to prove the equality of the corresponding characteristic functions (see no.6) for larger and larger F_q. Choose an F_q-rational structure on G (hence on $\mathfrak{g}$) with Frobenius map F such that $P, L, \sigma, \nu, \langle , \rangle$ are defined over F_q and such that there exists $\varphi: F^*K \tilde{\to} K$. Let $f_\circ = \chi_{K_\circ, \varphi_\circ}$ where $\varphi_\circ : F^*K_\circ \tilde{\to} K_\circ$ is defined by φ and let $f = \chi_{K, \varphi}$. We have

$$f(\xi) = \begin{cases} f_\circ(\mathrm{Ad}(g)\xi), & \text{if } \mathrm{Ad}(g)\xi_{ss} = \sigma \text{ for some } g \in G^F \\ 0, & \text{otherwise} \end{cases}$$

here ξ_{ss} is the semisimple part of ξ.

It is enough to show that

(b)
$$\hat{f}(\xi) = |\mathfrak{u}^F|\,|P^F|^{-1} \sum_{\substack{g \in G^F \\ \mathrm{Ad}(g)\xi \in \mathfrak{p}}} \hat{f}_\circ(\rho(\mathrm{Ad}(g)\xi)), \qquad \xi \in \mathfrak{g}^F.$$

We have

$$\hat{f}(\xi) = |L^F|^{-1} \sum_{\substack{g \in G^F \\ \eta \in \mathfrak{l}^F \\ \eta \text{ nilp.}}} \Psi\langle \xi, \mathrm{Ad}(g^{-1})(\sigma + \eta)\rangle f_\circ(\sigma + \eta).$$

Fix a coset $U^F g_\circ$, $g_\circ \in G^F$, and let g run only over this coset. Note that $u \to \mathrm{Ad}(u^{-1})(\sigma + \eta) - (\sigma + \eta)$ is a bijection $U^F \to \mathfrak{u}^F$, hence this part of the sum is

$$\sum_{\mu \in \mathfrak{u}^F} \Psi\langle \xi, \mathrm{Ad}(g_\circ^{-1})(\sigma + \eta + \mu)\rangle f_\circ(\sigma + \eta)$$

$$= \sum_{\mu \in \mathfrak{u}^F} \Psi\langle \mathrm{Ad}(g_\circ)\xi, \mu\rangle \Psi\langle \mathrm{Ad}(g_\circ)\xi, \sigma + \eta\rangle f_\circ(\sigma + \eta)$$

$$= 0, \quad \text{unless } \mathrm{Ad}(g_\circ)\xi \in \mathfrak{p}.$$

Hence

$$\hat{f}(\xi) = |L^F|^{-1} \sum_{\substack{g \in G^F \\ \mathrm{Ad}(g)\xi \in \mathfrak{p} \\ \eta \in \mathfrak{l}^F \\ \eta \text{ nilp.}}} \Psi\langle \mathrm{Ad}(g)\xi, \sigma + \eta\rangle f_\circ(\sigma + \eta)$$

and (b) (hence (a)) follows.

8. Let $L, P, U, \mathfrak{l}, \mathfrak{p}, \mathfrak{u}$ be as in 2(a) and let $A \in M_L(\mathfrak{l})$ be cuspidal. Write $A = h^*\xi_{\Psi_\circ} \boxtimes A_1$ as in 3(d) for $\mathfrak{l}$ instead of $\mathfrak{g}$. Here h is a linear form on the centre z_1 of l. Let $\sigma \in z_1$ be defined by $h(\xi) = -\langle \sigma, \xi\rangle$ for all $\xi \in z_1$. Assume that σ is in the centre of $\mathfrak{g}$.

Let $N_\sigma = \sigma + N$. Assuming also that 5(b) holds for L/Z_L^o, we shall prove that

(a) $F(i_L^G A) \cong i_L^G A|N_\sigma$ extended by 0 on $g - N_\sigma$, shifted by codim N_σ.

(The special case where L is a maximal torus and $A = \overline{\mathbb{Q}}_\ell$ appears in [B].)

We can easily reduce the general case to the case $\sigma = 0$. Assume now that $\sigma = 0$. Using 3(a),(b) we see that it is enough to prove instead of (a) the equality of the corresponding characteristic functions (see no.6) for larger and larger F_q. Choose an F_q-rational structure on G (hence on g) with Frobenius map F such that P, L, <, > are defined over F_q and such that there exists $\varphi = F^*A \tilde{\to} A$. Let $f_\circ = \chi_{A,\varphi}: \mathcal{Z}^F \to \overline{\mathbb{Q}}_\ell$. From 5(b) for L/Z_L^o it follows that

(b) $$\hat{f}_o(\xi_o) = \begin{cases} \gamma f_o(\xi_o), & \text{if } \xi_o \in \mathcal{Z}^F \text{ is nilpotent} \\ 0, & \text{otherwise} \end{cases}$$

where γ is a constant.

It is enough to prove that

$$|P^F|^{-1} \sum_{\xi' \in g^F} \Psi\langle \xi, \xi' \rangle \sum_{\substack{g \in G^F \\ Ad(g)\xi' \in p}} f_o(\rho(Ad(g)\xi'))$$

$$= \begin{cases} \gamma |u^F|\ |P^F|^{-1} \sum\limits_{\substack{g \in G^F \\ Ad(g)\xi \in p}} f_o(\rho(Ad(g)\xi)) & , \text{ if } \xi \in g^F \text{ is nilpotent} \\ 0 & , \text{ otherwise} \end{cases}$$

; this follows easily from (b).

9. Proof of Theorem 5. The theorem is obvious when G = {e}. We can assume that G ≠ {e} and that the theorem is already known for all Levi subgroups of proper parabolic subgroups of G. Moreover, using 3(d) we can assume that G is semisimple. We first show:

(a) If $A = \mathcal{F}K$ where $K \in M_G(g)$ is orbital and non-cuspidal, then A is admissible, non-cuspidal.

Let $\sigma + \nu \in C$, L, P, U, $\mathcal{Z}$, p, u be associated with K as in no.7. If P ≠ G, by the induction hypothesis and by 7(a) we have $A = \mathcal{F}K \cong i_L^G(\mathcal{F}K_o)$ where $K_o \in M_L(\mathcal{Z})$ is orbital; moreover, by the induction hypothesis, $\mathcal{F}K_o$ is admissible hence $i_L^G(\mathcal{F}K_o)$ is a direct sum of admissible non-cuspidal objects in $M_G(g)$, see 3(a). Since $i_L^G(\mathcal{F}K_o)$ is isomorphic to A, it follows that A is admissible, non-cuspidal. Therefore we may assume that L = P = G so that σ is in the centre of g. Since G is semisimple, it follows that $\sigma = 0$ so that the support of K is the closure of a single nilpotent class. Attach P, L, p, $\mathcal{Z}$, $A_\circ$ to K as in 3(f). (These P, L, p, $\mathcal{Z}$, are not the ones considered above). Since K is not cuspidal, we have P ≠ G. From 3(f) and 8(a) (which is applicable by the induction hypothesis) we see that K is a direct summand of $\mathcal{F}(i_L^G A_o)$ hence $A = \mathcal{F}K$ is a direct summand of $\mathcal{F}\mathcal{F}(i_L^G A_o) = j^*(i_L^G A_o)$, see 4(a). Then j^*A is a direct summand of $i_L^G A_o$ hence it is admissible, non-cuspidal, see 3(a).

But one checks easily that j^* permutes the admissible non-cuspidal objects in $\mathcal{M}_G(\mathfrak{g})$, so that A is admissible non-cuspidal, as asserted.

Next we show:

(b) <u>If</u> $A \in \mathcal{M}_G(\mathfrak{g})$ <u>is admissible, non-cuspidal, then</u> A <u>is anti-orbital</u>.

Let $L, P, \mathfrak{l}, \mathfrak{p}$ be as in 2(a) and let $A' \in \mathcal{M}_L(\mathfrak{l})$ be a cuspidal object such that A is a direct summand of $i_L^G A'$. Since A is not cuspidal we have $P \neq G$. By the induction hypothesis, we have $A' \cong \mathcal{F}K$ where $K \in \mathcal{M}_L(\mathfrak{l})$ is orbital. Let $\sigma + \nu \in C \subset \mathfrak{l}$ be attached to K as in no. 7 (for $\mathfrak{l}$ instead of $\mathfrak{g}$). Let L' be the centralizer of σ in G. Then $L' \supset L$ since σ is central in $\mathfrak{l}$. Let P' be a parabolic subgroup of G with Levi subgroup L' such that $P' \supset P$.

By 8(a), $\mathcal{F}(i_L^{L'} A')$ is a direct sum of orbital complexes on $\mathfrak{l}'$ with support contained in $\{\xi \in \mathfrak{l}' | \xi_{ss} = \sigma\}$. Hence $i_L^{L'} A' \cong \bigoplus_\alpha \mathcal{F}A_\alpha$ where $A_\alpha \in \mathcal{M}_{L'}(\mathfrak{l}')$ are orbital, supp $A_\alpha \subset \{\xi \in \mathfrak{l}' | \xi_{ss} = -\sigma\}$. We have $i_{L'}^G(\mathcal{F}A_\alpha) = \mathcal{F}\tilde{A}_\alpha$ where $\tilde{A}_\alpha \in \mathcal{M}_G(\mathfrak{g})$ is orbital with support contained in $\{\xi \in \mathfrak{g} | \xi_{ss} = -\sigma\}$. (See 7(a); this is applicable by the induction hypothesis if $L' \neq G$ and is trivial if $L' = G$.) Hence, using 3(c), we have $i_L^G A' = i_{L'}^G (i_L^{L'} A') = \bigoplus_\alpha i_{L'}^G(\mathcal{F}A_\alpha) = \bigoplus_\alpha \mathcal{F}\tilde{A}_\alpha$. Since A is a direct summand of $i_L^G A'$, we must have $A \cong \mathcal{F}\tilde{A}_\alpha$ for some α, so that A is anti-orbital, as asserted.

It remains to show that $\mathcal{F}K \cong K$ for any cuspidal $K \in \mathcal{M}_G(\mathfrak{g})$.

We now fix an F_q-rational structure on G (hence on $\mathfrak{g}$) with Frobenius map F such that $\langle , \rangle$ is defined over F_q and $F^*K \cong K$ for all cuspidal $K \in \mathcal{M}_G(\mathfrak{g})$. (This is possible since there are only finitely many cuspidal K).

Let I be the set of pairs $(C, \mathcal{L})$ where C is an F-stable G-orbit in $\mathfrak{g}$ and $\mathcal{L}$ is a G-equivariant irreducible local system on C (up to isomorphism) such that $F^*\mathcal{L}$ is isomorphic to $\mathcal{L}$.

For each $i = (C, \mathcal{L}) \in I$ we choose a definite $\varphi : F^*\mathcal{L} \xrightarrow{\sim} \mathcal{L}$ and define $f_i : \mathfrak{g}^F \to \bar{Q}_\ell^*$ by

$$f_i(\xi) = \begin{cases} \mathrm{Tr}(\varphi, \mathcal{L}_\xi), & \text{if } \xi \in C^F \\ 0, & \text{if } \xi \in \mathfrak{g}^F - C^F. \end{cases}$$

It is easy to see (cf. for example [L_2, V(24.2.7)]) that $f_i (i \in I)$ form a basis for the vector space $\mathcal{V}$ of all G^F-invariant functions $\mathfrak{g}^F \to \bar{Q}_\ell$. Now let $K_i \in \mathcal{M}_G(\mathfrak{g})$, $(i = (C, \mathcal{E}) \in I)$, be the orbital perverse object such that supp $K_i = \bar{C}$ and $K|C = \mathcal{E}$ up to shift.

Then $F^*K_i \xrightarrow{\sim} K_i$; we choose a definite isomorphism and we let f_i' be the characteristic function of K_i with respect to this isomorphism.

Then f_i' can be expressed in terms of the f_i by means of a triangular matrix with non-zero elements on the diagonal. In particular f_i' $(i \in I)$ form a basis of $\mathcal{V}$.

Let I_0 be the set of those $i \in I$ for which K_i is cuspidal (All cuspidal $K \in \mathcal{M}_G(\mathfrak{g})$ are among the K_i, $i \in I$).

Let (,) be the non-singular bilinear form on $\mathcal{V}$ defined by

$$(f, f') = \sum_{\xi \in \mathfrak{g}^F} f(\xi) f'(\xi).$$

We shall need the following properties.

(c) $f'_h = c_h f_h$ (c_h = constant) for $h \in I_0$.

(This follows from 3(g))

(d) If $i \in I - I_0$ and $h \in I_0$ then $(f_i, f_h) = 0$

(This follows from [L_2, V 24.4(d)] if $i = (C, E)$ with C nilpotent and is trivial, otherwise)

(e) If $i \in I - I_0$ and $h \in I_0$ then $(f'_i, f_h) = 0$.

(Indeed, from [L_2, V 24.4(d)] it follows that $f'_i|N^F$ is a linear combination of $f_t|N^F$ where $t \in I - I_0$, and it remains to use (d).)

We now choose for each $i \in I$ an isomorphism $F^* FK_i \xrightarrow{\sim} FK_i$ and denote by f''_i the corresponding characteristic function. We have

(f) $f''_i = d_i \hat{f}'_i$, (d_i = constant), $i \in I$.

(g) If $i \in I - I_0$ and $h \in I_0$ then $(f''_i, f_h) = 0$.

(Indeed from (a) and 3(h) it follows that $f''_i |N^F$ is a linear combination of $f_t|N^F$ where $t \in I - I_o$, and it remains to use (d).)

From (g) it follows that $(\hat{f}'_i, f_h) = 0$ for $i \in I - I_0$, $h \in I_0$. Since $f \to \hat{f}$ preserves (,) up to a scalar, it follows that $(f'_i, \hat{f}_h) = 0$ hence $(f'_i \circ j, \hat{f}_h) = 0$ for $i \in I - I_0$, $h \in I_0$. When i runs through $I - I_0$ the functions $\{f'_i \circ j\}$, $\{f'_i\}$ coincide up to order and up to scalars. Hence $(f'_i, \hat{f}_h) = 0$ for $i \in I - I_0$, $h \in I_0$. By computing dimensions, we deduce that $\{\hat{f}_h | h \in I_0\}$ span the subspace of V orthogonal to all f'_i, $(i \in I - I_0)$. Similarly from (e) we see that $\{f_h | h \in I_0\}$ span the subspace of V orthogonal to all f'_i $(i \in I - I_0)$. It follows that $\{\hat{f}_h | h \in I_0\}$ span the same subspace as $\{f_h | h \in I_0\}$. In particular each $\hat{f}_h (h \in I_0)$ is zero on $\mathfrak{g}^F - N^F$. Using (f) and (c) it follows that $f''_h (h \in I_0)$is zero on $\mathfrak{g}^F - N^F$. Letting now F_q become bigger and bigger, we deduce that $FK_h \mid \mathfrak{g} - N$ is zero, or in other words, supp $FK_h \subset N$ for $h \in I_0$. Since FK_h is G-equivariant, irreducible, and N consists of finitely many G-orbits, it follows that FK_h is orbital, $(h \in I_0)$.

Assume that FK_h is non-cuspidal, $h \in I_0$. Applying (a) to $K = FK_h$ we see that $F\, FK_h = j^* K_h$ is non-cuspidal hence K_h is non-cuspidal, a contradiction. Thus FK is cuspidal whenever K is cuspidal. Note that F preserves the character by which the centre of G acts on an irreducible object of $M_G(\mathfrak{g})$; hence using 3(e), we see that $FK \cong K$ whenever K is cuspidal. This completes the proof of the theorem.

10. In the following results we assume that we are given an F_q rational structure on G (hence on g) with Frobenius map F such that $\langle , \rangle$ is defined over F_q; we assume also that G is semisimple.

Corollary. Let $K \in M_G(g)$ be cuspidal. Assume that there exists $\varphi: F^*K \tilde{\to} K$ and let $f = \chi_{K,\varphi}: g^F \to \overline{Q}_\ell$. Then $\hat{f} = \gamma f$ where γ is a constant. In particular both f, $\hat{f}$ are concentrated on nilpotent elements.

his follows immediately from 5(b).

The corollary reminds us of a recent result of Kawanaka [Kw, (3.3.9)] stating that a certain function on a prehomogeneous vector space associated with the exceptional group G_2, F_4 or E_8 is essentially invariant under Fourier transform.

11. We now state the following result which is a complement to Cor. 10. (The setup is that in no. 10).

Theorem. Let $f: g^F \to \overline{Q}_\ell$ be a G^F-invariant function such that $f|g^F - N^F \equiv 0$, $\hat{f}|g^F - N^F \equiv 0$. Then there exist $K_i \in M_G(g)$, cuspidal, $(1 \leq i \leq m)$ and isomorphisms $\varphi_i: F^*K_i \tilde{\to} K_i$ such that f is a $\overline{Q}_\ell$-linear combination of the functions $\chi_{K_i,\varphi_i}: g^F \to \overline{Q}_\ell$, $(1 \leq i \leq m)$.

Proof. Let I be the set of all pairs (C, L) where C is a nilpotent orbit in g and L is a G-equivariant irreducible local system on C (up to isomorphism). For each $i \in I$ we denote by A_i the irreducible object in $M_G(g)$ with support $\overline{C}$ such that $A_i|C$ is L up to a shift; let $A_i^! \in M_G(g)$ be defined by $F A_i^! = A_i$.

Let J be the set of all pairs $(L, K_\circ)$ up to G-conjugacy where L is the Levi subgroup of a parabolic subgroup of G, l is its Lie algebra and $K_\circ \in M_L(l)$ is cuspidal. For each $j \in J$, $j = (L, K_\circ)$ we denote $K_j^! = i_L^G(K_\circ)$ and $K_j = K_j^!|N$, extended by zero on $g - N$, shifted by codim N. There is a unique map $\tau: I \to J$ with the following property: for any j, K_j is a direct sum of irreducible objects of $M_G(g)$, isomorphic to some A_i, $i \in \tau^{-1}(j)$; each A_i $(i \in \tau^{-1}(j))$ is isomorphic to a direct summand of K_j.

According to 5(b) and 8(a) we have $F(K_j^!) = K_j$, $(j \in J)$. Hence $K_j^!$ is a direct sum of irreducible objects of $M_G(g)$, isomorphic to some $A_i^!$, $i \in \tau^{-1}(j)$; each $A_i^!$ $(i \in \tau^{-1}(j))$ is isomorphic to a direct summand of $K_j^!$.

On the set I we have a natural action of the Frobius map F. If $i = (C, E) \in I^F$ then $F^*A_i \approx A_i$ and $F^*A_i^! \approx A_i^!$; we choose isomorphisms $\varphi_i: F^*A_i \to A_i$, $\varphi_i^!: F^*A_i^! \to A_i^!$.

Then f is a linear combination of the functions χ_{A_i,φ_i} (since $\hat{f} = 0$ on $g^F - N^F$) Hence f is a linear combination

$$f = \sum_{i \in I^F} c_i \chi_{A_i^!, \varphi_i} \qquad (c_i \in \overline{Q}_\ell).$$

Let $\lambda \to \overline{\lambda}$ be an automorphism of $\overline{Q}_\ell$ which corresponds to complex conjugation under some isomorphism $\overline{Q}_\ell \approx \mathbb{C}$.

We shall need the following fact.

(a) If $A', A'' \in \mathcal{M}_G(\mathfrak{g})$ are irreducible direct summands of $K^!_{j'}, K^!_{j''}$ respectively ($j', j'' \in J$) and $\varphi': F^*A' \tilde{\to} A'$, $\varphi'': F^*A'' \tilde{\to} A''$ are isomorphisms then

$$\sum_{\xi \in \mathfrak{g}^F} \chi_{A',\varphi'}(\xi)\overline{\chi_{A'', \varphi''}(\xi)} = q^{m(j')} \sum_{\xi \in \mathcal{N}^F} \chi_{A', \varphi'}(\xi) \overline{\chi_{A'', \varphi''}(\xi)}$$

and both sides are zero unless $j' = j''$; here $m:J \to \mathbf{N}$ is defined by $m(L,K_o) = \dim \mathring{\mathcal{Z}}_L$.

This can be proved by expressing $\chi_{A', \varphi'}(\xi)$, $\chi_{A'', \varphi''}(\xi)$ in terms of "generalized Green functions" as in [L_2, II 8.5, (10.4.5)] and then using the orthogonality relations [L_2, II 9.11, V (25.6.2)].

The formula (a) is applicable to $(A', \varphi') = (A^!_i, \varphi_i)$, $(A'', \varphi'') = (A^!_h, \varphi_h)$, $i, h \in I^F$. Hence we have

(b)
$$\begin{aligned}\sum_{\xi \in \mathfrak{g}^F} f(\xi)\overline{f(\xi)} &= \sum_{i,h \in I^F} c_i\bar{c}_h \sum_{\xi \in \mathfrak{g}^F} \chi_{A^!_i, \varphi_i}(\xi)\overline{\chi_{A^!_h, \varphi_h}(\xi)} \\ &= \sum_{\substack{i,h \in I^F \\ \tau(i) = \tau(h)}} c_i\bar{c}_h q^{m(\tau(i))} \sum_{\xi \in \mathcal{N}^F} \chi_{A^!_i, \varphi_i}(\xi)\overline{\chi_{A^!_h, \varphi_h}(\xi)} \\ &= \sum_{j \in J} q^{m(j)} \sum_{\xi \in \mathcal{N}^F} f_j(\xi)\overline{f_j(\xi)}\end{aligned}$$

$$\text{where } f_j(\xi) = \sum_{\substack{i \in \tau^{-1}(j) \\ F(i) = i}} c_i \chi_{A^!_i, \varphi_i}(\xi).$$

On the other hand, since $f = 0$ on $\mathfrak{g}^F - \mathcal{N}^F$, we have

(c)
$$\begin{aligned}\sum_{\xi \in \mathfrak{g}^F} f(\xi)\overline{f(\xi)} &= \sum_{\xi \in \mathcal{N}^F} f(\xi)\overline{f(\xi)} \\ &= \sum_{i,h \in I^F} c_i\bar{c}_h \sum_{\xi \in \mathcal{N}^F} \chi_{A^!_i, \varphi_i}(\xi)\overline{\chi_{A^!_h, \varphi_h}(\xi)} \\ &= \sum_{\substack{i,h \in I^F \\ \tau(i) = \tau(h)}} c_i\bar{c}_h \sum_{\xi \in \mathcal{N}^F} \chi_{A^!_i, \varphi_i}(\xi)\overline{\chi_{A^!_h, \varphi_h}(\xi)} \\ &= \sum_{j \in J} \sum_{\xi \in \mathcal{N}^F} f_j(\xi)\overline{f_j(\xi)}.\end{aligned}$$

Comparing (b) and (c) we obtain

$$\sum_{j \in J} (q^{m(j)} - 1) \sum_{\xi \in \mathcal{N}^F} f_j(\xi)\overline{f_j(\xi)} = 0$$

It follows that $f_j|\mathcal{N}^F$ is zero whenever $m(j) > 0$. Hence

$$f(\xi) = \sum_{\substack{i \in I^F \\ m(\tau(i)) = 0}} c_i \chi_{A^!_i, \varphi_i}(\xi) \qquad (\xi \in \mathcal{N}^F).$$

The condition $m(\tau(i)) = 0$ is equivalent to the condition that $A_i^!$ is cuspidal. The theorem follows.

12. We now state a variant of the previous theorem in which functions are replaced by perverse sheaves.

Theorem. Assume that G is semisimple. Let $K \in \mathcal{M}_G(g)$ be a semisimple object such that $\operatorname{supp} K \subset \mathcal{N}$, $\operatorname{supp} \mathcal{F}K \subset \mathcal{N}$. Then K is a direct sum of cuspidal objects in $\mathcal{M}_G(g)$.

Proof. We may assume that K is irreducible. We assume that K is not cuspidal and we shall find a contradiction. Since $\operatorname{supp} K \subset \mathcal{N}$, we see that K is orbital. From 9(a) we see that $\mathcal{F}K$ is admissible, non-cuspidal. From 3(a) we see that $\operatorname{supp} \mathcal{F}K \not\subset \mathcal{N}$. This contradicts our assumption that $\operatorname{supp} \mathcal{F}K \subset \mathcal{N}$, proving the proposition.

Acknowledgement. This research has been supported in part by an N.S.F. grant.

REFERENCES

[BR] P. Bardsley, R.W. Richardson: Étale slices for algebraic transformation groups in characteristic p, Proc. Lond. Math. Soc. 51(1985), 295 - 317.

[BBD] A. Beilinson, J. Bernstein, P. Deligne, Faisceaux pervers, Astérisque 100 (1982), Société Mathématique de France.

[B] J.L. Brylinski, Transformations canoniques, dualité projective, théorie de Lefschetz, transformation de Fourier et sommes trigonometriques, preprint

[HK] R. Hotta, M. Kashiwara, The invariant holonomic system on a semisimple Lie algebra, Inv. Math. 75(1984) 327 - 358.

[KLa] N. Katz, G. Laumon, to appear in Publ. Math. I.H.E.S.

[Kw] N. Kawanaka, Generalized Gelfand - Graev representations of exceptional simple algebraic groups over a finite field, preprint.

[Kz] D. Kazhdan, Proof of Springer's hypothesis. Israel J. Math. 28(1977) 272 - 286.

[L_1] G. Lusztig, Intersection cohomology complexes on a reductive group, Inv. Math. 75(1984), 205 - 272.

[L_2] G. Lusztig, Character sheaves, I Adv. in Math. 56(1985), 193 - 237, II Adv. in Math. 57(1985) 226 - 265, III Adv. in Math. 57(1985) 266 - 315, IV Adv. in Math. 59(1986), 1 -63, V to appear in Adv. in Math.

[S] T.A. Springer, Trigonometric sums, Green functions of finite groups and representations of Weyl groups, Inv. Math. 36(1976), 173 - 207.

COMMUTING DIFFERENTIAL OPERATORS AND ZONAL SPHERICAL FUNCTIONS

I.G. Macdonald
School of Mathematical Sciences
Queen Mary College
London E1 4NS.

1. Introduction

The subject of my talk is a class of polynomial symmetric functions

$$J_\lambda(x_1,\ldots,x_n;\alpha)$$

indexed by partitions $\lambda = (\lambda_1,\ldots,\lambda_n)$ of length $\leq n$, and involving a parameter α, which may be regarded as an indeterminate or (more usefully) as a positive real number.

For particular values of α these polynomials occur "in nature":

(i) when $\alpha = 1$ they are (up to a scalar factor) the Schur functions $s_\lambda(x_1,\ldots,x_n)$, i.e. the characters of the polynomial representations of $GL_n(\mathbb{C})$.

(ii) when $\alpha = 2$ they are the zonal polynomials familiar to the statisticians (see, e.g., [4]). These are zonal spherical functions on $SL_n(\mathbb{R})/SO(n)$ (or on $SU(n)/SO(n)$) defined by polynomial representations of $SL_n(\mathbb{R})$ (or $SU(n)$).

(iii) when $\alpha = \frac{1}{2}$ they can again be interpreted as zonal spherical functions, this time on the symmetric space $SL_n(\mathbb{H})/Sp(n)$ or its compact form $SU(2n)/Sp(n)$.

(iv) when $\alpha = 0$ we have $J_\lambda(x_1,\ldots,x_n;0) = e_{\lambda'}(x_1,\ldots,x_n)$, the product of elementary symmetric functions corresponding to the conjugate λ' of the partition λ ([10], Chapter I).

(v) Finally, when suitably normalized $J_\lambda(x_1,\ldots,x_n;\alpha)$ makes sense for $\alpha = \infty$, and reduces in this case to the monomial symmetric function $m_\lambda(x_1,\ldots,x_n) = \Sigma\ x_1^{\lambda_1}\ldots x_n^{\lambda_n}$.

The cases (i) - (iii) above may be described uniformly as follows. Let G be a connected non-compact semisimple Lie group, and K a maximal compact subgroup of G, such that the restricted root system of the symmetric space G/K is of type A_{n-1}. Let m denote the multiplicity of each restricted root. (If $n > 3$ the only possibilities are $G = SL_n(F)$, where $F = \mathbb{R}$, $\mathbb{C}$ or $\mathbb{H}$, and $m = \dim_{\mathbb{R}} F = 1,2$ or 4. If $n = 3$ there is another possibility with G of type E_6, K of type F_4, and $m = 8$ (type EIV in Cartan's classification [5]).) Then the polynomials $J_\lambda(x_1,\dots,x_n;\alpha)$ with $\alpha = 2/m$ can be viewed as zonal spherical functions on G/K (or also on U/K, where U is a compact form of G) arising from finite-dimensional polynomial representations of G.

2. The differential operators $D_r^{(\alpha)}$

The polynomials $J_\lambda(x_1,\dots,x_n\,;\alpha)$ were first introduced by Henry Jack [7], and we shall come back to his definition in §3. Here we shall obtain them as simultaneous eigenfunctions of a family of commuting differential operators introduced by Debiard [2] and Sekiguchi [12].

We shall use the notation and terminology of symmetric functions [10]. Let $x_1,\dots,x_n$ be independent variables; the symmetric group S_n acts on the ring $\mathbb{R}[x_1,\dots,x_n]$ by permuting the x_i, and we write

$$\Lambda_{n,\mathbb{R}} = \mathbb{R}[x_1,\dots,x_n]^{S_n}$$

for the subring of symmetric polynomials in $x_1,\dots,x_n$.

If $\alpha = (\alpha_1,\dots,\alpha_n) \in \mathbb{N}^n$, let x^α denote $x_1^{\alpha_1}\dots x_n^{\alpha_n}$. In particular, let $\delta = (n-1,n-2,\dots,1,0)$; then the Vandermonde determinant is

$$\Delta = \prod_{i<j}(x_i - x_j) = \sum_{w\in S_n} \varepsilon(w)x^{w\delta},$$

where $\varepsilon(w)$ is the sign of the permutation w.

Now let t be an indeterminate, and let

$$D(t;\alpha) = \Delta^{-1} \sum_{w\in S_n} \varepsilon(w)x^{w\delta} \prod_{i=1}^{n} (1 + t(\alpha D_i + (w\delta)_i)) \;,$$

where $D_i = x_i\partial/\partial x_i$. The coefficient of $t^r (0 \le r \le n)$ in $D(t,\alpha)$ is a linear differential operator which we denote by $D_r^{(\alpha)}$:

$$D(t,\alpha) = \sum_{r=0}^{n} t^r D_r^{(\alpha)} \;.$$

If $f \in \Lambda_n$ is homogeneous of degree d, we have

$$D_0^{(\alpha)} f = f, \quad D_1^{(\alpha)} f = (d\alpha + \tfrac{1}{2}n(n-1))f ,$$

$$D_2^{(\alpha)} = (C - \alpha(\alpha U + V))f$$

where C is a constant (depending on n, d and α) and

$$U = \tfrac{1}{2} \sum_{i=1}^{n} x_i^2 \frac{\partial^2}{\partial x_i^2} , \qquad V = \sum_{i \neq j} \frac{x_i^2}{x_i - x_j} \frac{\partial}{\partial x_i} .$$

Remark: When $\alpha = 1, 2$ or $\tfrac{1}{2}$ (cases (i) - (iii) in §1) the operators $D_r^{(\alpha)}$ generate the algebra of radial components of invariant differential operators on the corresponding symmetric space G/K (or U/K). In particular $D_2^{(\alpha)}$ (or $\alpha U + V$) is essentially the Laplace-Beltrami operator.

Proposition ([2], [12]): The operators $D_r^{(\alpha)}$ $(0 \leq r \leq n)$ commute with each other.

We shall sketch a proof, but we have no space for all the details. Define a scalar product $\langle u, v\rangle_\alpha'$ on $\Lambda_{n,\mathbb{R}}$ as follows (the reason for the prime in the notation is that we shall later define another scalar product):

$$\langle u, v\rangle_\alpha' = \int_{T^n} u(t)\overline{v(t)}|\Delta(t)|^{2/\alpha} \, dt$$

where T^n is the torus

$$T^n = \{t = (t_1,\ldots,t_n) \in \mathbb{C}^n : |t_i| = 1, \ 1 \leq i \leq n\}$$

and dt is the Haar measure, normalized so that $\langle 1,1\rangle_\alpha' = 1$. Alternatively, $\langle u,v\rangle_\alpha'$ is the constant term in

$$c^{-1} u(t)v(t^{-1}) \prod_{i \neq j} (1 - t_i t_j)^{1/\alpha} ,$$

where the constant c is determined by Dyson's conjecture ([3], [11]).

Then formal computations show that

(a) each operator $D_r^{(\alpha)}$ maps $\Lambda_{n,\mathbb{R}}$ into $\Lambda_{n,\mathbb{R}}$ and is degree preserving; moreover if $m_\lambda = \Sigma x^\lambda$ is the monomial symmetric function indexed by the partition λ ([10], Chapter I) then

$$D_r^{(\alpha)} m_\lambda = \sum_{\mu \le \lambda} a_{\lambda\mu} m_\mu$$

where $\mu \le \lambda$ is the usual dominance order on partitions (loc. cit.)

(b) $D_r^{(\alpha)}$ is self-adjoint for the scalar product $\langle u,v\rangle'_\alpha$.

Now fix a positive integer $k \le n$ and apply Gram-Schmidt orthogonalization to the basis $(m_\lambda)_{|\lambda|=k}$ of the space Λ_n^k , starting at the bottom (i.e., with $m_{(1^k)}$). We shall obtain an orthogonal basis $(\tilde{J}_\lambda)$ with each $\tilde{J}_\lambda$ of the form

$$\tilde{J}_\lambda = \sum_{\mu \le \lambda} \tilde{v}_{\lambda\mu}(\alpha) m_\mu \ . \qquad (1)$$

With respect to the basis $(\tilde{J}_\lambda)$, the matrix of each differential operator $D_r^{(\alpha)}$ is simultaneously triangular and hermitian, by virtue of (a) and (b) above, hence is diagonal. It follows that the $D_r^{(\alpha)}$ commute with each other and that the $\tilde{J}_\lambda$ are simultaneous eigenfunctions of each $D_r^{(\alpha)}$.

Remark: In fact the Laplace-Beltrami operator $D_2^{(\alpha)}$ is decisive: the $\tilde{J}_\lambda$ are uniquely determined (up to scalar factors) as the symmetric polynomials of the form (1) that are eigenfunctions for $D_2^{(\alpha)}$.

3. Jack's symmetric functions

In order to avoid extraneous scalar factors, it is convenient to renormalize the symmetric polynomials $\tilde{J}_\lambda$ defined in §2. For a partition λ of $k \le n$ we define $J_\lambda(x_1,\dots,x_n;\alpha)$ to be the scalar multiple of $\tilde{J}_\lambda$ in which the coefficient of $m_{(1^k)}$ (i.e. of the kth elementary symmetric function e_k) is $k!$

It is then not difficult to verify that, when we pass to $n+1$ variables, we have

$$J_\lambda(x_1,\dots,x_n 0;\alpha) = J_\lambda(x_1,\dots,x_n;\alpha)$$

(which would not be the case for the $\tilde{J}_\lambda$). Hence for each partition λ we have a well defined element $J_\lambda(x;\alpha)$ of the ring $\Lambda_{\mathbb{R}} = \varprojlim \Lambda_{n,\mathbb{R}}$ of symmetric functions with real coefficients ([10], Chapter I). The $J_\lambda(x;\alpha)$ are <u>Jack's symmetric functions</u>; $J_\lambda(x;\alpha)$ is homogenous of degree $|\lambda|$.

There is another and more combinatorial way of defining these symmetric functions. For each integer $r \geq 1$, let $p_r = \Sigma x_i^r$ be the rth power sum, and for each partition $\lambda = (\lambda_1, \lambda_2, \ldots)$ let $p_\lambda = p_{\lambda_1} p_{\lambda_2} \cdots$. The products p_λ for all partitions λ from an $\mathbb{R}$-basis of $\Lambda_{\mathbb{R}}$. The usual scalar product on Λ, for which the Schur functions form an orthonormal basis, is such that for any two partitions λ, μ we have

$$<p_\lambda, p_\mu> = \delta_{\lambda\mu} z_\lambda,$$

where $\delta_{\lambda\mu}$ is the Kronecker delta, and z_λ is the order of the centralizer of a permutation of cycle-type λ in the symmetric group $S_{|\lambda|}$.

We modify this scalar product as follows: define

$$<p_\lambda, p_\mu>_\alpha = \delta_{\lambda\mu} z_\lambda \alpha^{\ell(\lambda)}$$

where $\ell(\lambda)$ is the length of the partition λ (i.e. the number of nonzero parts λ_i). Then it can be shown that the $J_\lambda(x;\alpha)$ are pairwise orthogonal with respect to this scalar product, as well as with respect to the scalar product $<u,v>'_\alpha$ defined in §2. (When $\alpha = 1$, the two scalar products coincide; when $\alpha \neq 1$, they don't.)

In other words, the $J_\lambda(x;\alpha)$ can be constructed by Gram-Schmidt orthogonalization relative to the scalar product $<u,v>_\alpha$ on $\Lambda_{\mathbb{R}}$ from the monomial symmetric functions $m_\mu(x)$, the scalar factors being adjusted so as to ensure that the coefficient of $m_{(1^k)}$ in J_λ is equal to $k!$ (where $k = |\lambda|$). This is essentially Jack's original definition [7]. To show that this definition agrees with the previous one it is enough to verify that the J_λ so defined are eigenfunctions of the Laplace-Beltrami operator $D_2^{(\alpha)}$ with the appropriate eigenvalues.

In general the formal properties of the $J_\lambda(x;\alpha)$ appear to mimic those of the Schur functions $s_\lambda(x)$ in a very satisfactory way. (We should observe at this point that when $\alpha = 1$, $J_\lambda(x;\alpha)$ reduces not to $s_\lambda(x)$ but to $h(\lambda)s_\lambda(x)$, where $h(\lambda)$ is the product of the hook lengths of the diagram of λ.)

One example is the following. Recall [10] that there is an involution ω on the algebra $\Lambda_{\mathbb{R}}$ which may be defined by

$$\omega(p_r) = (-1)^{r-1} p_r \qquad (r \geq 1).$$

This involution has the property that $\omega(s_\lambda) = s_{\lambda'}$, for any partition λ, where as before λ' is the conjugate of λ. This property now generalizes as follows: if we define an automorphism ω_α of $\Lambda_{\mathbb{R}}$ by

$$\omega_\alpha(p_r) = (-1)^{r-1}\alpha p_r \qquad (r \geq 1)$$

we have the duality theorem

$$\omega_\alpha(J_\lambda(x;\alpha)) = \alpha^{|\lambda|}J_{\lambda'}(x;\alpha^{-1})$$

for any α and any partition λ. This can be proved again by verifying that the $\omega_\alpha(J_\lambda)$ are eigenfunctions of the operator $D_2^{(\alpha)}$ with the appropriate eigenvalues.

4. Some theorems and some conjectures

Let λ be a partition, s a square in the diagram of λ. Suppose that s lies in the ith row and jth column of λ. We define the arm-length $a(s)$ and arm-colength $a'(s)$ of s by

$$a(s) = \lambda_i - j, \qquad a'(s) = j - 1,$$

and likewise the leg-length $\ell(s)$ and leg-colength $\ell'(s)$ of s by

$$\ell(s) = \lambda'_j - i, \qquad \ell'(s) = i - 1.$$

The upper and lower hook-lengths $h^*_\lambda(s)$, $h^\lambda_*(s)$ are then defined by

$$h^*_\lambda(s) = \ell(s) + (1 + a(s))\alpha,$$

$$h^\lambda_*(s) = \ell(s) + 1 + a(s)\alpha.$$

When $\alpha = 1$, both of these are equal to the hook length $h_\lambda(s) = \ell(s) + a(s) + 1$. Finally define

$$h^*(\lambda) = \prod_{s\in\lambda} h^*_\lambda(s), \qquad h_*(\lambda) = \prod_{s\in\lambda} h^\lambda_*(s).$$

The following results, which confirm earlier conjectures of mine, are due to R. Stanley [14]:

(1) $\langle J_\lambda, J_\lambda\rangle_\alpha = h^*(\lambda)h_*(\lambda)$.

(2) The coefficient of m_λ in J_λ is equal to $h_*(\lambda)$.

(3) Let X be an indeterminate and define $\phi_X : \Lambda_{\mathbb{R}} \to \mathbb{R}[X]$ by $\phi_X(p_r) = X$ for all $r \geq 1$. Then

$$\phi_X(J_\lambda) = \prod_{s\in\lambda} (X + a'(s)\alpha - \ell'(s)) .$$

(When X is specialized to a positive integer n, $\phi_X(J_\lambda)$ is the value of $J_\lambda(x_1,\ldots,x_n;\alpha)$ at $x_1 = \ldots = x_n = 1$.)

In addition, Stanley has proved a Pieri formula for the J_λ:

(4) The coefficient of J_λ in $J_\mu J_{(n)}$ is zero unless $\lambda \supset \mu$ and $\lambda - \mu$ is a horizontal strip ([10] Chapter 1) of length n, and then it is equal to

$$n!\alpha^n \Big(\prod_{s\in\mu} \tilde{h}_\mu(s)\Big)\Big(\prod_{s\in\lambda} h_\lambda(s)\Big)^{-1}$$

where (for $\sigma = \lambda$ or μ)

$$\tilde{h}_\sigma(s) = \begin{cases} h^*_\sigma(s) & \text{if } \lambda - \mu \text{ contains a square in the same column as } s, \\ h^\sigma_*(s) & \text{otherwise.} \end{cases}$$

From this result it follows (as in the case of Schur functions or Hall-Littlewood functions) that $J_\lambda(x;\alpha)$ can be written explicitly as a sum of monomials:

(5) $\quad J_\lambda(x;\alpha) = \sum_T w_T(\alpha) x^T$

summed over all (column strict) tableaux T of shape λ. Here $x^T = x^\mu$ where μ is the weight of T (the terminology is that of [10]) and $w_T(\alpha)$ is an explicitly given rational function of α, in which both numerator and denominator are products of linear factors, namely upper and lower hook-lengths for the intermediate partitions defined by the tableau T. It should be remarked that, in general, different tableaux T, T' of the same wieght (i.e., such that $x^T = x^{T'}$) give rise to different coefficients $w_T(\alpha)$, $w_{T'}(\alpha)$.

We shall conclude this section with some conjectures. Some of these are due to R. Stanley, others to K. Kadell and the author. First of all, tables of the J_λ, as far as they have been computed, suggest that:

(C1) The coefficient $v_{\lambda\mu}(\alpha)$ of m_μ in J_λ is a polynomial in α with non-negative integral coefficients.

We may remark that (C1) does not obviously follow from (5) above, which gives $v_{\lambda\mu}(\alpha)$ only as a sum of rational functions of α.

(C2) When J_λ is expressed in terms of the power sum products p_μ, the coefficients are polynomials in α with integer coefficients.

In other words, $J_\lambda \in \mathbb{Z}[\alpha, p_1, p_2, \ldots]$. We may remark that (C2) would imply that each $v_{\lambda\mu}(\alpha) \in \mathbb{Z}[\alpha]$, with each coefficient of $v_{\lambda\mu}(\alpha)$ divisible by $\prod_{i\geq 1} m_i!$, if $\mu = (1^{m_1}2^{m_2}\ldots)$.

(C3) Let λ, μ, ν be partitions and let

$$c^{\nu}_{\lambda\mu}(\alpha) = \langle J_\lambda J_\mu, J_\nu\rangle_\alpha, \qquad c^{\nu}_{\lambda\mu} = \langle s_\lambda s_\mu, s_\nu\rangle$$

(so that $c^{\nu}_{\lambda\mu}$ is a non-negative integer, given by the Littlewood-Richardson rule ([10] Chapter 1)). Then

(i) $c^{\nu}_{\lambda\mu}(\alpha)$ is a polynomial in α with non-negative ingegral coefficients;

(ii) $c^{\nu}_{\lambda\mu}(\alpha) \neq 0$ iff $c^{\nu}_{\lambda\mu} \neq 0$;

(iii) If $c^{\nu}_{\lambda\mu} = 1$, then $c^{\nu}_{\lambda\mu}(\alpha)$ is of the form $\tilde{h}(\lambda)\tilde{h}(\mu)\tilde{h}(\nu)$, where (for $\sigma = \lambda, \mu, \nu$)

$$\tilde{h}_\sigma = \prod_{s\in\sigma} \tilde{h}_\sigma(s)$$

and each factor $\tilde{h}_\sigma(s)$ is either $h^*_\sigma(s)$ or $h^\sigma_*(s)$.

Exactly what should be expected to happen when $c^{\nu}_{\lambda\mu} > 1$ (i.e., when there is more than one LR-tableau of shape $\nu - \lambda$ and weight μ) is at present unclear, at any rate to the author.

The last two conjectures relate to the situation where the number of variables x_i is finite, say $x = (x_1,\ldots,x_n)$. Conjecture (C4) compares the two scalar products on $\Lambda_{n,\mathbb{R}}$:

(C4) $$\frac{\langle J_\lambda, J_\lambda\rangle'_\alpha}{\langle J_\lambda, J_\lambda\rangle_\alpha} = \prod_{s\in\lambda} \frac{n+a'(s)\alpha-\ell(s)}{n+(a'(s)+1)\alpha-(\ell'(s)+1)}.$$

This is true for $\alpha = 1$ (obviously) and $\alpha = 2$.

Finally, conjecture (C5) connects Jack's symmetric functions with Selberg integrals. Let

$$W(x) = \prod_{i=1}^{n} x_i^{a}(1 - x_i)^b |\Delta(x)|^{2c}$$

where for the sake of prudence a, b, c are real numbers ≥ 0, and $\Delta(x)$ is the Vandermonde determinant, as in §2. Let $\lambda = (\lambda_1, \ldots, \lambda_n)$ be a partition of length $\leq n$:-

(C5) $$\int_{[0,1]^n} J_\lambda(x;c^{-1}) W(x)\, dx_1 \ldots dx_n =$$

$$J_\lambda(1,\ldots,1;c^{-1}) \prod_{i=1}^{n} \frac{(\lambda_i+a+c(b-i))!\,(b+c(n-i))!\,(ci)!}{(\lambda_i+b+c(2n-i-1)+1)!\,c!}$$

where $x!$ means $\Gamma(x+1)$.

This is true when $c = 0$ or 1, for all a, b and λ; also when $\lambda = (1^k)$ $(0 \leq k \leq n)$ for all a, b, c (Selberg [13], Aomoto [1]).

5. Zonal polynomials

When $\alpha = 2$, the symmetric functions $J_\lambda(x;\alpha)$ coincide with the zonal polynomials Z_λ introduced by L - K. Hua [6] and A.T. James [8] around 1960. There is a large literature on these, mostly due to statisticians (see [4] and the bibliography there).

When the Schur functions are expressed in terms of the power sums, the coefficients involve the characters of the symmetric groups. If λ runs through the partitions of n, we have

$$S_\lambda = \sum_\mu z_\mu^{-1} \chi_\mu^\lambda \, p_\mu$$

where as before p_μ is the product of power sums corresponding to the partition μ; z_μ is the order of the centralizer of an element of cycle-type μ in S_n; and χ_μ^λ is the value of the irreducible character χ^λ of S_n at such an element.

Correspondingly, in the case of the zonal polynomials Z_λ we have

$$Z_\lambda = \sum_\mu z_{2\mu}^{-1} \omega_\mu^\lambda \, p_\mu \qquad (1)$$

where 2μ is the partition $(2\mu_1, 2\mu_2, \ldots)$ of $2n$, and the ω_μ^λ (like the χ_μ^λ) are integers, which arise from zonal spherical functions in place of characters.

In detail, let S_{2n} be the group of all permutations of $2n$ symbols $(1,2,\ldots,n,1',2',\ldots,n')$, and let H_n be the centralizer in S_{2n} of the element $(11')(22')\ldots(nn')$. (H_n is the hyper-octahedral group, or the Weyl group of type B_n.) The pair (S_{2n}, H_n) is a Gelfand pair, that is to say the induced character $1_{H_n}^{S_{2n}}$ is multiplicity-free; in fact,

$$1_{H_n}^{S_{2n}} = \sum_{|\lambda|=n} \chi^{2\lambda} .$$

Let ϕ be the characteristic function of H_n, and let

$$\omega^\lambda = \phi\chi^{2\lambda} ,$$

the product being convolution in the group algebra of S_{2n}. Then ω^λ is the zonal spherical function corresponding to the component $\chi^{2\lambda}$, and is constant on each double coset $H_n \times H_n$ in S_{2n}. Now these double cosets are naturally indexed by the partitions of n, as follows: each permutation $w \in S_{2n}$ defines a graph $\Gamma(w)$ with $2n$ vertices $1, 2, \ldots, n, 1', 2', \ldots n'$ and edges (rr'), (wr, wr') $(1 \le r \le n)$. The components of $\Gamma(w)$ are cycles of even lengths $2\mu_1, 2\mu_2, \ldots$. Thus w determines a partition μ of n, and one may verify that two permutations determine the same partition μ if and only if they lie in the same double coset of H_n. Then the coefficient ω_μ^λ in (1) above is the value of ω^λ at elements of the double coset indexed by μ.

On the other hand, the polynomial representation of $G = GL_n(\mathbb{R})$ corresponding to a partition ν of length $\le n$ has a fixed vector under the orthogonal group $K = O(n)$ if and only if all the parts ν_i of ν are even, i.e. $\nu = 2\lambda$ for some partition λ. Corresponding to each partition λ of length $\le n$ there is therefore a zonal spherical function ζ_λ on the double coset space $K\backslash G/K$. Now this space may be identified with the space of positive diagonal matrices diag $(x_1, \ldots, x_n)$ modulo the action of the symmetric group S_n (by permuting the x_i). Thus ζ_λ may be regarded as a symmetric function of $x_1, \ldots, x_n$, and (up to a scalar factor) it is just the zonal polynomial Z_λ.

Finally, in the case $\alpha = \frac{1}{2}$ of Jack's symmetric functions, there is a similar set-up, but with the trivial character of H_n replaced by the sign character ε; here again the induced character of S_{2n} is multiplicity-free, namely

$$\varepsilon_{H_n}^{S_{2n}} = \sum_{|\lambda|=n} \chi^{(2\lambda)'} .$$

In this situation we can define 'twisted' zonal spherical functions $\pi^{\lambda}(x) = \varepsilon(x)\omega^{\lambda'}(x)$, which occur in the formula

$$J_{\lambda}(x;\tfrac{1}{2}) = 2^{-n} \sum_{\mu} z_{\mu}^{-1} \pi_{\mu}^{\lambda} p_{\mu}(x)$$

obtained from (1) by use of duality (§3) .

References

[1] K. Aomoto, Jacobi polynomials associated with Selberg integrals, SIAM J. Math. Analysis, to appear.

[2] A. Debiard, Polynômes de Tchébychev et de Jacobi dans un espace euclidien de dimension p, C.R. Acad. Sc. Paris 296 (1983) Série I, 529-532.

[3] F.J. Dyson, Statistical theory of the energy levels of complex systems I, J. Math. Phys., 3(1962) 140-156.

[4] R.H. Farrell, Multivariate calculation, Springer-Verlag (1985).

[5] S. Helgason, Differential geometry, Lie groups and symmetric spaces, Academic Press (1978).

[6] L.-K. Hua, Harmonic analysis of functions of several complex variables in the classical domains, AMS Translations 6 (1963).

[7] H. Jack, A class of symmetric polynomials with a parameter, Proc. R.S. Edinburgh 69A (1970) 1-18.

[8] A.T. James, Zonal polynomials of the real positive definite symmetric matrices, Ann. Math. 74 (1961) 456-469.

[9] K. Kadell, A proof of some q-analogs of Selberg's integral for k = 1, SIAM J. Math. Analysis, to appear.

[10] I.G. Macdonald, Symmetric functions and Hall polynomials, Oxford University Press (1979).

[11] I.G. Macdonald, Some conjectures for root systems, SIAM J. Math. Analysis, 13 (1982) 988-1007.

[12] J. Sekiguchi, Zonal spherical functions on some symmetric spaces, Publ. RIMS, Kyoto University 12 (1977) 455-459.

[13] A. Selberg, Bemerkninger om et Multipelt Integral, Norsk. Mat. Tidsskrift 26 (1944) 71-78.

[14] R. Stanley, private communication.

Some Surfaces Covered by the Ball and
A Problem in Finite Groups

G.D. Mostow* and Stephen S.T. Yau

§1. Lattices in PU(1,n) defined by monodromy

Consider the multivalued function on $(\mathbb{P}^1)^n$

$$F_{st}(x_2,\ldots,x_{n+1}) = \int_s^t z^{-\mu_0}(z-1)^{-\mu_1}(z-x_2)^{-\mu_2}\ldots(z-x_{n+1})^{-\mu_{n+1}}dz$$

where

$s,t \in \{0,1,\infty,x_2,\ldots,x_{n+1}\}$

$x_2,\ldots,x_{n+1}$ are distinct elements in $\mathbb{P}^1 - \{0,1,\infty\}$

$0 < \mu_i < 1 \quad , \qquad i = 0,1,\ldots,n+1$

$1 < \sum_0^{n+1} \mu_i < 2.$

Locally, in a neighborhood of each point, F_{st} is holomorphic and globally it has many determinations. However, at any $x_2,\ldots,x_{n+1}$ the linear span of all its determinations is an $n+1$ dimensional space W -- for topological reasons (cf. [2], §1, §3.8).

An isotopy φ of $\mathbb{P}^1$ which returns the $n+3$ points $S = \{0,1,\infty,x_2,\ldots,x_{n+1}\}$ to their initial positions effects a linear transformation on the integrand and on the homology class in the punctured line $\mathbb{P}^1 - \{0,1,\infty,x_2,\ldots,x_{n+1}\}$ of the path of integration and hence effects a linear transformation φ_W of W. The group of equivalence classes of isotopies of the line $\mathbb{P}^1$ which return each point of S to its initial position is just the colored (or pure) braid group on $n+3$ strings in $\mathbb{P}^1$: $C_{n+3}(\mathbb{P}^1)$ or alternatively $\pi_1((\mathbb{P}^1)^{n+3}$ - all diagonals).

For each automorphism A of the vector space W, let Proj A denote the automorphism induced by A on the projective space Proj W of 1-dimensional subspaces of W.

The maps $\varphi \to \varphi_W$ and $\varphi \to \text{Proj}\ \varphi_W$ define homomorphisms

*Supported in part by NSF Grant DMS-8506130

$$\theta' : C_{n+3}(\mathbb{P}^1) \to \mathrm{Aut}\, W$$

$$\theta : C_{n+3}(\mathbb{P}^1) \to \mathrm{Aut\ Proj}\, W$$

which we call the monodromy actions of the colored braid group. Set

$$\Gamma'_\mu = \mathrm{Im}\ \theta' \ , \ \Gamma_\mu = \mathrm{Im}\ \theta$$

Set

$$\mu_\infty = 2 - \sum_0^{n+1} \mu_i .$$

In [2] we prove

1. (Corollary 2.21). The group Γ'_μ preserves a hermitian form ψ on W of signature $(1,n)$ [one plus sign and n minus signs].

Let $B_+ = \mathrm{Proj}\{w \in W;\ \psi(w,w) > 0\}$. B_+ may be identified with the unit ball in $\mathbb{C}_n$.

2. (Theorem 10.19). Assume condition INT: for all $s,t \in S$ with $\mu_s + \mu_t < 1$, one has $(1 - (\mu_s + \mu_t))^{-1}$ an integer. Then

a) Γ_μ is a lattice subgroup in $\mathrm{Aut}\ B_+ (\approx PU(1,n))$.

b) If $\mu_s + \mu_t < 1$ for all $s,t \in S$, then $\Gamma_\mu \backslash B_+$ is compact.

c) $\Gamma_\mu \backslash B_+$ is the set of μ-stable points in the Mumford μ-quotient variety of $\mathrm{Aut}\ \mathbb{P}^1 \backslash (\mathbb{P}^1)^{n+3}$, and coincides with the quotient variety if $\Gamma_\mu \backslash B_+$ is compact.

These results were proved by H.A. Schwarz for $n = 1$ and essentially stated but only partially proved by E. Picard for $n = 2$ ([6], [5]).

The integrality condition INT is satisfied only for $n \leq 5$. It is therefore of some interest to understand the varieties $\Gamma_0 \backslash B_+$ where Γ_0 is a torsion-free subgroup of finite index in the lattice Γ_μ.

The computation of the one dimensional $\beta_1(\Gamma_0 \backslash B_+)$ poses a challenge even for $n = 2$. In this paper we describe an attempt to compute $\beta_1(\Gamma_0 \backslash B_+)$ via Morse theory.

§2. The Surface Y

For convenience, we restrict ourselves to the case $\mu_s + \mu_t < 1$ for all $s,t \in S$. We assume $n = 2$, we write Γ for Γ_μ, B for B_+, and set

$$M = \Gamma\backslash B.$$

In this case, M arises from $\mathbb{P}^1 \times \mathbb{P}^1$ by blowing up the three points $(0,0)$, $(1,1)$ and (∞,∞). Alternatively, M may be described as arising from complex projective 2-space $\mathbb{P}^2$ by blowing up 4 points. M has 10 exceptional lines of self intersection -1, and schematically one can depict them with the diagram

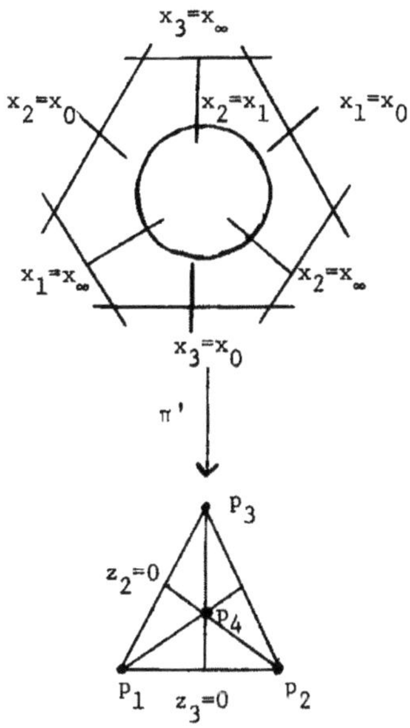

That is, if in $\mathbb{P}^2$ we blow up 4 points $\{p_1,p_2,p_3,p_4\}$ no three of which are colinear, we obtain a DelPezzo surface whose 10 exceptional lines lie above the 4 points and the 6 lines of their complete quadrangle.

Let Γ_0 denote a normal subgroup of finite index in Γ; set

$Y: = \Gamma_0\backslash B$

$\pi:\ Y \to \Gamma\backslash B = M$, $\pi':M \to \mathbb{P}^2$

L': = the set of 10 exceptional lines of M.

Choosing non-homogeneous coordinates (z_2,z_3) on $\mathbb{P}^2$, we can assume that the six lines of the complete quadrangle are

$$z_2 = \{^0_1 \quad , \quad z_3 = \{^0_1 \quad , \quad z_2 = z_3, \quad \text{the line at } \infty .$$

The 10 exceptional lines $\{L; L \in L'\}$ may be regarded equally well as the image of the 10 diagonals $x_s = x_t$ in $\mathrm{Aut}\, \mathbb{P}_1 \backslash (\mathbb{P}^1)^S_{\text{stable}}$ where $(\mathbb{P}^1)^S_{\text{stable}}$ denotes the subset of points $(x_s; s \in S)$ of $(\mathbb{P}^1)^S$ with $\sum_{x_s = z} \mu_s < 1$ for all $z \in \mathbb{P}^1$ (i.e. $\mu_{s_1} + \mu_{s_2} + \ldots + \mu_{s_k} < 1$ whenever $x_{s_1} = x_{s_2} = \ldots = x_{s_k}$).

We have the dictionary

$$\pi'^{-1}(z_2 = 0) \iff x_2 = x_0$$
$$\pi'^{-1}(z_2 = 1) \iff x_2 = x_1$$
$$\pi'^{-1}(z_3 = 0) \iff x_3 = x_0$$
$$\pi'^{-1}(z_3 = 1) \iff x_3 = x_1$$
$$\pi'^{-1}(z_2 = z_3) \iff x_2 = x_3$$
$$\pi'^{-1}(p_1) \iff x_1 = x_\infty$$
$$\pi'^{-1}(p_2) \iff x_2 = x_\infty$$
$$\pi'^{-1}(p_3) \iff x_3 = x_\infty$$
$$\pi'^{-1}(p_4) \iff x_0 = x_\infty$$

The map $\pi : Y \to M$ is a branched cover with ramification index $(1 - \mu_s - \mu_t)^{-1}$ over the line L_{st} in L' corresponding to the diagonal $x_s = x_t$ of $(\mathbb{P}^1)^S_{\text{stable}}$ where $s,t \in \{0,1,2,3,\infty\}$.

By hypothesis, Γ_0 is a normal subgroup of finite index in Γ. Set

$$G = \Gamma/\Gamma_0 .$$

Then G operates on Y and

$$G \backslash Y = M.$$

As mentioned above, the action of Γ on the ball B arises from linear transformations preserving the hermitian form; thus Γ preserves the Fubini-Study metric on B. The effect of an element γ_{st} of the colored braid group $C_5(\mathbb{P}^1)$ which loops s once around t is an isometry of B which fixes each point of a complex line lying above the line $L_{st} \in L'$, and has order

$(1 - \mu_s - \mu_t)^{-1}$; thus each γ_{st} acts as a complex-reflection on B ([2], Prop. 9.2).

For suitable choice of γ_{st} in its conjugacy class and for any set L'' of 5 lines of L' no four of which are disjoint, the five complex reflections $\{\gamma_{st};\ L_{st} \in L''\}$ generate Γ.

We give Y the metric induced from B. Correspondingly, on Y, the group G acts by isometries and is generated by complex reflections in the connected one-dimensional complex subvarieties lying over the lines of L''.

If two elements γ_{st} and γ_{uv} of the braid group $C_5(P^1)$ commute, it follows at once that their fixed point sets in the ball B and in the surface Y are orthogonal. For any point $q \in Y$, the order of its stabilizer G_q in G is

$$1 \quad \text{if} \quad \pi(q) \notin \coprod_{L \in L'} L$$

$$(1 - \mu_s - \mu_t)^{-1} \quad \text{if} \quad \pi(q) \in L_{st} \text{ and no other line}$$

$$(1 - \mu_s - \mu_t)^{-1}(1 - \mu_u - \mu_v)^{-1} \quad \text{if} \quad \pi(q) \in L_{st} \cap L_{uv}\,, \quad (st) \neq (uv)\ .$$

§3. Morse Theory on Y

On $\mathbb{P}^2$, let f denote the meromorphic function

$$f = z_1(z_1-1)z_2(z_2-1)(z_1-z_2)$$

and define the real valued function φ by

$$\varphi = \log f\bar{f},$$

where (z_1, z_2) denote non-homogeneous coordinates.

Set $z_i = x_i + \sqrt{-1}y_i$ $(i = 1,2)$. The critical points of φ, at which the gradient of φ vanishes, are obtained by solving the simultaneous equations

$$\begin{cases} 0 = \partial_{z_1}\varphi = \dfrac{1}{z_1} + \dfrac{1}{z_1-1} + \dfrac{1}{z_1-z_2} \\ 0 = \partial_{z_2}\varphi = \dfrac{1}{z_2} + \dfrac{1}{z_2-1} - \dfrac{1}{z_1-z_2} \end{cases}$$

One finds that there are exactly two critical points

$$p = (a,a') \quad \text{and} \quad p' = (a',a)$$

where $a = \dfrac{5+\sqrt{5}}{10}$ and $a' = \dfrac{5-\sqrt{5}}{10}$ are the two solutions of the quadratic equation $a^2 - a + \frac{1}{5} = 0$. We note also that

$$\varphi(p) = \varphi(p') = 2\log aa'aa'(a-a') = 2\log \frac{1}{5}\cdot\frac{1}{5}\frac{1}{\sqrt{5}} = -5\log 5$$
$$:= c$$

The computation of the Hessian at p is straightforward.

$$\partial^2_{x_1}\varphi = (\partial_{z_1} + \partial_{\bar{z}_1})^2\varphi = (\partial^2_{z_1} + \partial^2_{\bar{z}_1})\varphi$$

$$\partial^2_{z_1}\varphi\big|_p = -\frac{1}{z_1^2} - \frac{1}{(z_1-1)^2} - \frac{1}{(z_1-z_2)^2}\bigg|_p = -\frac{1}{a^2} - \frac{1}{a'^2} - \frac{1}{\frac{1}{5}} = -[5 + \frac{a^2+a'^2}{a^2\cdot a'^2}]$$

$$= -[5 + \frac{a+a' - 2/5}{\frac{1}{25}}] = -[5 + \frac{3/5}{1/25}] = -[5+15] = -20.$$

Hence

$$\frac{\partial^2\varphi}{\partial x_1^2}\bigg|_p = -20 -20 = -40$$

Similarly

$$\frac{\partial^2}{\partial x_1\, x_2}\Big|_p = (\partial_{z_1} + \partial_{\bar z_1})(\partial_{z_2} + \partial_{\bar z_2})\varphi\Big|_p = 2\mathrm{Re}\ \partial_{z_1}\partial_{z_2}\varphi\Big|_p$$

$$= 2\,\frac{1}{(a-a')^2} = 2\,\frac{1}{\frac{1}{5}} = 10.$$

Thus the Hessian at p, and at p' too, is

$$10\begin{pmatrix} -4 & 1 & 0 & 0 \\ 1 & -4 & 0 & 0 \\ 0 & 0 & 4 & -1 \\ 0 & 0 & -1 & 4 \end{pmatrix}$$

which is conjugate to

$$10\begin{pmatrix} -3 & & & \bigcirc \\ & -5 & & \\ & & 3 & \\ \bigcirc & & & 5 \end{pmatrix}$$

In particular, the critical points are non-degenerate.

Let V denote the tangent space to $\mathbb{P}^2$ at the point p. Then

$$V = V_+ + V_- \quad \text{(direct)}$$

where V_+ and V_- are the eigenspaces of the Hessian corresponding to positive and negative eigenvalues respectively. We have

$$V_- = \mathbb{R}\partial_{x_1}\big|_p + \mathbb{R}\partial_{x_2}\big|_p \quad , \quad V_+ = \sqrt{-1}\ V_-$$

and an identical decomposition of the tangent space to $\mathbb{P}^2$ at the other critical point p'.

We shall apply the following result from Morse theory ([4], Theorem 5.3) :

Let M be a smooth manifold and $\psi: M \to \mathbb{R} \cup -\infty \cup \infty$ a continuous function. For any b with $-\infty \le b \le \infty$, set

$$M^b = \psi^{-1}(b), \quad M^{b]} = \psi^{-1}[-\infty, b].$$

Assume that

(1) ψ is smooth on $M - (M^{-\infty} \cup M^{\infty})$.

(2) c is a critical value such that all critical points in $\psi^{-1}(c)$ are non-degenerate.

(3) c is the only critical value of ψ in $[c-\varepsilon, c+\varepsilon]$.

(4) $\psi^{-1}[c-\varepsilon, c+\varepsilon]$ is compact.

Then homotopically

$$M^{c+\varepsilon]} \sim M^{c-\varepsilon]} \cup \coprod D^{\lambda}$$

where in the disjoint union $\coprod D^{\lambda}$, one has one λ-dimensional disc for each critical point of index λ(: = the sum of the dimensions of the eigenspaces of the Hessian with negative eigenvalues).

We apply this result to the function $\psi = \varphi \circ \pi' \circ \pi$ on $Y - Y^{\infty}$; it yields for any $\varepsilon > 0$:

(MT) $$Y^{c+\varepsilon]} \sim Y^{c-\varepsilon]} \cup \coprod_{g \in G} gE^{\lambda} \cup \coprod_{g \in G} gE'^{\lambda}$$

where E^{λ} is a connected component in $\pi'^{-1}(D^{\lambda})$ and similarly for E'^{λ}. Thus E^{λ} (resp. E'^{λ}) is a 2-disc containing a point over p (resp. p').

The tangent subspace V_- to $\mathbb{P}^2$ at p may be identified in a natural way with the tangent space at p to the real projective subspace $\mathbb{RP}^2$. Under downward gradient flow of the function φ on $\mathbb{P}^2$, a small disc about p in V_- flows down to fill out the triangle Δ in $\mathbb{RP}^2 - \infty$ lying between $x_1 = 0$, $x_2 = 1$, and $x_1 = x_2$

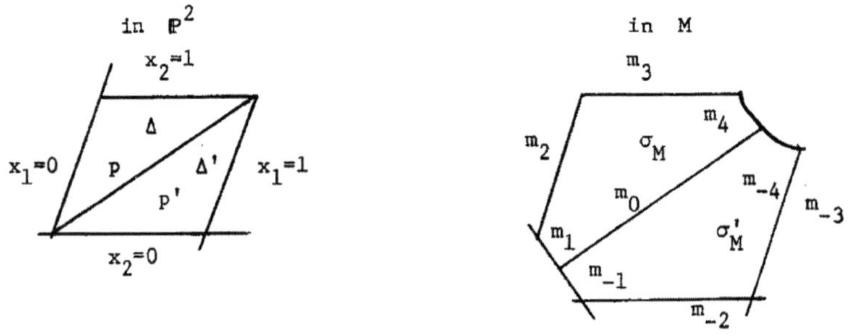

Correspondingly in M, the 2-disc D^{λ} flows down to a pentagon σ_M under the gradient flow of φ and similarly the 2-disc E^{λ} flows down to a pentagon σ in $Y - Y^{\infty}$ mapping into Δ under $\pi' \circ \pi$. Similarly, D'^{λ} flows down to a pentagon σ'_M containing the point p'. We orient the 2-cells σ_M and σ'_M and label the edges of their boundaries $\partial\sigma_M$ and $\partial\sigma'_M$ so as to have

$$\partial\sigma_M = m_0 + m_1 + m_2 + m_3 + m_4$$

$$\partial\sigma'_M = m_0 + m_{-1} + m_{-2} + m_{-3} + m_{-4}$$

and we index the 7 lines of L so that L_i is the line containing the edge m_i, $-4 \le i \le 4$. (Note that $L_{-4} = L_4$ and $L_{-1} = L_1$.) We choose a connected component σ' of $\pi^{-1}(\sigma'_M)$ so as to satisfy

$$\pi(\text{support } \sigma \cap \text{support } \sigma') = \text{support } \sigma_M \cap \text{support } \sigma'_M .$$

We label the oriented edge of $\partial\sigma$ and $\partial\sigma'$ which lie over m_i by y_i, $-4 \le i \le 4$. Thus

$$\partial\sigma = \sum_0^4 y_i$$

$$\partial\sigma' = \sum_0^4 y_{-i}$$

Inasmuch as the complex conjugation of $\mathbb{P}^2$ which fixes each point of $\mathbb{RP}^2$ can be lifted to M and in turn to a complex conjugation of the ball B, the 2-cells σ and σ' can be taken to lie below a real geodesic 2-plane $\mathbb{R}^2 \cap B$ and indeed to be geodesic pentagons in Y.

§4. The small cell complex ${}^{m}Y^{-\infty}$

As above

$$Y^{-\infty} = \bigcup_{L \in \mathcal{L}} \pi^{-1}(L) := \pi^{-1}(M^{-\infty})$$

with $\mathcal{L} = \{L_i;\ -3 \le i \le 4,\ i \ne -1\}$ and

$$M^{\infty} = \bigcup_{L \in \mathcal{L}' - \mathcal{L}} L$$

with $\pi'(M^{\infty})$ lying at ∞ in $\mathbb{P}^2$. We note that each line $L \in \mathcal{L}$ which meets a line in $\mathcal{L}' - \mathcal{L}$ meets a unique such line; we denote this line by L' (for example, $L_2' = L_{-3}'$).

We give $Y^{-\infty}$ the structure of a cell complex ${}^{m}Y^{-\infty}$ which is the pull-back of the following cell complex structure ${}^{m}M^{-\infty}$ on $M^{-\infty}$.

0. The vertices of $M^{-\infty}$ are the points of intersection of the seven lines

$$L_i \cap L_j \qquad \text{with} \qquad L_i, L_j \in \mathcal{L}.$$

1. The 1 cell of a line $L \in \mathcal{L}$ consist of a single slit joining the two vertices of L if $L \cap M^{\infty}$ is not empty; and consists of 2 slits joining the vertices of L if $L \cap M^{\infty}$ is empty, i.e. if L has three vertices.

2. The open 2-cells of $L \in \mathcal{L}$ consist of the complement of the union of its 1-cells.

We take as open k-cells σ^k in ${}^{m}Y^{-\infty}$ the connected components of inverse images of open k-cells in ${}^{m}M^{-\infty}$, $0 \le k \le 2$. It is clear that π is a homeomorphism on each open k-cell σ^k of ${}^{m}Y^{-\infty}$ for $k = 0,1$, and also for $k = 2$ if $\pi(\sigma^2)$ lies in a line L which does not meet M^{∞}. However, if $\pi(\sigma^2) \subset L$ with $L \cap M^{\infty}$ not empty, then

$$\pi: \sigma^2 \to \pi(\sigma^2)$$

has degree c', the order of the complex reflection in the line $L' \in \mathcal{L}' - \mathcal{L}$ which L meets.

It is clear that the one cells in ${}^{m}M^{-\infty}$ can be taken to coincide with the one-cells

$$\{m_i;\ -4 \le i \le 4\}$$

described in §3.

Thus the 1-cells of ${}^{m}Y^{-\infty}$ are $\{G\, y_i;\ -4 \le i \le 4\}$ where $\pi(y_i) = m_i$, $-4 \le i \le 4$, and as above $G = \Gamma/\Gamma_0$. The closed 2-cells in $\pi^{-1}(L)$ consist of quadrilaterals if $L \cap M^{\infty}$ is empty and of $2c'$ - gons if $L \cap M^{\infty}$ is not empty, where c' is the ramification index of the line L'. Moreover, the closed cells of ${}^{m}Y^{-\infty}$ may be taken to be closed convex geodesic subsets in Y with respect to the metric on Y induced from the ball B.

The group G permutes the cells of the cell complex ${}^{m}Y^{-\infty}$. For each i between -4 and 4, we let L_i denote the line in $\mathcal{L}$ which contains the 1-cell m_i of ${}^{m}M^{-\infty}$. The seven lines of $\mathcal{L}$ are $\{L_{-3},L_{-2},L_0,L_1,L_2,L_3,L_4\}$ with $L_{-4} = L_4$, $L_{-1} = L_1$. It will be convenient sometimes to write L_5 for L_0.

The stabilizer in G of a 2-cell in $\pi^{-1}(L_i)$ is a cyclic group of order c_i if $\pm i = 1, 4$, since it fixes each point of the 2-cell. However, it is a product of two cyclic groups of order $c_i c_i'$ if $\pm i = 0, 2, 3$.

§ 5. $H_1(Y)$ as a quotient of $H_1(Y^{-\infty})$.

We continue the notation of §3.

Lemma 5.1. $H_1(Y^{-\infty} - (Y^{-\infty} \cap Y^{\infty}) \to H_1(Y)$ is surjective.

Proof. Let Y_+ (resp. Y_-) denote the set of points in $Y - Y^{\infty}$ which flow into $\pi^{-1}(p)$ (resp. $\pi^{-1}(p'))$ under the downward flow of the function $\varphi \circ \pi' \circ \pi$. Then each connected component of $Y_{\pm}$ is a 2-cell. Given any path in Y representing an element h in $H_1(Y)$, it can be deformed in Y so as to avoid the 2-dimensional set $Y^{\infty} \cup Y_+ \cup Y_-$, and thus h can be represented by a path which flows downward under the gradient flow into an arbitrary small neighborhood of $Y^{-\infty} - (Y^{-\infty} \cap Y^{\infty})$. Since the latter is an absolute neighborhood retract, the Lemma follows.

Lemma 5.2. In the exact homology sequence with coefficients in $\mathbb{Q}$,

$$H_2(Y - Y^{\infty}, Y^{-\infty} - (Y^{\infty} \cap Y^{-\infty})) \xrightarrow{\partial} H_1(Y^{-\infty} - (Y^{\infty} \cap Y^{-\infty})) \longrightarrow H_1(Y - Y^{\infty})$$

the image of ∂ is the subgroup represented by the group of 1-cycles $\mathbb{Q}[G\partial\sigma, G\partial\sigma']$.

Proof. By (MT) of §3 and the fact that gradient flow moves $Y^{c-\varepsilon]}$ downward arbitrarily close to $Y^{-\infty} - (Y^{\infty} \cap Y^{-\infty})$, we see that homotopically

$$Y - Y^{\infty} \sim (Y^{-\infty} - (Y^{\infty} \cap Y^{-\infty})) \cup G\sigma \cup G\sigma'$$

where, by abuse of notation, we denote the support of σ, σ' by σ, σ' respectively. From this, the lemma follows at once.

Lemma 5.3. $H_2(Y, (Y - Y^{\infty}) \cup Y^{-\infty}) = 0$.

Proof. Any 2-chain in Y can be deformed in Y so as to meet Y^{∞} transversally at only a finite number of points. Since any small disc meeting Y^{∞} transversally can be deformed into a disc in $Y^{-\infty}$ meeting Y^{∞} in a point of $Y^{\infty} \cap Y^{-\infty}$, we can deform τ, near each intersection point with Y^{∞} in turn, so as to meet Y^{∞} only at points of $Y^{\infty} \cap Y^{-\infty}$ and to lie in $(Y-Y^{\infty}) \cup Y^{-\infty}$. Thus the inclusion $i:(Y-Y^{\infty}) \cup Y^{-\infty} \to Y$ induces an injection $i_*:H_i((Y-Y^{\infty}) \cup Y^{-\infty}) \to H_i(Y)$ for $i = 1$ and a surjection for $i = 2$. From this the Lemma follows.

<u>Proposition 5.4</u>. $H_1(Y) = H_1(Y^{-\infty})/\mathrm{Im}\ i_*\partial$ where ∂ is as in Lemma 5.2 and i denotes the inclusion $Y^{-\infty} - (Y^{\infty} \cap Y^{-\infty}) \to Y^{-\infty}$.

<u>Proof</u>. Consider the diagram of exact homology sequences

$$
\begin{array}{ccccccc}
 & & & & H_2(Y,(Y-Y^{\infty}) \cup Y^{-\infty}) & & \\
 & & & & \downarrow & & \\
H_2((Y-Y^{\infty}) \cup Y^{-\infty}, Y^{-\infty}) & \xrightarrow{\partial'} & H_1(Y^{-\infty}) & \longrightarrow & H_1((Y-Y^{\infty}) \cup Y^{-\infty}) & \longrightarrow & 0 \\
\downarrow & & \| & & \downarrow & & \\
H_2(Y, Y^{-\infty}) & \longrightarrow & H_1(Y^{-\infty}) & \longrightarrow & H_1(Y) & \longrightarrow & 0 \\
\downarrow & & & & \downarrow & & \\
H_2(Y, (Y-Y^{\infty}) \cup Y^{-\infty}) & & & & 0 & &
\end{array}
$$

the zero on the third row following from Lemma 5.1.

From Lemma 5.3, we infer $H_1(Y) = H_1(Y^{-\infty})/\mathrm{Im}\ \partial'$. On the other hand, using a deformation retraction of a neighborhood $U(Y^{\infty})$ onto Y^{∞} followed by excision, we have

$$
\begin{aligned}
((Y-Y^{\infty}) \cup Y^{-\infty}, Y^{-\infty}) &\sim ((Y-U(Y^{\infty})) \cup Y^{-\infty}, Y^{-\infty}) \\
&\approx ((Y-U(Y^{\infty})) \cup Y^{-\infty} - [Y^{-\infty} \cap U(Y^{\infty})], Y^{-\infty} - [Y^{\infty} \cap U(Y^{\infty})] \\
&= (Y-U(Y^{\infty}), Y^{-\infty} - [Y^{-\infty} \cap U(Y^{\infty})]) \\
&\sim (Y-Y^{\infty}, Y^{-\infty} - [Y^{-\infty} \cap Y^{\infty}])
\end{aligned}
$$

where $\sim$ denotes homotopy equivalence. From this we deduce that $\mathrm{Im}\ \partial'$ can be identified with $\mathrm{Im}\ \partial$. Proof of the lemma is now complete.

§6. $H_1(Y^{-\infty})$.

In view of Proposition 5.4, we take a closer look at $H_1(Y^{-\infty})$. We define

$$H_1^{hor}(Y^{-\infty}) = \sum_{L \in \mathcal{L}} i_* H_1(\pi^{-1}L)$$

where i denotes the inclusion map $\pi^{-1}L \to Y$. Next we define the quotient

$$H_1^{loop}(Y^{-\infty}) = H_1(Y^{-\infty})/H_1^{hor}(Y^{-\infty}).$$

Lemma 6.1. Let $L_i \in \mathcal{L}'$ and let L_j, L_k, L_ℓ be the other 3 lines of $\mathcal{L}'$ which intersect L_i. Set c_i = the ramification index of L_i. Then

$$\frac{1}{c_j} + \frac{1}{c_k} + \frac{1}{c_\ell} = 1 - \frac{2}{c_i} .$$

Proof. Relabelling the indices of the set S, we can assume that $L_i = L_{04}$, $L_j = L_{12}$, $L_k = L_{23}$, $L_\ell = L_{31}$. Since the ramification index of L_{st} is $(1 - \mu_s - \mu_t)^{-1}$, we find

$$1 - c_j^{-1} = \mu_1 + \mu_2$$

$$1 - c_k^{-1} = \mu_2 + \mu_3$$

$$1 - c_\ell^{-1} = \mu_3 + \mu_1$$

$$2(1 - c_i^{-1}) = 2(\mu_0 + \mu_4)$$

Adding yields $5 - (\frac{1}{c_j} + \frac{1}{c_k} + \frac{1}{c_\ell} + \frac{2}{c_i}) = 2 \sum_0^4 \mu_i = 4$, which implies the lemma.

Notation. For any finite set G, $|G|$ = cardinal of G. For any elements $g, h, \ldots$ in a group G, $\langle g,h,\ldots \rangle$ denotes the group generated by $g,h, \ldots$.

Lemma 6.2. Let $L \in \mathcal{L}'$, let X denote a connected component of $\pi^{-1}(L)$, and let c_L denote the ramification index of L. Then the first Betti number of X is given by

$$\beta_1(X) = \frac{2|G_X|}{c_L^2} + 2 \tag{2}$$

where G_X denotes the stabilizer of X in G.

Proof. G_X acts on X with fundamental domain a convex geodesic quadrilateral τ having three vertices lying above $L \cap L_j$, $L \cap L_k$, $L \cap L_\ell$. Computing the Euler characteristic $\chi(X)$ from the cell complex on X with 2-cells $G_X \tau$, we find

$$\beta_2(X) - \beta_1(X) + \beta_0(X) = \frac{|G_X|}{c_L} \left(1 - 2 + \frac{1}{c_j} + \frac{1}{c_k} + \frac{1}{c_\ell}\right)$$

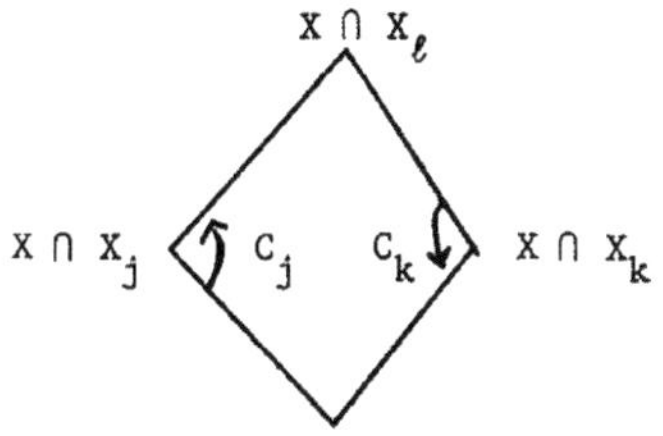

where $c_L c_j = |G_{X \cap X_j}|$, $G_{X \cap X_j} = \langle C_X, C_j \rangle$ etc.

C_X (resp. C_j, C_k, C_ℓ) denoting the complex reflection in the subvarieties X (resp. X_j, X_k, X_ℓ) lying over L (resp. L_i, L_j, L_k). By Lemma 6.1, using $\beta_0(X) = \beta_2(X) = 1$, we get

$$\beta_1(X) - 2 = \frac{|G_X|}{c_L} \left(\frac{2}{c_L} \right)$$

as required.

<u>Lemma 6.3</u>. Let C^* denote the one-dimensional simplicial complex whose vertices correspond to the connected components of $\pi^{-1}(L)$ as L varies over the set of 7 lines $\mathcal{L}$, and whose one cells correspond to points of $\pi^{-1}(L_i \cap L_j)$ with $L_i, L_j \in \mathcal{L}$. Then

$$H_1^{\text{loop}}(Y^{-\infty}) = H_1(C^*).$$

<u>Proof</u>. Modulo $H_1^{\text{hor}}(Y^{-\infty})$, each closed path in $Y^{-\infty}$ is determined by a sequence

$$L_{i_1}, L_{i_1} \cap L_{i_2}, L_{i_2}, L_{i_2} \cap L_{i_3}, \ldots, L_{i_n} = L_{i_1}$$

with homotopy corresponding to homotopy in C^*. This implies the result.

<u>Lemma 6.4</u>. $\beta_0(C^*) = 1$.

The lemma is equivalent to the assertion that $\coprod_{L \in \mathcal{L}} \pi^{-1}(L)$ is connected in Y. This will follow at once from the stronger assertion:

Let $\nu: B \to M = \Gamma \backslash B$ denote the natural projection, and let L_i, L_j be two distinct lines of $\mathcal{L}'$ with $L_i \cap L_j$ not empty. Then $\nu^{-1}(L_i \cup L_j)$ is connected.

Proof. Let $q \in \nu^{-1}(L_i \cap L_j)$, and let Z, X_i, X_j denote the connected component containing q of $\nu^{-1}(L_i \cup L_j)$, L_i, L_j respectively. Clearly Z is stable under Γ_{X_i} and Γ_{X_j}, the stabilizers in Γ of X_i and X_j. Moreover, Γ_{X_i} contains the complex reflections in all the complex lines over elements of L' which meet it orthogonally. Hence $\Gamma_{X_i} \cup \Gamma_{X_j}$ contain the complex reflections in at least 6 complex lines lying over 6 distinct elements of L' for which $< \Gamma_{X_i}, \Gamma_{X_j} > = \Gamma$. Hence $Z = \Gamma Z = \nu^{-1}(L_i \cup L_j)$. Proof of the lemma is now complete.

Lemma 6.5. $\dim H_1^{loop}(Y^{-\infty}) = |G| \left(\sum_{\substack{i=-4 \\ i \neq -1}}^{4} \frac{1}{c_i c_{i+1}} - \sum_{\substack{-3 \\ i \neq -1}}^{4} \frac{1}{|G_i|} \right) + 1$

where c_i is the ramification index of the line $L_i \in L$ and $|G_i|$ is the order of $|G_{X_i}|$, where X_i is a connected component of $\pi^{-1}(L_i)$, $-3 \leq i \leq 4$.

Proof. The Euler characteristic of C^* is given by

$$\begin{aligned}
\chi(C^*) &= |\{0\text{-cells in } C^*\}| - |\{1\text{-cells in } C^*\}| \\
&= \sum_{L \in L} |\text{connected components of } \pi^{-1}(L) - \sum_{\substack{L_i \neq L_j \\ L_i, L_j \in L, L_i \cap L_j \neq \phi}} |\text{conn.comp. } \pi^{-1}(L_i \cap L_j)| \\
&= |G| \left(\sum_{\substack{-3 \\ i \neq -1}}^{4} \frac{1}{|G_i|} - \sum_{\substack{-4 \\ i \neq -1}}^{4} \frac{1}{c_i c_{i+1}} \right).
\end{aligned}$$

Hence

$$\begin{aligned}
\beta_1(C^*) &= \beta_0(C^*) - \chi(C^*) \\
&= 1 - \chi(C^*)
\end{aligned}$$

which implies the lemma.

Corollary 6.6. $\beta_1(Y^{-\infty}) = |G| \left[\sum_{\substack{-3 \\ i \neq -1}}^{4} \left(\frac{2}{c_i^2} + \frac{1}{|G_i|} \right) + \sum_{\substack{-4 \\ i \neq -1}}^{4} \frac{1}{c_i c_{i+1}} \right] + 1$

This follows immediately from Lemmas 6.2, 6.4, and definitions.

We close this section with some additional identities that will be used below.

Lemma 6.7. $\displaystyle\sum_{L \in \mathcal{L}'} \frac{1}{c_L} = 2.$

Proof.
$$\sum_{L \in \mathcal{L}'} \frac{1}{c_L} = \sum_{0 \le i < j \le 4} 1 - (\mu_i + \mu_j)$$
$$= 10 - 4 \sum_{0}^{4} \mu_i$$
$$= 10 - 4 \cdot 2$$
$$= 2.$$

Lemma 6.8. Let c_i be the ramification index of the line $L_i \in \mathcal{L}'$ $(-4 \le i \le 4)$. Then

$$\sum_{0}^{4} \frac{1}{c_i} = 1 = \sum_{0}^{4} \frac{1}{c_{-i}}$$

Proof. We can assume that the five lines $L_0, L_1, \ldots, L_4$ are the lines L_{st} with $(st) = (01)(12)(23)(34)(40)$. Then writing $\mu_5 = \mu_0$,

$$\sum_{0}^{4} \frac{1}{c_i} = \sum_{i=0}^{4} 1 - (\mu_i + \mu_{i+1}) = 5 - 2\sum_{0}^{4} \mu_i = 5 - 4 = 1.$$

Similarly

$$\sum_{0}^{4} \frac{1}{c_{-i}} = 1.$$

Lemma 6.9.
$$2 \sum_{\substack{i<j \\ L_i \cap L_j \ne \phi \\ L_i, L_j \in \mathcal{L}'}} \frac{1}{c_i c_j} = \sum_{L \in \mathcal{L}'} \frac{1}{c_L} - 2 \sum_{L \in \mathcal{L}'} \frac{1}{c_L^2} .$$

Proof.
$$2 \sum_{\substack{i<j \\ L_i \cap L_j \ne \phi}} \frac{1}{c_i c_j} = \sum_{L_i \cap L_j \ne \phi} \frac{1}{c_i c_j} = \sum_{L \in \mathcal{L}'} \frac{1}{c_L}\left(1 - \frac{2}{c_L}\right)$$

by Lemma 6.1.

§7. The Boundary Operator

Set $V = \mathbb{Q}[G]$, the group algebra of G over $\mathbb{Q}$. Introduce the inner product on V

$$<\alpha,\beta> = \frac{1}{|G|} \operatorname{Tr} \rho(\alpha)\rho(\beta^t)$$

where $\beta \to \beta^t$ is the involution of V carrying each g to g^{-1} for $g \in G$, and ρ denotes the regular representation of V

$$\rho(\alpha): v \to \alpha v.$$

Then for any $g,h \in G$

$$<g,h> = \begin{cases} 0 & \text{if } g \neq h \\ 1 & \text{if } g = h \end{cases}$$

Thus for $\alpha = \sum_{g \in G} \alpha_g g$ and $\beta = \sum_{g \in G} \beta_g g$,

$$<\alpha,\beta> = \Sigma\, \alpha_g \beta_g$$

This inner product is positive definite and G bi-invariant. On $V \oplus V$ we take the induced inner product; it is bi-invariant under $G \times G$.

We define two G-module maps of V to the one dimensional chain group $C_1(^m Y^{-\infty})$ of the cell complex $^m Y^{-\infty}$

$$\partial : \sum_{g \in G} \alpha_g g \to \sum_{g \in G} \alpha_g g\, \partial \sigma$$

$$\partial' : \sum_{g \in G} \beta_g g \to \sum_{g \in G} \beta_g g\, \partial \sigma'$$

where $\alpha_g, \beta_g \in \mathbb{Q}$ for all $g \in G$, and the 2 cells σ, σ' are the geodesic pentagons described at the end of §3. Define $D: V \oplus V \to C_1(^m Y^{-\infty})$ as

$$D = \partial \oplus \partial'$$

For any subgroup H of G set

$$V^H = \{v \in V;\ vH = v\}$$

V^H is a left $\mathbb{Q}[G]$-module. The G-module $C_1(^m Y^{-\infty})$ is isomorphic to the direct sum of the nine G-modules

$$\bigoplus_{-4\le i\le 4} V^{G_{y_i}}$$

where G_{y_i} denotes the stabilizer of the 1-cell y_i of §3.

The group G_{y_i} is cyclic (cf. §2); let C_i denote a generator of G_{y_i} $-4 \le i \le 4$ with $C_{-i} = C_i$ for $i = 1,4$. Set

$$V_i = V^{<C_i>} \qquad -4 \le i \le 4$$

$$V_{0,0} = \{v \oplus v \ ; \ v \in V_0\}$$

<u>Lemma 7.1.</u> $(\mathrm{Ker}\ D)^{\perp} = V_{0,0} + \sum_{1}^{4} V_i \oplus V_{-i}$.

<u>Proof</u>. The element $\alpha \oplus \beta \in \mathrm{Ker}\ D$ if and only if in $C_1(^{m}Y^{-\infty})$

$$\sum_{g\in G} \alpha_g g(y_0+y_1+y_2+y_3+y_4) + \sum_{g\in G} \beta_g g(y_0+y_{-1}+y_{-2}+y_{-3}+y_{-4}) = 0$$

Equating the coefficient of each gy_i to zero, yields

(0) $$\sum_{h,h' \in <C_0>} \alpha_{gh} + \beta_{gh'} = 0 \qquad \text{for all} \quad g \in G/<C_0>$$

(i) $$\sum_{h \in <C_i>} \alpha_{gh} = 0 \qquad \text{for all} \quad g \in G/<C_i>$$

(-i) $$\sum_{h \in <C_{-i}>} \beta_{gh} = 0 \qquad \text{for all} \quad g \in G/<C_{-i}>$$

Let $\delta_{g<C_i>} = \sum_{h\in <C_i>} \delta_{gh} h$, where $\delta_{gh} = 0$ or 1 according as $g \ne h$ or $g = h$, $-4 \le i \le 4$. Then the conditions above are equivalent to

(0) $$\alpha \oplus \beta \perp \delta_{g<C_0>} \oplus \delta_{g<C_0>} \quad , \text{ for all } \quad g$$

(i) $$\alpha \perp \delta_{g<C_i>}, \ 1 \le i \le 4 \quad , \text{ for all } \quad g$$

(-i) $$\beta \perp \delta_{g<C_{-i}>}, \ 1 \le i \le 4 \quad , \text{ for all } \quad g$$

From this the lemma follows.

The following elementary observation is used repeatedly in the estimates we are about to make.

Lemma 7.2. Let A_1, A, B_1, B be vector spaces with $A_1 \subset A$ and $B_1 \subset B_1$. Then $\dim A \cap B - \dim A_1 \cap B_1 \le \dim(A/A_1) + \dim(B/B_1)$.

Proof. $$\frac{A \cap B}{A_1 \cap B_1} \approx \frac{A \cap B}{A_1 \cap B} \times \frac{A_1 \cap B}{A_1 \cap B_1} \hookrightarrow \frac{A}{A_1} \times \frac{B}{B_1} .$$

Set

$$\mathcal{O}(G) = \sup_{\substack{0 \le i < j \le 4 \\ |i-j| = 2 \text{ or } 3}} \{\dim V_i \cap V_j,\ \dim V_{-i} \cap V_{-j}\} . \tag{7.2.1}$$

(Note that $\dim V_i \cap V_j = |G|/|\langle C_i, C_j\rangle|$.

We have therefore

$$\dim \sum_1^4 V_0 \cap V_i - \dim(V_0 \cap V_1 + V_0 \cap V_4) \le 2\mathcal{O}(G)$$

and

$$\dim \sum_1^4 V_0 \cap V_{-i} - \dim(V_0 \cap V_1 + V_0 \cap V_4) \le 2\mathcal{O}(G).$$

Define ${}_0\mathcal{O}_1^4$ and ${}_0\mathcal{O}_{-1}^{-4}$ via

$$\dim V_0 \cap \sum_1^4 V_i - \dim \sum_1^4 (V_0 \cap V_i) = {}_0\mathcal{O}_1^4$$

$$\dim V_0 \cap \sum_1^4 V_{-i} - \dim \sum_1^4 (V_0 \cap V_{-i}) = {}_0\mathcal{O}_{-1}^{-4} .$$

Similarly define ${}_4\mathcal{O}_1^3$ and ${}_4\mathcal{O}_{-1}^{-3}$ via

$$\dim V_4 \cap \sum_1^3 V_i = \dim \sum_1^3 (V_4 \cap V_i) + {}_4\mathcal{O}_1^3$$

$$\dim V_{-4} \cap \sum_1^3 V_{-i} = \dim \sum_1^3 (V_{-4} \cap V_{-i}) + {}_4\mathcal{O}_{-1}^{-3} .$$

Lemma 7.3. Let G be a finite group and G_1, G_2, G_3 subgroups with G_2 centralizing G_1 and G_3. Set $V = \mathbb{Q}[G]$ and $V_i = V^{G_i}$ $(i=1,2,3)$. Then

$$V_2 \cap (V_1 + V_3) = V_2 \cap V_1 + V_2 \cap V_3.$$

Proof. Let $\sigma = \frac{1}{|G_2|} \sum_{x \in G_2} x$. Then $v \to v\sigma$ is a projection of V onto V_2 which stabilizes V_1 and V_3. Given $f_2 = f_1 + f_3$ with $f_i \in V_i$ $(i=1,2,3)$, we get

$$f_2 = f_2 \cdot \sigma = f_1 \cdot \sigma + f_3 \cdot \sigma \in (V_2 \cap V_1) + (V_2 \cap V_3).$$

<u>Lemma 7.4.</u>

$$\dim \operatorname{Im} D = \sum_{-4}^{4} \dim V_i - \sum_{\substack{-4 \\ i \neq -1}}^{4} \dim V_i \cap V_{i+1} - \varepsilon$$

where $\varepsilon \le {}_0\mathcal{O}_1^4 + {}_0\mathcal{O}_{-1}^{-4} + {}_4\mathcal{O}_1^3 + {}_{-4}\mathcal{O}_{-1}^{-3} + 8\mathcal{O}(G)$.

<u>Proof.</u> $\dim \operatorname{Im} D = \dim(\operatorname{Ker} D)^{\perp}$ – in fact, since our inner product on $V \oplus V$ is positive definite, D maps $(\operatorname{Ker} D)^{\perp}$ isomorphically onto $\operatorname{Im} D$. Thus

$$\begin{aligned}\dim \operatorname{Im} D &= \dim V_{0,0} + \dim \sum_1^4 V_i \oplus V_{-i} - \dim V_{0,0} \cap \sum_1^4 V_i \oplus \sum_1^4 V_{-i} \\ &= \dim V_0 + \dim \sum_1^4 V_i + \dim \sum_1^4 V_{-i} - \dim(V_0 \cap \sum_1^4 V_i) \cap (V_0 \cap \sum_1^4 V_{-i})\end{aligned}$$

We have, by Lemma 7.2

$$\dim(V_0 \cap \sum_1^4 V_i) \cap (V_0 \cap \sum_1^4 V_{-i}) = \dim[\sum_1^4 (V_0 \cap V_i) \cap \sum_1^4 (V_0 \cap V_{-i})] + \varepsilon_0$$

with $\varepsilon_0 \le {}_0\mathcal{O}_1^4 + {}_0\mathcal{O}_{-1}^{-4}$. Moreover,

$$\dim \sum_1^4 V_i = \dim V_4 + \dim \sum_1^3 V_i - \dim V_4 \cap \sum_1^3 V_i.$$

Substituting $\dim \sum_1^3 V_i = \dim V_2 + \dim(V_1 + V_3) - \dim V_2 \cap (V_1 + V_3)$ we get, using Lemma 7.3,

$$\begin{aligned}\dim \sum_1^4 V_i = \sum_1^4 \dim V_i &- \dim(V_1 \cap V_3) - \dim(V_2 \cap V_1) - \dim(V_2 \cap V_3) \\ &- \sum_1^3 \dim(V_4 \cap V_i) - {}_4\mathcal{O}_1^3\end{aligned}$$

and a similar expression for $\dim \sum_1^4 V_{-i}$. Therefore

$$\begin{aligned}\dim \operatorname{Im} D = \sum_{-4}^{4} \dim V_i &- \sum_{\substack{-4 \\ i \neq -1,0}}^{3} \dim(V_i \cap V_{i+1}) - \dim \sum_1^4 (V_0 \cap V_i) \cap \sum_1^4 (V_0 \cap V_{-i}) \\ &- 6\mathcal{O}(G) - {}_4\mathcal{O}_1^3 - {}_4\mathcal{O}_{-1}^{-3} - \varepsilon_0.\end{aligned}$$

Finally,

$$\dim \sum_1^4 (V_0 \cap V_i) = \dim[(V_0 \cap V_1) + (V_0 \cap V_4)] + \varepsilon_1, \quad \varepsilon_1 \le 2O(G)$$

$$\dim \sum_1^4 (V_0 \cap V_{-i}) = \dim[(V_0 \cap V_1) + (V_0 \cap V_4)] + \varepsilon_{-1}, \quad \varepsilon_{-1} \le 2O(G).$$

Consequently $\dim \sum_1^4 V_0 \cap V_i \cap \sum_1^4 V_0 \cap V_{-i} = \dim V_0 \cap V_1 + V_0 \cap V_4 + \varepsilon_2$ where $\varepsilon_2 \le \varepsilon_1 + \varepsilon_{-1}$ by Lemma 7. 2, and indeed we can take $\varepsilon_2 = \varepsilon_1$ as seen from the proof of Lemma 7.2 in the special case $A_1 = B_1$. This yields

$$\dim \operatorname{Im}D = \sum_{-4}^{4} \dim V_i - \sum_{\substack{-4 \\ i \neq -1}}^{4} \dim(V_i \cap V_{i+1}) - \varepsilon$$

where $V_5 = V_0$ and $\varepsilon \le {}_4O_1^3 + {}_4O_{-1}^{-3} + \varepsilon_0 + 8O(G)$, i.e. $\varepsilon \le {}_0O_1^4 + {}_0O_{-1}^{-4} + {}_4O_1^3 + {}_4O_{-1}^{-3} + 8O(G)$, as required.

Set

$$P_D = \sum_{-4}^{4} \frac{1}{c_i} - \sum_{\substack{-4 \\ i \neq -1}}^{4} \frac{1}{c_i c_{i+1}}, \quad \varepsilon_D = \frac{\varepsilon}{|G|}.$$

We can restate Lemma 7.4 as:

(7.4)' $$\dim \operatorname{Im} D = |G| \, (P_D - \varepsilon_D).$$

§8. $\beta_1(Y)$

Let $C_i({}^mY^{-\infty})$ denote the group of i-chains of ${}^mY^{-\infty}$ with coefficients in Q, let ∂_i denote the boundary operator in $C_i({}^mY^{-\infty})$, $Z_i = \operatorname{Ker} \partial_i$, $B_i = \operatorname{Im} \partial_{i+1}$ $(i=0,1,2)$.

Lemma 8.1. $\dim Z_1 = |G| \left[\frac{1}{c_1} + \frac{1}{c_4} + \sum_{\substack{-3\\ i\neq\pm1}}^{3} \frac{1}{c_i c_i'} + \sum_{\substack{-3\\ i\neq-1}}^{4} \frac{2}{c_i^2} + \sum_{\substack{-4\\ i\neq-1}}^{4} \frac{1}{c_i c_{i+1}}\right] + 1$

where G_i is stabilizer in G of a connected component of $\pi^{-1}(L_i)$.

Proof. The kernel of the boundary map ∂_2 is spanned by the fundamental 2-cycles of each connected component of $\pi^{-1}(L_i)$, $-3 \leq i \leq 4$, $i \neq -1$. Hence

$$\dim \operatorname{Im} \partial_2 = \dim C_2({}^mY^{-\infty}) - \sum_{\substack{-3\\ i\neq-1}}^{4} \frac{|G|}{|G_i|}$$

$$= |G| \left[\frac{1}{c_1} + \frac{1}{c_4} + \sum_{\substack{i=-3\\ i\neq\pm1}}^{3} \frac{1}{c_i c_i'} - \sum_{\substack{-3\\ i\neq-1}}^{4} \frac{1}{|G_i|}\right]$$

by the result of the end of §2. Inasmuch as $\dim Z_1 = \dim H_1(Y^{-\infty}) + \dim B_1$, the result follows at once from Corollary 6.6.

Set

$$P_Z = \sum_{\substack{-3\\ i\neq-1}}^{4} \frac{2}{c_i^2} + \sum_{\substack{-4\\ i\neq-1}}^{4} \frac{1}{c_i c_{i+1}} + \sum_{\substack{-3\\ i\neq\pm1}}^{3} \frac{1}{c_i c_i'} + \frac{1}{c_1} + \frac{1}{c_4}.$$

From Lemma 8.1, we have

$$\frac{1}{|G|} \dim Z_1 = P_Z + \varepsilon_Z \tag{8.1)'}$$

where

$$\varepsilon_Z = \frac{1}{|G|}$$

Lemma 8.2. $P_Z = P_D$.

Proof. The identity

$$\sum_{-4}^{4} \frac{1}{c_i} - \sum_{\substack{-4\\ i\neq-1}}^{4} \frac{1}{c_i c_{i+1}} = 2 \sum_{\substack{-3\\ i\neq-1}}^{4} \frac{1}{c_i^2} + \sum_{\substack{-4\\ i\neq-1}}^{4} \frac{1}{c_i c_{i+1}} + \sum_{\substack{-3\\ i\neq\pm1}}^{3} \frac{1}{c_i c_i'} + \frac{1}{c_1} + \frac{1}{c_4}$$

is equivalent to

$$2\sum_{\substack{-3\\ i\neq -1}}^{4}\frac{1}{c_i^2}+2\sum_{\substack{-4\\ i\neq -1}}^{4}\frac{1}{c_ic_{i+1}}+\sum_{\substack{-3\\ i\neq \pm 1}}^{3}\frac{1}{c_ic_i'}=\sum_{\substack{-3\\ i\neq -1}}^{4}\frac{1}{c_i}$$

Set $c_0' = c_7$, $c_{-3}' = c_6 = c_2'$, $c_3' = c_8 = c_{-2}'$.

We use the fact that the fifteen points

$\{L \cap L';\ L, L' \in \mathcal{L}'\}$ consist of eight of the form

$\{L \cap L'; L, L' \in \mathcal{L}\}$ and five of the form $\{L \cap L';\ L \in \mathcal{L},\ L' \notin \mathcal{L}\}$ and two of the form $\{L \cap L', L, L' \notin \mathcal{L}\}$.

Thus
$$\sum_{\substack{L_i\cap L_j\neq\phi\\ i<j}}\frac{1}{c_ic_j}=\sum_{\substack{-4\\ i\neq -1}}^{4}\frac{1}{c_ic_{i+1}}+\sum_{\substack{-3\\ i\neq \pm 1}}^{3}\frac{1}{c_ic_i'}+\frac{1}{c_6c_7}+\frac{1}{c_7c_8}$$

and the identity can be rewritten, using Lemma 6.7

$$2\sum_{\substack{-3\\ i\neq -1}}^{4}\frac{1}{c_i^2}-\sum_{\substack{-3\\ i\neq \pm 1}}^{3}\frac{1}{c_ic_i'}+2\Big(\sum_{\substack{L_i\cap L_j\neq\phi\\ i<j}}\frac{1}{c_ic_j}-\frac{1}{c_6c_7}-\frac{1}{c_7c_8}\Big)$$
$$=2-\frac{1}{c_6}-\frac{1}{c_7}-\frac{1}{c_8}\ .$$

By Lemma 6.9, this is equivalent to

$$2\Big(-\frac{1}{c_6^2}-\frac{1}{c_7^2}-\frac{1}{c_8^2}\Big)-\sum_{\substack{-3\\ i\neq \pm 1}}^{3}\frac{1}{c_ic_{i+1}'}+2\Big(-\frac{1}{c_6c_7}-\frac{1}{c_7c_8}\Big)=-\frac{1}{c_6}-\frac{1}{c_7}-\frac{1}{c_8}$$

that is

$$\frac{1}{c_6}+\frac{1}{c_7}+\frac{1}{c_8}-\frac{1}{c_{-3}c_{-6}}-\frac{1}{c_{-2}c_8}-\frac{1}{c_0c_7}-\frac{1}{c_2c_6}-\frac{1}{c_3c_8}=$$
$$2\Big(\frac{1}{c_6^2}+\frac{1}{c_7^2}+\frac{1}{c_8^2}+\frac{1}{c_6c_7}+\frac{1}{c_7c_8}\Big)$$

The left side is

$$\frac{1}{c_6}\Big(1-\frac{1}{c_{-3}}-\frac{1}{c_2}\Big)+\frac{1}{c_8}\Big(1-\frac{1}{c_3}-\frac{1}{c_{-2}}\Big)+\frac{1}{c_7}\Big(1-\frac{1}{c_0}\Big)=$$
$$\frac{1}{c_6}\Big(\frac{2}{c_6}+\frac{1}{c_7}\Big)+\frac{1}{c_8}\Big(\frac{2}{c_8}+\frac{1}{c_7}\Big)+\frac{1}{c_7}\Big(1-\frac{1}{c_0}\Big)\quad \text{by Lemma 6.1.}$$

Hence the identity reduces to

$$\frac{1}{c_7}\left(1-\frac{1}{c_0}\right)=\frac{2}{c_7^2}+\frac{1}{c_6c_7}+\frac{1}{c_7c_8}$$

i.e.
$$1-\frac{1}{c_0}=\frac{2}{c_7}+\frac{1}{c_6}+\frac{1}{c_8},$$

which follows from Lemma 6.1, applied to the line L_7 in M^∞ meeting L_0.

<u>Theorem 8.3</u>.

$$\frac{1}{|G|}\ \beta_1(Y) \le \varepsilon_Z + \varepsilon_D \le \frac{1}{|G|}\left(8\mathcal{O}(G) + {}_0\mathcal{O}_1^4 + {}_0\mathcal{O}_{-1}^{-4} + {}_4\mathcal{O}_1^3 + {}_4\mathcal{O}_{-1}^{-3} + 1\right).$$

<u>Proof</u>. Combining (7.4)', (8.1)', and Lemma 8.2 yields

$$\beta_1(Y) = \dim \frac{Z_1(Y^{-\infty})}{\operatorname{Im} D+\operatorname{Im} \partial_2} \le \dim \frac{Z_1(Y^{-\infty})}{\operatorname{Im} D} = |G|\ (\varepsilon_Z + \varepsilon_D).$$

The remaining inequality of Theorem 8.3 now follows from the definition of ε_Z and ε_D.

§9. A problem in finite groups.

Let G be a finite group and let $G_0, G_1, G_2, \ldots, G_k$ be subgroups of G. Set $V = \mathbb{Q}[G]$, the group algebra of G over the field $\mathbb{Q}$ of rational numbers and set

$$V_i = V^{G_i} = \{v \in V;\ v\cdot G_i = v\}\ .$$

Set

$$ {}_0\mathcal{O}_1^k(G) = \dim V_0 \cap \sum_1^k V_i - \dim \sum_1^k V_0 \cap V_i .$$

Problem: Estimate ${}_0\mathcal{O}_1^k(G)$.

If the group G is abelian, then ${}_0\mathcal{O}_1^k(G) = 0$. However, one can find groups with ${}_0\mathcal{O}_1^k(G) \neq 0$. If $G \subset H$ and one considers $G_0, G_1, \ldots, G_k$ as subgroups of H, then one has

$$ {}_0\mathcal{O}_1^k(H) = {}_0\mathcal{O}_1^k(G) \cdot \frac{|H|}{|G|} .$$

Conjecture: If $\sum_0^k \frac{1}{|G_i|} \leq 1$, and $G_0, \ldots, G_k$ generate G and are cyclic and no element $\neq 1$ of G_i is conjugate to an element of G_j if $i \neq j$, then

(9.1) $$ {}_0\mathcal{O}_1^k(G) \leq |G|^\alpha \quad , \alpha < 1.$$

In the situation of §8, the group G is a homomorphic image of an infinite group with the presentation given by the Coxeter diagram:

(9.2) 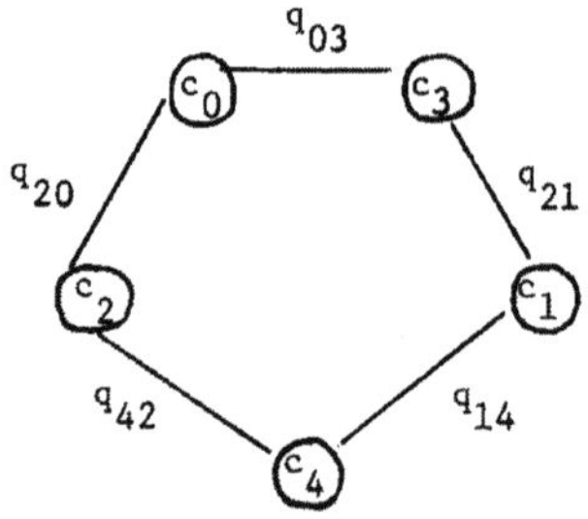

$$, \quad \sum_0^4 \frac{1}{c_i} = 1$$

Each node represents a generator C_i of order c_i with

$$(C_iC_j)^{q_{ij}} = 1, \quad \text{if the nodes are joined}$$

$$C_iC_j = C_jC_i , \quad \text{if the nodes are not joined.}$$

Each G_i is cyclic, generated by the image of C_i $(i=0,1,2,3,4)$.

The lattice Γ'_μ of $U(2,1)$ defined in §1 has such a presentation. When we take Γ'_0 to be a congruence subgroup modulo a prime ideal, then the resulting Γ'_μ/Γ'_0 becomes a subgroup of $GL_3(\mathbb{F}_q)$ of 3×3 matrices with coefficients in the finite field $\mathbb{F}_q$ and is a central extension of G, and the problem for G can be deduced from Γ'_μ/Γ'_0. Thus we may take $G = \Gamma'_\mu/\Gamma'_0$. For each non-commuting pair of groups G_i, G_j, the subgroup $< G_i, G_j >$ contains a subgroup isomorphic to $SL_2(F)$ where F is a subfield of $\mathbb{F}_q$. The group G is a large subgroup of $GL_3(\mathbb{F}_q)$ and we conjecture for the $\mathcal{O}(G)$ defined in (7.2.1).

$$\mathcal{O}(G) \leq |G|^\alpha \quad , \alpha < 1 \tag{9.3}$$

for the group $G = \Gamma_\mu/\Gamma_0$.

Discussion. As a consequence of conjectures (9.1) and (9.3), Theorem 8.3 would imply

$$\frac{\beta_1(Y)}{|G|} \leq 12\ |G|^{-a} \quad , a > 0. \tag{9.4}$$

This conjecture can be compared with the known result of DeGeorge and Wallach [3]:

Given a nested sequence of normal subgroups Γ_j in a cocompact discrete subgroup Γ of a linear connected semi-simple group Lie group G with $\bigcap_1^\infty \Gamma_j = \{1\}$, and given $\omega \in \hat{G}$, the set of all equivalence classes of irreducible unitary representations of G, then

$$\lim_{j \to \infty} \operatorname{vol}(\Gamma_j \backslash G)^{-1} N(\Gamma_j, \omega) = 0, \text{ if } \omega \text{ is not square integrable;}$$

here $N(\Gamma,\omega)$ is the multiplicity of ω in $L^2(\Gamma\backslash G)$.

Set $Y_j = \Gamma_j \backslash B$.

If as $j \to \infty$

$$\frac{\beta_i(Y_j)}{|\Gamma/\Gamma_j|} \geq \text{constant} \quad c > 0$$

then one could find an element $\omega \in \hat{G}$ with $\dfrac{N(\Gamma_j,\omega)}{\operatorname{vol}(\Gamma_j\backslash G)} \geq c > 0$, as $j \to \infty$, and hence ω would be in the discrete series (cf. DeGeorge-Wallach [3] Theorem 5.4, also [1], pg. 214). But it is known that the discrete series contributes to cohomology only in $\frac{1}{2}$ dimension G/K, K being a maximal compact subgroup of G; in our case this is $\frac{1}{2}\dim B = 2$. Thus

$$\frac{\beta_1(Y_j)}{|\Gamma/\Gamma_j|} \to 0 \quad \text{as} \quad j \to \infty .$$

Conjecture (9.4) offers a more precise estimate for $\frac{\beta_1(Y_j)}{|\Gamma/\Gamma_j|}$.

Bibliography

[1] Borel, A., and Wallach, N., Continuous Cohomology, Discrete Subgroups, and Representations of Reductive Groups, Ann. of Math. Studies, 94, (1980), Princeton Univ. Press.

[2] Deligne, P., and Mostow, G. D., Monodromy of Hypergeometric Functions and Non-Lattice Integral Monodromy, Publ. I.H.E.S., 1986.

[3] DeGeorge, D.L., and Wallach, N., Limit Formulas for Multiplicities in $L^2(G/\Gamma)$, Annals of Math., 107, (1978), pp. 133-150.

[4] Milnor, J., Morse Theory, Annals of Math. Studies, v. 51, (1951), Princeton Univ. Press.

[5] Picard, E., Sur les Fonctions Hyperfuchsiennes Provenant des Series Hypergeometriques de Deux Variables, Ann. ENS, III 2 (1885), pp. 357-384.

[6] Schwarz, H.A., Uber Diejenigen Fälle in Welchen die Gaussische Hypergeometrische Reihe Eine Algebraische Function Ihres Viertes Elementes Darstellt, J. Reine u. Angew. Math., 75, (1873), pp. 292-335.

* Dept. of Math., Yale University, New Haven Conn. 06520, U.S.A.

INVARIANT THEORY AND KLOOSTERMAN SUMS

I. Piatetski-Shapiro

1. Introduction.

A. Selberg [1] introduced the following type of series

$$P(z,s) = \Sigma \frac{y^s e^{2\pi i k \frac{az+b}{cz+d}}}{|cz+d|^{2s}} \qquad (z \text{ in the upper half plane})$$

where the summation is over representatives of cosets of $\Gamma_0 \backslash SL(2,\mathbb{Z})$. $\Gamma_0 = \{\begin{pmatrix}1 & x\\ 0 & 1\end{pmatrix} | x \in \mathbb{Z}\}$. k is a positive integer. A. Selberg was able to prove the following remarkable property: The poles of $P(z,s)$ correspond to Maass wave forms. More precisely, if $\varphi(z)$ is a Maass wave form satisfying $\Delta\varphi = \lambda\varphi$, $(\Delta = -y^2(\frac{\partial^2}{\partial x^2} + \frac{\partial^2}{\partial y^2}))$, then for s_0 such that $\lambda = s_0(1 - s_0)$ there is a pole of $P(z,s)$ at $s = s_0$ and the residue of $P(z,s)$ at s_0 is $\varphi(z)$. Another remarkable property found by A. Selberg was that the Fourier coefficient of $P(z,s)$ is Dirichlet series $Z(s)$ whose coefficients are Kloosterman sums

$$Z(s) = \sum_{n=1}^{\infty} \frac{S(k,n)}{n^{2s}}$$

$$S(k,n) = \sum_{xy \equiv 1 (\text{mod } n)} e^{2\pi i k \frac{x+y}{n}}$$

The aim of this note is to generalize the Selberg construction to arbitrary split reductive group G.

Different generalizations of Selberg's construction were considered by D. Bump, S. Friedberg, D. Goldfeld [3], [4], and G. Stevens [5]. They are dealing with Poincaré series which, like Eisenstein series, depends on many complex variables. The Poincaré series, which we introduce here, depends only on one complex parameter. Their main property is that they produce global L-functions which were introduced by R. P. Langlands. It is possible also to define Poincaré series which produce local Langlands L-factors. More precisely, we prove the following result.

Denote by ${}^LG$ the L group of G. In our situation it is a reductive group over $\mathbb{C}$. Let $\pi = \otimes \pi p$ is a cuspidal automorphic representation of G. Let ρ be an irreducible finite dimensional representation of ${}^L G$. Let Σ be a finite set of places of k which includes all the archimedean places in case k is a number field.

We assume that πp is an unramified representation for $p \notin \Sigma$. R. Langlands introduced

$$L^{\Sigma}(\pi,\rho,s) = \prod_{p\notin\Sigma} L_p(\pi p,\rho,s).$$

We construct a Poincaré series $P(g,s)$ such that

$$\int_{G_k\backslash G_A} P(g,s)\varphi(g)dg = cL^{\Sigma}(\pi,\rho,s)$$

where $c \neq 0$, and φ is a cusp form with Whittaker model lying in automorphic representation π such that φ is right invariant under K_p for all $p \notin \Sigma$. K_p is the standard maximal compact subgroup of G_p. $P(g,s)$ of course depends on Σ.

Our construction can be modified in such a way that it gives a local L-function at a given nonarchimedean place. In all cases our Poincaré series converges absolutely in some right half plane. In the case where the series produces a local L-factor at a given place, we can easily get the meromorphic continuation of the series using spectral theory. In the case of series which produce global L-functions $L^{\Sigma}(\pi,\rho,s)$, the problem of meromorphic continuation is equivalent to the problem of meromorphic continuation of $L^{\Sigma}(\pi,\rho,s)$ which is part of Langlands' conjectures.

It is interesting to observe that the Whittaker Fourier coefficient of our Poincaré-Selberg series is related to some sorts of generalized Kloosterman sums. We prove that in the general case these sums can be expressed through sums of the type

$$S(n) = \sum_{x_1\cdots x_\ell \equiv a(n)} \psi(x_1+\cdots+x_\ell)$$

and ψ is a character of $n\mathbb{Z}\backslash\mathbb{Z}$. Usually Weil's estimation does not produce very good results. However, one can always expect that some version of Linnik's conjecture implies Ramanujan's conjecture in general.

2. Construction of Poincaré-Selberg Series.

First we recall some known facts from invariant theory. Let H be a reductive group defined over $\mathbb{C}$, and ρ an irreducible finite dimensional representation. In our application $H = {}^LG$. Consider $\mathrm{Sym}^n\rho$. It is usually reducible. Write

$$\mathrm{Sym}^n\rho = \oplus a(n,\tau)\tau$$

where τ are irreducible finite dimensional representations of H. $a(n,\tau)$ is the multiplicity of τ. Consider the Poincaré-Molien series

$$M_\tau(t) = \sum_{n=0}^{\infty} a(n,\tau)t^n \tag{2.1}$$

$M_\tau(t)$ can be presented in the form

$$M_\tau(t) = \frac{P_\tau(t)}{Q(t)} \tag{2.2}$$

where $P_\tau(t)$ and $Q(t)$ are polynomials and $Q(t)$ has the form

$$Q(t) = \prod_{i=1}^{r} (1 - t^{d_i}).$$

One can choose $Q(t)$, which does not depend on τ. [2]

Let G be a split semisimple group defined over a global field k. Let C_p be the Cartan subgroup of $G(k_p)$, (p - a place of k). Let B_p be the Borel subgroup of $G(k_p)$, and K_p - the maximal compact subgroup. Denote by ψ_p the non-degenerate normalized character of the maximal unipotent subgroup X_p of $G(k_p)$. Consider any function f on G satisfying

$$f(xgk) = \psi_p(x)f(g) \qquad \forall x \in X_p,\ k \in K_p \tag{2.3}$$

Such a function f is uniquely defined by its restriction to C_p, and more: Let $E = C \cap K$ and L be the lattice C/E. Denote by j the projection $C \to L$. An element of L can be naturally interpreted as a character of the maximal torus T of $H = {}^LG$. It is easy to see that $f(c) = 0$ for c such that $j(c)$ does not lie in the corresponding Weyl chamber. Denote by $\Delta(c)$ the jacobian of the map $x \to cxc^{-1}$ (from X_p to X_p). Let f be as in (2.3) such that

$$f(c,s) = P_{\tau(c)}(q_p^{-s})\Delta^{\frac{1}{2}}(c) \tag{2.4}$$

where $\tau(c)$ is an irreducible representation of H with highest weight equal to $j(c)$. q_p is the number of elements in the residue field of k_p. Let Σ be a finite set of places containing all the Archimedean places. For $p \notin \Sigma$, let $f_p(g,s)$ be a function on G_p satisfying (2.3), (2.4). For $p \in \Sigma$, let $f_p(g)$ be a function on G_p satisfying

$$f_p(xg) = \psi_p(x)f_p(g) \tag{2.5}$$

Usually, but not always, we assume that for $p \in \Sigma$, $f_p(g)$ does not depend on s. Now for $g \in G_A$ define

$$f(g) = \prod_p f_p(g_p,s)$$

We assume that $\psi = \Pi\psi_p$ is a nondegenerate character of $X_k\backslash X_A$. Then

$$f(\delta g,s) = f(g,s) \qquad \forall\delta \in X_k$$

Put

$$P(g,s) = \sum_{\gamma\in X_k\backslash G_k} f(\gamma g,s) \tag{2.6}$$

This series converges absolutely in some half plane.

Denote $K = \prod_p K_p$. Let $\sigma = \otimes\sigma p$ be a decomposable finite dimensional representation of K such that $\sigma|K_p = 1$ for $p \notin \Sigma$. We say that an automorphic form φ on G(A) is of type (K,σ), if under right translation by K of φ, it generates the representation, which is isomorphic to σ.

THEOREM. Let φ be a cusp form on G(A) which generates an irreducible automorphic representation π, of type (K,σ), and assume that the Whittaker functional takes the value 1 on φ, then there exists a Poincaré series (2.6) which depends only on Σ and (K,σ) (and not on φ) such that

$$\prod_{i=1}^{r} \zeta_k^{\Sigma}(d_i s)\int_{G_k\backslash G_A} P(g,s)\varphi(g)dg = c(\varphi)L^{\Sigma}(\pi,\rho,s) \tag{2.7}$$

Here ζ_k^{Σ} is the restricted ζ-function of k, d_i appear in Q(t) in (2.2). $c(\varphi) \neq 0$.

First we prove two lemmas.

LEMMA 1. Let π_p be an unramified representation of G_p. Let W(g) be the Whittaker function with respect to ψ_p^{-1} which corresponds to an unramified vector of π_p, then

$$\prod_{i=1}^{r} (1 - q_p^{-d_i s})^{-1}\int_{X_p\backslash G_p} f_p(g,s)W(g)dg = L(\pi_p,\rho,s) \tag{2.8}$$

PROOF: It is well known that π_p corresponds to a semisimple conjugacy class A_{π_p} in $H = {}^LG$, such that

$$L(\pi_p,\rho,s) = \det(I - \rho(A_{\pi_p})q_p^{-s})^{-1}$$

Let us recall the Shintani-Casselman-Shalika formula for W(g):

$$W(c) = \operatorname{tr} \tau(A_{\pi_p})\Delta(c)^{\frac{1}{2}} \tag{2.9}$$

where τ is the representation of H with highest weight $j(c)$, ($c \in C_p$ with $j(c)$ in the corresponding Weyl chamber). Recall that the Haar measure on G_p is $dg = \Delta^{-1}(c)dxdcdk$. Using this, it remains to prove

$$\int_C M_{r(c)}(q_p^{-s})W(c)dc = L(\pi_p,\rho,s) \tag{2.10}$$

Using (2.1) and (2.9) and changing the order of integration we get that (2.10) is equivalent to

$$\sum_{m=1}^{\infty} \operatorname{tr}(\operatorname{sym}_n\rho(A_{\pi_p}))q_p^{-ns} = \det(I - \rho(A_{\pi_p})q_p^{-s})^{-1}$$

which is an identity.

Proof is easy.

LEMMA 2. Let $p \in \sigma$ and W_p in the Whittaker model of π_p such that $W_p(I) \neq 0$, then there is a function $f_p(g)$ satisfying (2.5), which depends only on $(K,\hat{\sigma}_p)$ such that

$$\int_{X_p\backslash G_p} f_p(g)W_p(g)dg \neq 0.$$

PROOF OF THE THEOREM: Pick f satisfying Lemma 1 and Lemma 2, then

$$\int_{C_k\backslash G_A} P(g,s)\varphi(g)ds = \int_{X_k\backslash G_A} f(g,s)\varphi(g)dg =$$

$$\int_{X_A\backslash G_A} f(g,s)W(g)dg =$$

$$\prod_p \int_{X_p\backslash G_p} f_p(g,s)W_p(g)dg$$

Now (2.7) it follows from Lemmas 1 and 2.

3. Fourier Coefficients as Kloosterman Zeta Functions.

Let G be as before. Consider the Fourier coefficient

$$Z(g,s) = \int_{X_k\backslash X_A} \psi^{-1}(x)P(xg,g)dx \tag{3.1}$$

We have

$$Z(g,s) = \sum_{\delta\in X_k\backslash G_{K/X_k}} \int_{X_k^\delta\backslash X_A} f(\delta xs,s)\psi^{-1}(x)dx$$

where $X^\delta = X \cap \delta^{-1}X\delta$ (since $\delta \in G_k$, X^δ is an algebraic group)

$$= \sum_{\delta\in X_k\backslash G_k/X_k} \int_{X_k^\delta\backslash X_A} f(\delta xg,s)\psi^{-1}(x)dx$$

Choose representatives δ in the normalizer N of the Cartan subgroup C,

(Bruhat's lemma). Let W be the Weyl group of G, then $C\backslash N \simeq W$. Put for $w \in W$

$$Z_w(g,s) = \sum_{\delta \text{ projects to } w} \int_{X_A^\delta \backslash X_A} f(\delta x g, s)\psi^{-1}(x)\,dx$$

It is easy to show that $Z_w(g,s) \equiv 0$ unless w satisfies the following assumption: Iff α is a simple root then $w\alpha$ is either negative or a simple root. For instance, if $G = GL(n)$ then $W \simeq S_n$ the permutations on n elements. W is isomorphic to the subgroup of permutation matrices in G, i.e. those which have in each row and each column only one nonzero element which is one. Then the elements w satisfying the above assumption are of the form $\begin{pmatrix} 0 & \cdots & I_{k_1} \\ \vdots & & \vdots \\ \vdots & \cdot & \vdots \\ \cdot & \cdot & \vdots \\ I_{k_r} & \cdots & 0 \end{pmatrix}$. Put

$$I_\delta(g,s) = \int_{X_A^\delta \backslash X_A} f(\delta x g, s)\psi^{-1}(x)\,dx$$

Let Y^δ be the subgroup of X generated by all the root subgroups X_α such that $\alpha > 0$ and $w\alpha < 0$, then

$$I_\delta(g,s) = \int_{Y_A^\delta} f(\delta y g, s)\psi^{-1}(y)\,dy = \prod_p \int_{Y_p^\delta} f_p(\delta y g, s)\psi_p^{-1}(y)\,dy$$

Put

$$I_\delta^{(p)}(g,s) = \int_{Y_p^\delta} f_p(\delta y g, s)\psi_p^{-1}(y)\,dy$$

We now consider the case $G = PGL(2)$, then we have the following result:

Write $\delta = \begin{pmatrix} \alpha & 0 \\ 0 & 1 \end{pmatrix} \begin{pmatrix} 0 & -1 \\ 1 & 0 \end{pmatrix}$. Assume that $g = I$, and write $I_p(\alpha,s)$ for $I_\delta^{(p)}(I,s)$, then we have:

Assume that $|\alpha|_p > q_p$ then

$$I_p(\alpha,s) = \begin{cases} 0 & \text{if } \operatorname{val}_p(\alpha) \text{ is odd} \\ \int_{|x|=q_p^\ell} \psi^{-1}\left(\frac{\alpha}{x} + x\right)dx & \text{if } |\alpha|_p = q_p^{2\ell} \end{cases}$$

We see that the integral does not depend on s and that it equals a Kloosterman sum.

Assume that $|\alpha| = q^{-m}$, $m \geq 0$ then

$$I_p(\alpha,s) = P_m(q^{-s})q_p^{\frac{m}{2}} - P_{m+2}(q^{-s})q^{-\frac{m}{2}-1}$$

Assume that $|\alpha| = q$, then

$$I_p(\alpha,s) = -P_1(q^{-s})q^{-\frac{1}{2}}$$

We see that for $|\alpha|_p \le q_p$ the integral $I_p(\alpha,p)$ depends on s and on ρ. (ρ enters in the definition of the polynomials $P_n(t)$).

Analogous properties are true for a general semisimple group G, and in general, the part of $I_\delta(g,\rho)$ which depends on trigonometric sums can be expressed through sums of the form

$$\int\limits_{\substack{|\alpha|=q^m \\ |x_1|=\cdots=|x_{m-1}|=q^\ell}} \psi(x_1+x_2+\cdots+x_{m-1}+\frac{\alpha}{x_1x_2\cdots x_{m-1}})dx_1\cdots dx_m.$$

where ℓ is a positive integer.

References

(1) A. Selberg. On estimation of Fourier coefficients of modular forms. Proc. of Sym. in Pure Math., vol. III (1915).

(2) T.A. Springer. Invariant theory. Lecture Notes 585, Springer-Verlag.

(3) D. Bump, S. Friedberg, D. Goldfeld. Poincare series and Kloosterman sums for SL(3,Z). Preprint.

(4) S. Friedberg. Poincare series for GL(2). Preprint.

(5) G. Stevens. Poincare series on GL(n) and Kloosterman sums. Preprint

Department of Mathematics
Yale University
12 Hillhouse Avenue
New Haven, Connecticut 06520
and
School of Mathematical Sciences
Tel-Aviv University
Tel-Aviv, Israel

ON ACTIONS OF $\mathbb{G}_a$ ON $\mathbb{A}^n$

V.L. Popov

MehMat, MGU

Moscow, USSR

To T.A.Springer on his 60th birthday

1. Let k be an algebraically closed field of characteristic zero. We identify the k-algebra $F_n = k[x_1,\dots,x_n]$ of polynomials in the indeterminates $x_1,\dots,x_n$ with the algebra of regular functions on n-dimensional affine space $\mathbb{A}^n$ by means of the isomorphism which sends x_i to i-th standard coordinate function. If $\sigma:\mathbb{A}^n \to \mathbb{A}^n$ is an isomorphism, we denote by σ^* the automorphism $F_n \to F_n$ given by $(\sigma^*f)(a) = f(\sigma(a))$, $f \in F_n$, $a \in \mathbb{A}^n$. The map $\operatorname{Aut}\mathbb{A}^n \to \operatorname{Aut}_k F_n$, $\sigma \mapsto \sigma^*$, is an anti-isomorphism. We identify σ with the set of polynomials $(\sigma^*x_1,\dots,\sigma^*x_n)$ (which defines σ by the formula $\sigma\, a = ((\sigma^*x_1)(a),\dots,(\sigma^*x_n)(a))$, $a \in \mathbb{A}^n$).

2. The group $\operatorname{Aut}\mathbb{A}^n$ (or, which is the same, $\operatorname{Aut}_k F_n$) has the structure of an infinite dimensional algebraic group, [2]. At present a satisfactory description of its structure is known only for $n \leqslant 2$, see [2,3,4,5,...]. For $n \geqslant 3$ a number of key problems remains open, [1,2,5]. One of these is the problem of the structure of finite dimensional algebraic subgroups of $\operatorname{Aut}\mathbb{A}^n$ or, which is the same, the structure of (regular algebraic) actions of finite dimensional algebraic groups on $\mathbb{A}^n$. The precise formulation of this problem is connected with the consideration of two subgroups of $\operatorname{Aut}\mathbb{A}^n$: the affine subgroup

$$A f_n = \{\sigma = (f_1,\dots,f_n) \in \operatorname{Aut}\mathbb{A}^n \mid \deg f_i \leqslant 1 \text{ for each } i\}$$

and the triangular "Borel" subgroup

$$B_n = \{(f_1,\dots f_n) \in \mathrm{Aut}\,\mathbb{A}^n \mid f_i = c_i x_i + h_i,\ c_i \in k,\ c_i \neq 0,$$
$$h_i \in k[x_1,\dots,x_{i-1}] \text{ for each } i\ \}$$

(we assume that $h_1 = 0$). If $n \leqslant 2$ then every finite dimensional algebraic subgroup G of Aut $\mathbb{A}^n$ is conjugate to either a subgroup of A f_n or to a subgroup of B_n, see [2,3,4,5,...]. The problem (posed in [2] by I.R.Shafarevich) is : can one generalize this to the case of an arbitrary n ? If G is unipotent then this reduces (because of the Lie - Kolchin theorem) to the following question : is G conjugate to a subgroup of B_n ?

3. In his recent paper [1] H. Bass constructed an example of a one parameter unipotent algebraic subgroup of Aut $\mathbb{A}^2$ which is not conjugate to a subgroup of B_3 (or, in other words, an example of an action of the additive group $\mathbb{G}_a$ on $\mathbb{A}^3$ which cannot be triangularized). This construction is based on the automorphism

$$\tau = (x_1, x_2 + x_1 U,\ x_3 - 2x_2 U - x_1 U^2) \in \mathrm{Aut}\,\mathbb{A}^3,\quad U = x_1 x_3 + x_2^2,$$

proposed earlier by M. Nagata [6] as a conjectural example of an element of Aut $\mathbb{A}^3$ which is not a product of elements of A f_3 and B_3. H. Bass observed that one can include τ into the one parameter unipotent algebraic subgroup

$$\{\sigma_t = (x_1, x_2 + t x_1 U,\ x_3 - 2t x_2 U - t^2 x_1 U^2) \in \mathrm{Aut}\,\mathbb{A}^3 \mid t \in k\};$$

this subgroup furnishes the desired example (the latter is proved by means of an investigation of the ideal in F_3 defining the variety Fixσ_t of fixed points of the automorphism σ_t, $t \neq 0$).

I shall show here that this example of H. Bass is a special case of a simple general construction which furnishes in a unified way examples of non-triangular actions of $\mathbb{G}_a$ on A^n for arbitrary n.

4. Let D be a locally nilpotent k-derivation of F_n (i.e. for each $f \in F_n$ there exists an s such that $D^s f = 0$). Then, for $t \in k$, one has a well-defined endomorphism exp tD of the k-algebra F_n given by the formula

$$(\exp tD)(f) = \sum_{m \geqslant 0} \frac{t^m}{m!} D^m f, \qquad f \in F_n,$$

(the sum is finite because D is locally nilpotent, and the property $(\exp tD)(fg) = (\exp tD)(f).(\exp tD)(g)$ follows from Leibniz' formula). We write

$$\exp tD = \sum_{m \geq 0} \frac{t^m}{m!} D^m .$$

It follows from formally verifiable properties of exponentials, to wit $(\exp tD)(\exp sD) = \exp (t+s)D$ for each $t,s \in k$, and $\exp 0D = \mathrm{Id}$, that in fact $\exp tD$ is an automorphism of the k-algebra F_n and that $\{\exp tD | t \in k\}$ is a subgroup of $\mathrm{Aut}_k F_n$. This subgroup is a one parameter algebraic unipotent subgroup, i.e. for each $f \in F_n$ the linear span of $\{(\exp tD)(f) | t \in k\}$ over k is finite dimensional and the action of $\mathbb{G}_a$ on this span induced by $\{\exp tD | t \in k\} \subset \mathrm{Aut}_k F_n$ is given by a rational linear representation of $\mathbb{G}_a$. Or, in other words, if $\sigma_{tD} \in \mathrm{Aut}\,\mathbb{A}^n$ is an element such that $\sigma^*_{tD} = \exp tD$ then the formula $t(a) = \sigma_{tD}a$, $t \in k$, $a \in \mathbb{A}^n$, defines a regular algebraic action of $\mathbb{G}_a$ on $\mathbb{A}^n$ (i.e. $\mathbb{G}_a \times \mathbb{A}^n \to \mathbb{A}^n$, $(t,a) \mapsto t(a)$, is a morphism).

<u>LEMMA 1</u>. <u>If</u> $\{\exp tD | t \in k\}$ <u>is conjugate in</u> $\mathrm{Aut}_k F_n$ <u>to a subgroup of</u> B_n <u>then</u> $\mathrm{Fix}\sigma_{tD}$ <u>is for each</u> $t \in K$ <u>a cylindrical variety, i.e. is isomorphic to</u> $\mathbb{A}^1 \times Z$ <u>for a certain variety</u> Z <u>(depending on</u> t).

<u>PROOF</u>. Let $\gamma \in \mathrm{Aut}\,\mathbb{A}^n$ be an element such that $\delta_t = (\gamma^*)^{-1}.\ \exp tD.\ \gamma^* \in B_n$ for each $t \in k$. We have $\delta_t x_i = c_{it}x_i + h_{it}$, $c_{it} \in k$, $c_{it} \neq 0$, $h_{it} \in k[x_1,\dots,x_{i-1}]$, $h_{1t} = 0$ for each $t \in k$ and $i = 1,\dots,n$. Since $\{\delta_t | t \in k\}$ is an algebraic subgroup of $\mathrm{Aut}_k F_n$, the map $t \mapsto c_{it}$ is a regular (polynomial) function on $\mathbb{A}^1$. It follows from this and from the conditions $c_{it} \neq 0$ and $c_{i0} = 1$ that $c_{it} = 1$ for each i and t. Using the equality $\delta_t = (\gamma\sigma_{tD}\gamma^{-1})^*$ we see now that $\mathrm{Fix}\gamma\sigma_{tD}\gamma^{-1} = \{a \in \mathbb{A}^n | h_{it}(a) = 0$ for each $i\}$. Therefore $\mathrm{Fix}\gamma\sigma_{tD}\gamma^{-1}$ is a cylindrical variety in the coordinates $x_1,\dots,x_n$ (because $h_{it} \in k[x_1,\dots,x_{n-1}]$ for each i and t). The assertion of the lemma follows now from the equality $\gamma\,\mathrm{Fix}\sigma_{tD} = \mathrm{Fix}\gamma\sigma_{tD}\gamma^{-1}$. □

5. Let us point out now two simple ways to construct locally nilpotent k-derivations of F_n.

Each k-derivation Δ of F_n is completely defined by the elements Δx_i, $i = 1,\dots,n$, and for an arbitrary set of elements $f_1,\dots,f_n \in F_n$ there exists a k-derivation Δ such that $\Delta x_i = f_i$, $i = 1,\dots,n$. It follows from Leibniz' formula that Δ is locally nilpotent iff there exists an s such that $\Delta^s x_i = 0$, $i = 1,\dots,n$.

Let V be the linear span of $x_1,\dots,x_n$ over k. We shall say that Δ is linearized (in the coordinates $x_1,\dots,x_n$) if V is invariant with respect to Δ. In this case one can consider the restriction of Δ to V; this is a linear operator $\Delta|_V$ on V which completely defines Δ. The k-derivation Δ is locally nilpotent iff $\Delta|_V$ is nilpotent. If $\Delta|_V$ is nilpotent we can, without loss of generality, assume that $x_1,\dots,x_n$ is a Jordan basis for $\Delta|_V$, i.e.

$$\Delta x_i = \begin{cases} x_{i+1} & \text{for } i \neq p_1,\dots,p_s, \\ 0 & \text{for } i = p_1,\dots,p_s \end{cases} \tag{$*$}$$

for a certain set of integers $1 \leqslant p_1 < \dots < p_s \leqslant n$. Therefore with each such set of integers $p_1,\dots,p_s$ we can associate a locally nilpotent k-derivation of F_n defined by ($*$); we denote this derivation by ${}_n\Delta_{p_1\dots p_s}$.

Another way to construct (a lot of) locally nilpotent k-derivations of F_n is as follows. Let D be such a k-derivation and $h \in F_n$ be one of its invariants, i.e. $Dh = 0$. Then it is easy to see by induction that $(hD)^m = h^m D^m$ for an arbitrary integer $m \geqslant 0$. Hence hD is also a locally nilpotent k-derivation of F_n and

$$\{\exp thD = \sum_{m \geqslant 0} \frac{t^m}{m!} h^m D^m \mid t \in k\} \quad \text{is a one parameter unipotent algebraic}$$

subgroup of Aut F_n.

<u>LEMMA 2. Consider a nonzero linearized k-derivation</u> $\Delta = {}_n\Delta_{p_1\dots p_s}$ <u>of</u> F_n, <u>a non-constant invariant</u> h <u>of</u> Δ <u>and the k-derivation</u> $D = h\Delta$. <u>Then for each</u> $t \neq 0$ <u>the hypersurface</u> $\Gamma_h = \{a \in A^n | h(a) = 0\}$ <u>is a union of certain irreducible components of the variety</u> Fix σ_{tD}.

<u>PROOF</u>. Let i be an integer, $1 \leqslant i \leqslant n$. If i is equal to one of $p_1,\dots,p_s$ then $(\exp tD)x_i = x_i$; if not - and, say, $p_{r-1} < i < p_r$ - then

$$(\exp tD)x_i = \sum_{j=0}^{p_r - i} \frac{t^j}{j!} h^j x_{i+j}.$$

It follows from $\Delta \neq 0$ that there exists at least one i of the second kind. But Fix σ_{tD} is defined by the system of equations $-x_i + (\exp tD)x_i = 0$, $i = 1,\dots,n$.

Hence we have that: 1) h divides each polynomial $(\exp tD)x_i - x_i$, $i = 1,\dots,n$; 2) if $t \neq o$ then at least one of these polynomials is not equal to zero. It follows from 1) that $\Gamma_h \subset \mathrm{Fix}\,\sigma_{tD}$, and from 2) that $\dim \mathrm{Fix}\,\sigma_{tD} \leqslant n-1$. The assertion of the lemma follows now from the fact that each irreducible component of Γ_h has dimension n-1. □

REMARK. It follows from the proof that $\mathrm{Fix}\,\sigma_{tD}$, $t \neq 0$, is the union of Γ_h and of several linear subspaces which are defined by the vanishing of some x_i's.

6. Since the union of some irreducible components of a cylindrical variety is itself cylindrical, Lemmas 1 and 2 imply a way to construct non-triangular actions of $\mathbb{G}_a$ on $\mathbb{A}^n$: if (using the notations of Lemma 2) the hypersurface Γ_h is not cylindrical then $t \mapsto \delta_{tD}$ is an action of such type. Therefore we come to the problem: how can one construct such invariants h that the hypersurface Γ_h is not cylindrical ? I do not know a general criterion for Γ_h to be cylindrical; apparently, "in general" Γ_h is not cylindrical. Nevertheless, having in mind the furnishing of examples of non-triangular actions of $\mathbb{G}_a$ on $\mathbb{A}^n$ the following observation will be sufficient for us : if h is a nondegerate quadratic form in $x_1,\dots,x_n$ then Γ_h is not a cylindrical hypersurface (indeed, such Γ_h has only one singular point; on the other hand it is clear that the dimension of the singular locus of a singular cylindrical variety is positive). There are in principle no difficulties in solving when ${}_n\Delta_{p_1\dots p_s}$ has a nondegenerate quadratic invariant. We shall show that one can furnish the desired examples in this way for an arbitrary n.

It is convenient to use some facts of the representation theory of sl_2. Let L be a finite dimensional sl_2-module. This module is selfdual, [7], hence there exists a nondegenerate sl_2-invariant in the symmetric square of the sl_2-module $L \oplus L$. On the other hand it is known, [7], that if L is simple then there exists a non-zero sl_2-invariant (automatically nondegenerate) in the symmetric square of L iff dim L is odd.

Let now Δ be a linearized locally nilpotent k-derivation of F_n. The operator $\Delta|_V$ being nilpotent, it follows from the Jacobson - Morozov theorem that one can include $\Delta|_V$ into a sl_2-triple. So one has a structure of sl_2-module on V and it is clear that each sl_2-invariant in the symmetric algebra of V is also an invariant of Δ. This sl_2-module is simple if $\Delta = {}_n\Delta_n$ and is the sum of two isomorphic simple sl_2-modules if n is even and $\Delta = {}_n\Delta_{n/2,n}$. It follows from this that Δ definitely has a nondegenerate quadratic invariant h if either n is odd and $\Delta = {}_n\Delta_n$ or n is even and $\Delta = {}_n\Delta_{n/2,n}$. It is not difficult to point out this invariant explicitly:

$$h = \sum_{i=1}^{d} (-1)^i x_i x_{n+1-i},$$ where d = n if n is odd and d = n/2 if n is even.

Therefore the above proves the following

THEOREM. *The action of* $\mathbb{G}_a$ *on* $\mathbb{A}^n$, $t \mapsto (f_1,\ldots,f_n)$, *given by the formulas*

$$f_s = \sum_{i=0}^{d-s} \frac{t^i}{i!} x_{i+s} h^i, \qquad h = \sum_{i=1}^{r} (-1)^i x_i x_{n+1-i}$$

where $d = r = n$ *if* n *is odd, and* $d = n/2$ *for* $1 \leq s \leq n/2$, $d = n$ *for* $n/2 < s \leq n$, $r = n/2$ *if* n *is even, is non-triangular.*

It is easy to see that for $n = 3$ the action given in this theorem is conjugate by the automorphism $(-x_3/2, x_2, x_1)$ with the action of $\mathbb{G}_a$ on $\mathbb{A}^3$ given in the example of H.Bass [1].

REFERENCES

[1] H.Bass, A non-triangular action of $\mathbb{G}_a$ on $\mathbb{A}^3$, Journal of Pure and Applied Algebra 33, 1984, 1 - 5.

[2] I.R.Shafarevich, On some infinite dimensional groups, Rendiconti di Matematica e della sue applicazioni, Ser.5, Vol.25, 1966, 208 - 212.

[3] D.Wright, Abelian subgroups of $\mathrm{Aut}_k(k[X,Y])$ and applications to actions on the affine plane, Illinois J.Math. 23, 1979, 579 - 634.

[4] R.Rentschler, Operations du groupe additif sur le plan affine, C.R.Acad.Sci. Paris, Series A, t. 267, 384 - 387.

[5] T. Kambayashi, Automorphism group of a polynomial ring and algebraic group action on an affine space, J.Algebra 60, 1979, 439 - 451.

[6] M.Nagata, On automorphism group of k[x,y], Lectures in Math., Kyoto Univ. n.5, 1972, Kinokuniya - Tokio.

[7] T.A.Springer, Invariant Theory, Lect.Notes Math. v. 585, 1977.

NORMALITY OF G-STABLE SUBVARIETIES OF A SEMISIMPLE LIE ALGEBRA

R. W. Richardson

Department of Mathematics
Research School of Physical Sciences
Australian National University
Canberra ACT
Australia

To T. A. Springer

§0. Introduction

Let $\underline{g}$ be a semisimple Lie algebra over an algebraically closed field k of characteristic zero and let G be the adjoint group of $\underline{g}$. Let $\underline{t}$ be a Cartan subalgebra of $\underline{g}$ and let W be the Weyl group of $\underline{g}$ with respect to $\underline{t}$. Let X be a closed G-stable subvariety of $\underline{g}$ and let $k[X]^G$ denote the algebra of G-invariant regular functions on X . Let D be an irreducible component of the intersection $X \cap \underline{t}$ and let $W_0 = N_W(D)/Z_W(D)$. For the purposes of this Introduction, we assume that D is a normal variety. Let $k[\underline{t}]^W$ (resp. $k[D]^{W_0}$) denote the k-algebra of W-invariant polynomial functions (resp. W_0-invariant regular functions) on $\underline{t}$ (resp. D). In this paper we will prove the following result, which gives an elementary necessary condition for X to be a normal variety:

Let the notation be as above. Then the following two conditions are equivalent:
(N1) *the homomorphism* $k[\underline{t}]^W \to k[D]^{W_0}$ *given by restriction is surjective; and*
(N2) $k[X]^G$ *is an integrally closed* k-*algebra. In particular, if* X *is a normal variety then condition* (N1) *holds.*

The condition (N1) is our elementary necessary condition for the normality of X . This seems to be a very useful condition since, in a number of concrete examples, it can be easily checked by using the detailed information available on Weyl group invariants.

We were led to formulate and prove the above result by the following question of De Concini and Procesi [10, p.8]: *Let* $\theta: \underline{g} \to \underline{g}$ *be an involutive automorphism of* $\underline{g}$ *and let* $\underline{p}$ *denote the* -1 *eigenspace of* θ *on* $\underline{g}$. *Let* Z *denote the closure of the orbit* $G\cdot\underline{p}$. *Is* Z *a normal variety?* In this case, one can choose the Cartan subalgebra $\underline{t}$ such that $\underline{p} \cap \underline{t} = \underline{a}$ is an irreducible component of $Z \cap \underline{t}$; $\underline{a}$ is a "Cartan subspace" of $\underline{p}$ and $W_0 = N_W(\underline{a})/Z_W(\underline{a})$ is the "little Weyl group". It is known that $k[\underline{t}]^W$ and $k[\underline{a}]^{W_0}$ are graded polynomial algebras and one has explicit information on the degrees of the homogeneous generators of these polynomial algebras. Using this information and condition (N1), we give several examples of pairs $(\underline{g},\theta)$ such that the corresponding variety Z is not a normal variety, thus giving a negative answer to De Concini and Procesi's question. In all of our examples, $\underline{g}$ is an exceptional simple Lie algebra of type E_ℓ .

We also consider the situation in which X is the closure of a "decomposition class" ("Zerlegungsklasse" in the terminology of Borho and Kraft [3]) in $\underline{g}$. In this case, easy computations allow us to show that, in a large number of cases, X is not a normal variety. A number of examples are given in §7 - §9. We mention one particular family of examples: let $\underline{g}$ be a simple Lie algebra of type A_ℓ $(\ell>2)$, D_ℓ , or E_ℓ and let X be either (i) the closure of the subregular sheet of $\underline{g}$ or (ii) the complement of the set of regular semisimple elements of $\underline{g}$; then X is not a normal variety.

In §5 we prove a theorem which states that, for certain G-stable cones X in $\underline{g}$, condition (N1) is a necessary and sufficient condition for the normality of X .

§1. Preliminaries

Our basic reference for algebraic groups is [1] and our basic reference for algebraic geometry is [8]. All algebraic varieties are taken over an algebraically closed field k of characteristic zero.

Let G be a group and let X be a G-set. If $g \in G$ and $x \in X$, then $g\cdot x$ denotes the result of g acting on x , $G\cdot x$ is the G-orbit of x and G_x is the stabilizer, or isotropy subgroup, of G at x . Let Y be a subset of X . Then $G\cdot Y$ is the G-orbit of Y . The subgroup $Z_G(Y) = \{g \in G \mid g\cdot y = y(y \in Y)\}$ is the *centralizer* of Y in G and $N_G(Y) = \{g \in G \mid g\cdot Y = Y\}$ is the *normalizer* of Y in G; $Z_G(Y)$ is a normal subgroup of $N_G(Y)$.

If the (affine) algebraic group G acts morphically on the affine algebraic variety X , then we say that X is an *affine G-variety*. Let G be a linearly reductive algebraic group and let X be an affine G-variety. By Hilbert's theorem, the algebra $k[X]^G$ of G-invariant regular functions on X is a finitely generated k-algebra. We let X/G denote the affine algebraic variety such that $k[X/G] = k[X]^G$ and let $\pi_X : X \to X/G$ be the morphism of algebraic varieties corresponding to the inclusion homomorphism $k[X]^G \to k[X]$; we say that X/G is the "quotient" of X by G and π_X is the "quotient morphism". If reference to G is necessary, we write $\pi_{X,G}$ instead of π_X .

Let G be a linearly reductive algebraic group and let X be an affine G-variety.

Then the following results are known:

1.1. *π_X is a surjective map and each fibre $\pi_X^{-1}(y)$, $y \in X/G$, contains a unique closed G-orbit.*

1.2. *Let Y be a closed G-stable subset of X. Then $\pi_X(Y)$ is closed in X/G and the homomorphism $k[X]^G \to k[Y]^G$ given by restriction is surjective. The restriction of π_X to Y induces an isomorphism of Y/G onto $\pi_X(Y)$.*

1.3. *Assume that G is finite. Then π_X is a finite morphism and, for every $x \in X$, the fibre $\pi^{-1}(\pi(x))$ is just the orbit $G \cdot x$.*

It follows from 1.1 that π_X determines a bijection from the set of closed G-orbits on X to the points of X/G. If G is finite, then every G-orbit is closed and the points of X/G correspond to the orbits of G on X. The proof of 1.2 uses the Reynold's operator and requires the hypothesis that characteristic(k) = 0.

Let X be an irreducible affine G-variety and let Y be a subvariety of X. We let $m = \sup_{x \in Y} \dim G \cdot x$ and we set $Y^{reg} = \{y \in Y \mid \dim G \cdot y = m\}$. Then Y^{reg} is a non-empty relatively open subset of Y.

If $\underline{a}$ is a Lie subalgebra of a Lie algebra $\underline{g}$ (resp. $x \in \underline{g}$), then $\underline{z}_{\underline{g}}(\underline{a})$ (resp. $\underline{z}_{\underline{g}}(x)$) denotes the centralizer of $\underline{a}$ (resp. x) in $\underline{g}$. We sometimes write $\underline{g}^{\underline{a}}$ (resp. $\underline{g}^x$) instead of $\underline{z}_{\underline{g}}(\underline{a})$ (resp. $\underline{z}_{\underline{g}}(x)$).

The following notation will be used for the rest of the paper: $\underline{g}$ will always denote a semisimple Lie algebra with adjoint group G; T is a maximal torus of G and $\underline{t}$ = Lie (T) is the corresponding Cartan subalgebra of $\underline{g}$; $W = N_G(T)/T$ is the Weyl group of G with respect to T; R is the set of roots of $\underline{g}$ with respect to $\underline{t}$ and B is a base of the root system R; if $\underline{g}$ is simple then we label the roots in B as in Bourbaki [4, Planches I-IX].

§2. Action of a finite group on an affine variety

In this section H denotes a finite group and V is an irreducible affine H-variety. We let $\pi = \pi_V : V \to V/H$. First we prove an elementary lemma.

Lemma 2.1. *Let E be a closed H-stable subvariety of V such that $\pi(E)$ is a (closed) irreducible subvariety of V/H. Then H acts transitively on the set of irreducible components of E.*

Proof. By 1.3, π is a finite morphism and $E = \pi^{-1}(\pi(E))$. In particular, $\pi(E)$ is closed in V/H. Let $E_1, \ldots, E_r$ be the irreducible components of E. Then $\pi(E)$ is the union of the closed irreducible subsets $\pi(E_i)$, $i = 1, \ldots, r$. Since $\pi(E)$

is irreducible, we see that $\pi(E) = \pi(E_j)$ for some index j. Hence each H-orbit on E meets E_j. It follows easily from this that the irreducible components of E are just the translates $h \cdot E_j$, $h \in H$.

The following lemma is the key to the proof of our main theorem:

Lemma 2.2. *Let* D *be a closed irreducible subvariety of* V *and let* $K = N_H(D)/Z_H(D)$. *Consider the following two conditions on* D: (i) *the homomorphism* $\varphi : k[V]^H \to k[D]^K$ *given by restriction is surjective; and* (ii) $\pi(D)$ *is a normal subvariety of* V/H. *Then condition* (ii) *implies condition* (i). *If* D *is a normal variety, then conditions* (i) *and* (ii) *are equivalent.*

Proof. Let $\tau : D/K \to \pi(D)$ be the surjective morphism determined by the restriction of π to D and let $i : \pi(D) \to V/H$ be the inclusion map. The homomorphism φ admits the factorization

$$K[V]^H \xrightarrow{i^*} k[\pi(D)] \xrightarrow{\tau^*} k[D]^K$$

where i^* and τ^* are the comorphisms of i and τ. Now i^* is surjective since $\pi(D)$ is closed in V/H and τ^* is injective since τ is dominant. Thus φ is surjective if and only if τ^* is an isomorphism. Consequently we have proved:

2.2.1. *Condition* (i) *holds if and only if* $\tau: D/K \to \pi(D)$ *is an isomorphism of varieties.*

We need the following result:

2.2.2. τ *is a birational morphism.*

Proof. Let $E = H \cdot D = \pi^{-1}(\pi(D))$. Then E is a closed H-stable subvariety of D and $\pi(D) = \pi(E)$. Let E_0 be the set of all points x of E such that x belongs to exactly one irreducible component of E. Let $D_0 = D \cap E_0$. Then the following results are immediate: (a) E_0 is a dense, open H-stable subset of E; (b) D_0 is a dense, open K-stable subset of D; (c) $\pi(D_0) = \pi(E_0)$; (d) $\pi(E_0)$ is a (dense) open subset of $\pi(D) = \pi(E)$; and (e) $\pi_D(D_0)$ is a (dense) open subset of D/K. Now $\pi(D_0) = \tau(\pi_D(D_0))$. If x and y are points of D_0 such that $\pi(x) = \pi(y)$, then it follows immediately from the definition of D_0 that $y \in K \cdot x$, hence that $\pi_D(x) = \pi_D(y)$. But this shows that τ maps the open subset $\pi_D(D_0)$ of D/K bijectively onto the open subset $\pi(E_0)$ of $\pi(E)$. Since we are in characteristic zero, this implies that τ is birational. This proves 2.2.2.

Assume now that $\pi(D)$ is a normal variety. It is clear that all fibres of τ are finite. Therefore it follows from Zariski's Main Theorem [8, p. 137, Cor. 2] that τ is an isomorphism of varieties. Thus condition (ii) implies condition (i).

Assume now that D is normal and that condition (i) holds. Then D/K is normal and τ is an isomorphism of varieties. Thus condition (ii) holds. This proves Lemma 2.2.

§3. Proof of the necessary condition for normality

Let $\pi : \underline{g} \to \underline{g}/G$ and $\pi_1 : \underline{t} \to \underline{t}/W$ denote the quotient morphisms. The following theorem is the main result of this paper:

Theorem A. *Let* X *be a closed irreducible* G*-stable subvariety of* $\underline{g}$ *, let* D *be an irreducible component of the intersection* $X \cap \underline{t}$ *and let* $W_0 = N_W(D)/Z_W(D)$ *. Consider the following three conditions:*

(N1) *The homomorphism* $k[\underline{t}]^W \to k[D]^{W_0}$ *given by restriction is surjective.*

(N2) $k[X]^G$ *is integrally closed.*

(N3) $\pi_1(D)$ *is a normal subvariety of* $\underline{t}/W$ *.*

Then (N2) *is equivalent to* (N3) *and* (N3) *implies* (N1). *If, in addition,* D *is a normal variety, then* (N1) *implies* (N3) *and the three conditions are equivalent.*

Proof. The following two results are well known:

3.1. *Let* $x \in \underline{g}$ *. Then the orbit* $G \cdot x$ *is closed in* $\underline{g}$ *if and only if* $G \cdot x$ *meets* $\underline{t}$ *. If* $x \in \underline{t}$ *, then* $G \cdot x \cap \underline{t} = W \cdot x$ *.*

3.2. *The homomorphism* $k[\underline{g}]^G \to k[\underline{t}]^W$ *given by restriction is an isomorphism. Hence the corresponding morphism* $\mu : \underline{t}/W \to \underline{g}/G$ *of affine varieties is an isomorphism.*

Let $E = X \cap \underline{t}$.

Lemma 3.3. $\pi(X) = \pi(E)$.

Proof. This follows from 3.1 and 1.1.

Now we can prove Theorem A. It follows immediately from Lemma 2.2 that condition (N3) of Theorem A implies condition (N1). It also follows that, if D is a normal variety, then conditions (N1) and (N3) are equivalent. By Lemma 2.1, $\pi_1(D) = \pi_1(E)$ and, by 1.2, $\pi(X) \cong X/G$. Thus $\pi_1(D) \cong X/G$. Therefore we see that (N2) and (N3) are equivalent.

§4. A lemma on graded polynomial algebras.

A graded commutative k-algebra $A = \oplus_{n \geq 0} A_n$ with $A_0 = k$ is a *graded polynomial algebra* if there exists a finite family $(a_1, \ldots, a_q)$ of homogeneous elements of A which are algebraically independent and generate A.

Let A be a graded polynomial algebra. Then a family $(a_1, \ldots, a_q)$ as above is called a *Hilbert basis* of A. Let $A^+ = \oplus_{n \geq 0} A_n$. Then a family $(b_1, \ldots, b_r)$ of homogeneous elements of A^+ is a Hilbert basis of A if and only if the images of $b_1, \ldots, b_r$ in $A^+/(A^+)^2$ form a basis of $A^+/(A^+)^2$.

Lemma 4.1. *Let* $\varphi : A \to B$ *be a surjective homomorphism of degree zero of graded polynomial algebras. Then there exists a Hilbert basis* $(a_1, \ldots, a_q)$ *of* A *and an integer* $s \leq q$ *such that* $(\varphi(a_1), \ldots, \varphi(a_s))$ *is a Hilbert basis of* B *and* $\varphi(a_j) = 0$, $j > s$.

Proof. Let $J = \mathrm{Kernel}(\varphi)$ and let $I = A^+$. Let E be a graded subspace of J such that $I^2 + J = I^2 \oplus E$ and let D be a graded subspace of I such that $I = D \oplus I^2 \oplus E$. Let $(a_1, \ldots, a_s)$ (resp. $(a_{s+1}, \ldots, a_q)$) be a basis of D (resp. E) which consists of homogeneous elements. It follows from the characterization of Hilbert bases that the family $(a_1, \ldots, a_q)$ satisfies the conclusions of Lemma 4.1.

§5. A necessary and sufficient condition for normality

Let $\pi = \pi_{\underline{g}}$. The following theorem gives a necessary and sufficient condition for the normality of certain G-stable cones in $\underline{g}$.

Theorem B. *Let* $\underline{c}$ *be a linear subspace of* $\underline{t}$, *let* $W_0 = N_W(\underline{c})/Z_W(\underline{c})$ *and let* X *denote the closed irreducible* G-*stable cone* $\pi^{-1}(\pi(\underline{c}))$. *Assume that the algebra of invariants* $k[\underline{c}]^{W_0}$ *is a graded polynomial algebra. Then the following three conditions are equivalent:* (N1) *the homomorphism* $k[\underline{t}]^W \to k[\underline{c}]^{W_0}$ *given by restriction is surjective;* (N3) $\pi(\underline{c})$ *is a normal variety; and* (N4) X *is a normal Cohen-Macaulay variety.*

Proof. Assume that condition (N1) holds. Let $\eta : \underline{c}/W_0 \to \underline{t}/W$ and $\mu : \underline{t}/W \to \underline{g}/G$ be the morphisms determined by the inclusion maps $\underline{c} \to \underline{t}$ and $\underline{t} \to \underline{g}$ and let $\nu : \underline{c}/W_0 \to \underline{g}/G$ be the composition of μ and η. By 3.2, μ is an isomorphism of varieties. It is known that $k[\underline{g}]^G$ is a graded polynomial algebra. Thus the comorphism $\nu^* : k[\underline{g}]^G \to k[\underline{c}]^{W_0}$ is a surjective homorphism of graded polynomial algebras. Let $s = \dim \underline{c} = \dim \underline{c}/W_0$. By Lemma 4.1 there exists a family $P_1, \ldots, P_\ell$ of algebraically independent homogeneous elements of $k[\underline{g}]^G$ which generate $k[\underline{g}]^G$ and which satisfy the following two conditions: (i) $\nu^*(P_1), \ldots, \nu^*(P_s)$ are algebraically independent and generate $k[\underline{c}]^{W_0}$; and (ii) $\nu^*(P_i) = 0$, $i = s+1, \ldots, \ell$. Let $P : \underline{g} \to k^\ell$ be defined by $P(x) = (P_1(x), \ldots, P_\ell(x))$. Then P is constant on G-orbits and determines an isomorphism $\tau : \underline{g}/G \to k^\ell$. It follows easily from (i) and (ii) above that

$$X = \{x \in \underline{g} \mid P_i(x) = 0 \quad (i = s+1, \ldots, \ell)\} . \tag{5.1}$$

We need the following results of Kostant [12]:

5.2. *Let* $a \in \underline{g}/G$. *Then the fibre* $\pi^{-1}(a)$ *is an irreducible normal subvariety of* $\underline{g}$ *of codimension* ℓ. *There exists a dense* G-*orbit* O *in* $\pi^{-1}(a)$ *and* $O = \underline{g}^{reg} \cap \pi^{-1}(a)$. *Let* C *be the complement of* O *in* $\pi^{-1}(a)$. *Then the codimension of* C *in* O *is at least two. Let* $x \in \underline{g}^{reg}$. *Then the differentials* $(dP_i)_x$, $i = 1, \ldots, \ell$ *are linearly independent.*

We also need the following elementary lemma:

Lemma 5.3. X *is an irreducible subvariety of* $\underline{g}$.

Proof. Let $X_1, \ldots, X_r$ be the irreducible components of X. Then each X_i is closed irreducible G-stable subvariety of $\underline{g}$. By 1.2, each $\pi(X_i)$ is a closed irreducible subset of $\pi(X) = \pi(\underline{c})$. Since $\pi(\underline{c})$ is irreducible, it follows that $\pi(X_i) = \pi(\underline{c})$ for at least one index i. After renumbering, we may assume that $\pi(X_i) = \pi(\underline{c})$ for $i = 1, \ldots, m$ and that $\pi(X_i) \neq \pi(\underline{c})$ for $i = m+1, \ldots, r$. Let d denote the common dimension of the fibres of π. For $i = 1, \ldots, m$, let $d_i = \dim X_i$ and let $\pi_i : X_i \to \pi(\underline{c})$ denote the restriction of π to X_i. For each $i = 1, \ldots, m$, the generic fibre of π_i is of dimension $d_i - s \leqq d$.

I claim that there exists an index $j \in \{1, \ldots, m\}$ such that $d_i = d + s$. If not, then for each $i = 1, \ldots, m$, there exists a non-empty open subset U_i of $\pi(\underline{c})$ such that $\dim \pi_i^{-1}(a) < d$ for $a \in U_i$ and such that $U_i \cap \pi(X_p) = \emptyset$ for $p > m$. Let $U = \bigcap_i U_i$. Then U is a non-empty open subset of $\pi(\underline{c})$. If $a \in U$, then $\pi^{-1}(a) = \bigcup_i \pi_i^{-1}(a)$ is of dimension less than d, which gives a contradiction.

Thus we may assume that $d_1 = \dim X_1 = d+s$. Since π_1 is surjective and the and the generic fibre of π_1 is of dimension d, each fibre of π_1 has dimension at least d. Let $a \in \pi(\underline{c})$. Then $\dim \pi_1^{-1}(a) = d$ and $\pi_1^{-1}(a)$ is contained in $\pi^{-1}(a)$, which is irreducible and of dimension d. Thus $\pi^{-1}(a) = \pi_1^{-1}(a) \subset X_1$. Thus $X = \pi^{-1}(\pi(\underline{c})) = X_1$, and X is irreducible. This proves Lemma 5.3.

Since X is irreducible,

$$I(X) = \{f \in k[\underline{g}] \mid f(x) = 0 \ (x \in X)\}$$

is a prime ideal. It follows from 5.2(a) that $X^{reg} = X \cap \underline{g}^{reg}$ is a dense open subset of X, that the complement of X^{reg} in X has codimension $\geqq 2$ in X and that each fibre of $\pi|_X$ meets X^{reg}. If $x \in X^{reg}$, then by 5.2(b) the differentials $(dP_i)_x$, $i = s+1, \ldots, \ell$, are linearly independent. It follows from a standard result in commutative algebra (see e.g. [17, p. 345]) that the ideal $I(X)$ is generated by $P_{s+1}, \ldots, P_\ell$. Since X is of codimension $\ell - s$ in $\underline{g}$, the sequence $P_{s+1}, \ldots, P_\ell$ is an R-sequence in $k[\underline{g}]$. Therefore $k[\underline{g}]/(P_{s+1}, \ldots, P_\ell) = k[\underline{g}]/I(X) \cong k[X]$ is a Cohen-Macaulay algebra and X is an irreducible Cohen-Macaulay variety.

To show that X is normal, it will suffice to show that X is non-singular in codimension one. It is an easy consequence of 5.2(b) and (5.1) that each point of X^{reg} is a smooth point of X and, as noted above, the complement of X^{reg} in X has codimension in X at least two. Thus the set of singular points of X is of codimension at least two. Therefore X is a normal variety. Consequently we have shown that condition (N1) of Theorem B implies condition (N4). The other conclusions of Theorem B follow from Theorem A.

6. Application to the De Concini-Procesi question

Let $\theta : \underline{g} \to \underline{g}$ be an involutive automorphism and let $\underline{p}$ denote the -1 eigenspace of θ on $\underline{g}$. A linear subspace $\underline{a}$ of $\underline{p}$ is a *Cartan subspace* of $\underline{p}$ if (i) $\underline{a}$ is a maximal abelian subalgebra of $\underline{p}$ and (ii) all elements of $\underline{a}$ are semisimple. Let $\underline{a}$ be a Cartan subspace of $\underline{p}$ and let $\underline{t}$ be a Cartan subalgebra of $\underline{g}$ which contains $\underline{a}$. Let $W_0 = N_G(\underline{a})/Z_G(\underline{a})$; the group W_0 is often called "the little Weyl group". Let $K = \{g \in G \mid g\theta = \theta g\}$. Then $\underline{p}$ is a K-stable subspace of $\underline{g}$. Let Z denote the closure of the orbit $G\cdot\underline{p}$ in $\underline{g}$. The following results are well known [9, 13, 22]:

6.1.(a) $K\cdot\underline{a}$ *is dense in* $\underline{p}$. *Hence* $G\cdot\underline{a}$ *is dense in* Z. (b) *Let* $W_1 = N_W(\underline{a})/Z_W(\underline{a})$. *Then* W_1 *is canonically isomorphic to* W_0. (c) W_0 *is a finite subgroup of* $GL(\underline{a})$ *generated by reflections. Hence* $k[\underline{a}]^{W_0}$ *is graded polynomial algebra.*

Let $\pi : \underline{g} \to \underline{g}/G$ and $\pi_1 : \underline{t} \to \underline{t}/W$ be the quotient morphisms. Let $\mu : \underline{t}/W \to \underline{g}/G$ be the isomorphism of varieties determined by the inclusion $\underline{t} \to \underline{g}$.

Lemma 6.2. $\pi(Z) = \pi(\underline{a})$.

Proof. Clearly $\pi(\underline{a}) \subset \pi(Z)$. Since π_1 is a finite morphism, $\pi_1(\underline{a})$ is closed in $\underline{t}/W$ and consequently $\pi(\underline{a}) = \mu(\pi_1(\underline{a}))$ is closed in $\underline{g}/G$. Now $\pi(G\cdot\underline{a}) = \pi(\underline{a})$ and therefore $G\cdot\underline{a} \subset \pi^{-1}(\pi(\underline{a}))$. Since $G\cdot\underline{a}$ is dense in Z , we have $Z \subset \pi^{-1}(\pi(\underline{a}))$. Therefore $\pi(Z) \subset \pi(\underline{a})$.

Lemma 6.3. $\underline{a}$ *is an irreducible component of* $Z \cap \underline{t}$.

Proof. Clearly $\underline{a} \subset Z \cap \underline{t}$. Let $\underline{c}$ be an irreducible component of $Z \cap \underline{t}$ which contains $\underline{a}$. Then

$$\dim \underline{c} = \dim \pi_1(\underline{c}) = \dim \pi(\underline{c}) \leq \dim \pi(Z) = \dim \pi(\underline{a}) .$$

Since $\dim \underline{a} = \dim \pi_1(\underline{a}) = \dim \pi(\underline{a})$, we see that $\dim \underline{c} = \dim \underline{a}$, thus that $\underline{c} = \underline{a}$. This proves 6.3.

Now if Z is a normal variety, it follows from Theorem A that the restriction homomorphism $k[\underline{t}] \to k[\underline{a}]$ maps $k[\underline{t}]^W$ onto $k[\underline{a}]^{W_0}$. The invariant algebras $k[\underline{t}]^W$ and $k[\underline{a}]^{W_0}$ are graded polynomial algebras. For $n \geq 0$, let $k[\underline{t}]^W_n$ (resp. $k[\underline{a}]^{W_0}_n$) denote the homogeneous component of $k[\underline{t}]^W$ (resp. $k[\underline{a}]^{W_0}$) of degree n . Then as an easy consequence of the remarks above, we have:

Lemma 6.4. *If* Z *is a normal variety, then*

$$\dim k[\underline{t}]^W_n \geq \dim k[\underline{a}]^{W_0}_n \quad \textit{for every } n . \tag{6.4.1}$$

A classification of (conjugacy classes of) involutive automorphisms of simple Lie algebras is given in Helgason's book [9] (see in particular the tables on pp. 518-520). For each pair $(\underline{g},\theta)$, he also gives the type of the root system corresponding to $(\underline{a},W_0)$. Thus for each pair $(\underline{g},\theta)$, with $\underline{g}$ simple, we have precise information on $\dim k[\underline{t}]^W_n$ and $\dim k[\underline{a}]^{W_0}_n$. For exactly four classes of pairs $(\underline{g},\theta)$ the condition (6.4.1) above is not satisfied. These are:

(a) $(\underline{g},\theta)$ of type EIII. Here $(\underline{t},W)$ is of type E_6 and $(\underline{a},W_0)$ is of type B_2 . Thus $\dim k[\underline{t}]^W_4 = 1$ and $\dim k[\underline{a}]^{W_0}_4 = 2$.

(b) $(\underline{g},\theta)$ of type EIV. Then $(\underline{t},W)$ is of type E_6 and $(\underline{a},W_0)$ is of type A_2 . Thus $\dim k[\underline{t}]^W_3 = 0$ and $\dim k[\underline{a}]^{W_0}_3$.

(c) $(\underline{g},\theta)$ of type EVII. Then $(\underline{t},W)$ is of type E_7 and $(\underline{a},W_0)$ is of type C_3. Therefore $\dim k[\underline{t}]_4^W = 1$ and $\dim k[\underline{a}]_4^{W_0} = 2$.

(d) $(\underline{g},\theta)$ of type EIX. Here $(\underline{t},W)$ is of type E_8 and $(\underline{a},W_0)$ is of type F_4. In this case $\dim k[\underline{t}]_6^W = 1$ and $\dim k[\underline{a}]_6^{W_0} = 2$.

Thus we see that if $(\underline{g},\theta)$ is of type EIII, EIV, EVII, or EIX, the corresponding variety Z is not normal.

For all of the other classes of involution of simple Lie algebras, the condition (6.4.1) is satisfied. In these cases, one can probably show that $k[\underline{t}]^W \to k[\underline{a}]^{W_0}$ is surjective and hence, by Theorem A, that Z/G is a normal variety. However, we have not checked the details. In any case, the normality of Z/G does not directly imply the normality of Z.

§7. Decomposition classes and sheets

The concept of a "decomposition class" ("Zerlegungsklasse") in a semisimple Lie algebra $\underline{g}$ was introduced by Borho and Kraft [3] in their study of "sheets" ("Schichten") in $\underline{g}$. For a very clear and detailed discussion of decomposition classes and sheets in $\underline{g}$, we refer the reader to [2]. Roughly speaking, two elements of $\underline{g}$ are in the same decomposition class if they have "similar" Jordan decompositions. More precisely, we have:

Definition 7.1. Let $x_1 \in \underline{g}$ have Jordan decomposition $x_1 = h_1 + y_1$ (with h_1 semisimple and y_1 nilpotent) and let $x_2 \in \underline{g}$ have Jordan decomposition $x_2 = h_2 + y_2$. Then x_1 and x_2 are in the same decomposition class if there exists $g \in G$ such that, letting $g\cdot x_2 = h_3 + y_3$ be the Jordan decomposition of $g\cdot x_2$, the following conditions hold:

(i) $G_{h_1} = G_{h_3}$; and (ii) if $M = G_{h_1} = G_{h_3}$, then $M\cdot y_1 = M\cdot y_3$.

If $x \in \underline{g}$, we let $\mathcal{D}(x)$ denote the decomposition class of x. The decomposition classes $\mathcal{D}(x)$, $x \in \underline{g}$, give a partition of $\underline{g}$ into disjoint, G-stable locally closed irreducible subvarieties. The set of decomposition classes is finite.

Let $x \in \underline{g}$ have Jordan decomposition $x = h + y$ and let $\underline{z} = \underline{z}(\underline{g}^h)$ be the centre of $\underline{g}^h$; the (commutative) subalgebra $\underline{z}$ is the "double centralizer" subalgebra of h. It is easy to see that

$$\mathcal{D}(x) = G\cdot(\underline{z}^{reg} + M\cdot y) = G\cdot(\underline{z}^{reg} + y) . \qquad (7.2)$$

We let π denote the quotient map $\pi_{\underline{g}}$.

For each root $\alpha \in R$, we let $s_\alpha \in W$ be the reflection corresponding to α. If J is a subset of the base B we set $\underline{z}_J = \{h \in \underline{t} \mid \alpha(h) = 0 \ (\alpha \in J)\}$ and we let W_J be the subgroup of W generated by $\{s_\alpha \mid \alpha \in J\}$. We let $N_J = N_W(\underline{z}_J)$; then N_J is the normalizer of W_J in W. We set $M_J = N_J/W_J$ and consider M_J as a subgroup of $GL(\underline{z}_J)$.

The following standard result characterizes double centralizer subalgebras of semisimple elements in $\underline{g}$.

7.3.(a) *Let* $h \in \underline{g}$ *be semisimple and let* $\underline{z} = \underline{z}(\underline{g}^h)$ *be the double centralizer of* h. *Then there exists* $g \in G$ *and* $J \subset B$ *such that* $g \cdot \underline{z} = \underline{z}_J$. *Hence* $\pi(\underline{z}) = \pi(\underline{z}_J)$.

(b) *Let* $h \in \underline{z}_J^{reg}$. *Then* $\underline{z}_J = \underline{z}(\underline{g}^h)$.

(c) *Let* J *and* K *be subsets of* B. *Then the subalgebras* $\underline{z}_J$ *and* $\underline{z}_K$ *are conjugate under* G *if and only if there exists* $w \in W$ *such that* $w(J) = K$.

If $h \in \underline{g}$ is semisimple and if $\underline{z}$ is the double centralizer of h, then it follows from 7.3(a), 1.3, and 3.2 that $\pi(\underline{z})$ is closed in $\underline{g}/G$.

Lemma 7.4. *Let* $x \in \underline{g}$ *have Jordan decomposition* $x = h + y$, *let* $\underline{z} = \underline{z}(\underline{g}^h)$ *and let* $\mathcal{D} = \mathcal{D}(x)$. *Then* $\pi(\overline{\mathcal{D}}) = \pi(\underline{z})$.

Proof. Since $\pi(\underline{z})$ is closed, this follows from (7.2).

For each $n \geqq 0$, let $\underline{g}^{(n)} = \{x \in \underline{g} \mid \dim G \cdot x = n\}$. The sets $\underline{g}^{(n)}$ are locally closed subsets of $\underline{g}$. A *sheet* in $\underline{g}$ is an irreducible component of some $\underline{g}^{(n)}$. Each sheet S of $\underline{g}$ is a finite union of decomposition classes in $\underline{g}$. In particular, each sheet contains a dense decomposition class. If S is a sheet and $\mathcal{D}$ is the dense decomposition class in S, then clearly $\overline{\mathcal{D}} = \overline{S}$.

We wish to apply the necessary condition for normality given by Theorem A to the closures of decomposition classes in $\underline{g}$. The following proposition is an easy consequence of 7.3, Lemma 7.4 and Theorems A and B.

Proposition 7.5.(a) *Let* $x \in \underline{g}$ *have Jordan decomposition* $x = h + y$, *let* $\underline{z} = \underline{z}(\underline{g}^h)$ *and let* $J \subset B$ *be such that* $\underline{z}$ *is G-conjugate to* $\underline{z}_J$. *If the closure of the decomposition class* $\mathcal{D}(x)$ *is a normal variety, then the following condition holds:*

$(N1)_J$ *the homomorphism* $\rho_J : k[\underline{t}]^W \to k[\underline{z}_J]^{M_J}$ *given by restriction is surjective.*

(b) *Let the notation be as above and assume further that* (i) x *is a regular element of* $\underline{g}$ *and* (ii) $k[\underline{z}_J]^{M_J}$ *is a graded polynomial algebra. Then the closure of* $\mathcal{D}(x)$ *is a normal variety if and only if condition* $(N1)_J$ *above holds.*

In [11], Howlett has given an explicit description of the normalizer N_J of W_J in W and of the representation of $M_J = N_J/W_J$ on $\underline{z}_J$. In most cases, it is an easy matter to check whether condition $(N1)_J$ above holds. If J is a proper non-empty subset of B, it turns out that, in most cases, condition $(N1)_J$ does not hold, so that the closure of the corresponding decomposition classes is not normal. In §8, we check condition $(N1)_J$ for all subsets J of B when $\underline{g}$ is a simple Lie algebra of type A_ℓ, B_ℓ, C_ℓ or D_ℓ and in §9 we check the condition in cases related to the subregular sheet.

Remark 7.6. Perhaps the most interesting case of the closure of a decomposition class is the closure of a nilpotent conjugacy class. In this case, $J = \emptyset$, condition $(N1)_J$ is trivially satisfied and Proposition 7.5 gives no information. A great deal of detailed information on the closures of nilpotent classes in the classical Lie algebras has been obtained by Kraft and Procesi [14,15].

§8. Condition $(N1)_J$ for the classical Lie algebras

In §8 we shall use the results of Howlett [11] without explicit reference.

8.1. *$\underline{g}$ of type A_ℓ.* Let $\underline{g} = \underline{s\ell}_{\ell+1}(k)$, let $\underline{d}$ be the space of diagonal $(\ell+1) \times (\ell+1)$ matrices and let $\underline{t} = \underline{g} \cap \underline{d}$. We shall identify $\underline{d}$ with $k^{\ell+1}$ in the obvious manner. Thus $\underline{t}$ is identified with $\{(x_1, \ldots, x_{\ell+1}) \mid \Sigma x_i = 0\}$. The Weyl group W is identified with the symmetric group $S_{\ell+1}$ acting on $k^{\ell+1}$ by permutation of the coordinates.

Let $J \subset B$. Assume that (the subdiagram of the Dynkin diagram corresponding to) J has n_i components of type A_{d_i-1}, $i = 1, \ldots, s$, and let $n_0 = \ell + 1 - \Sigma_{i=1}^{s} n_i d_i$. Then $M_J = N_J/W_J$ is isomorphic to $S_J = S_{n_0} \times \ldots \times S_{n_s}$.

Each root $\alpha \in R$ can be considered as a linear function on $\underline{d}$. Let $\underline{d}_J = \{x \in \underline{d} \mid \alpha(x) = 0 \ (\alpha \in J)\}$. We will need the following elementary lemma:

Lemma 8.1.1. *Let $\varphi_J : k[\underline{d}]^W \to k[\underline{d}_J]^{M_J}$ be the homomorphism determined by restriction. Then ρ_J is surjective if and only if φ_J is surjective.*

We omit the proof, which is easy.

For each positive integer m, let A_m be the algebra $k[X_1, \ldots, X_m]^{S_m}$ of symmetric polynomials in the indeterminates $X_1, \ldots, X_m$. Then $k[\underline{d}]^W$ is canonically isomorphic (as a graded algebra) to $A_{\ell+1}$. It follows from Howlett's result (or by an easy direct argument) that $k[\underline{d}_J]^{M_J}$ is isomorphic to the tensor product $A_{n_0} \otimes \dots \otimes A_{n_s}$. By comparing the dimensions of the graded components of degree two of $A_{\ell+1}$ and $A_{n_0} \otimes \dots \otimes A_{n_s}$, one sees that if either (i) $s > 1$ or (ii) $s = 1$ and $n_0 > 0$, then φ_J cannot be surjective. If $s = 0$, then $\underline{d}_J = \underline{d}$ and φ_J is surjective. If $s = 1$ and $n_0 = 0$, then J has n_1 components, each of type A_{d_1-1} , where $n_1 d_1 = \ell + 1$. In this case, an easy direct argument shows that φ_J is surjective. Thus we obtain:

Proposition 8.1.2. *Let* $\underline{g}$ *be simple of type* A_ℓ . *Let* J *be a non-empty subset of* B . *Then the condition* $(N1)_J$ *of Proposition 7.5 holds if and only if* J *has* m *connected components, each of type* A_{d-1} , *where* $md = \ell + 1$.

As a consequence of this, we see tha the number of non-empty subsets J of B such that condition $(N1)_J$ holds is equal to the number of divisors of $\ell + 1$.

The following proposition is an amusing consequence of Proposition 8.1.2.

Proposition 8.1.3. *Let* $\underline{g}$ *be simple of type* A_ℓ *and assume that* $\ell + 1$ *is prime. Let* $x \in \underline{g}$. *Then the closure of* $\mathcal{D}(x)$ *is a normal variety if and only if* (i) x *is nilpotent or* (ii) x *is a regular semisimple element.*

Proof. If x is nilpotent, then the decomposition class $\mathcal{D}(x) = \mathcal{D}$ is just $C_{\underline{g}}(x)$, the conjugacy class of x in $\underline{g}$, and it follows from a result of Kraft and Procesi [14] that the closure of $C_{\underline{g}}(x)$ is a normal variety. If x is regular semisimple, then the closure of $\mathcal{D}(x)$ is equal to $\underline{g}$. In all other cases $\pi(\overline{\mathcal{D}}) = \pi(z_J)$, where J is a proper non-empty subset of B , and it follows from Propositions 8.1.2 and 7.5 that $\overline{\mathcal{D}}$ is not a normal variety.

Remark 8.1.4. Let $\underline{g}$ be of type A_ℓ , let $x \in \underline{g}$ be semisimple and let $\mathcal{D} = \mathcal{D}(x)$. Then the set $\overline{\mathcal{D}}^{reg}$ is a "Dixmier sheet" in $\underline{g}$. It has been shown by Peterson [16] that every Dixmier sheet in $\underline{g}$ is non-singular and it is easy to show that the complement of $\overline{\mathcal{D}}^{reg}$ in $\overline{\mathcal{D}}$ is of codimension ≥ 2 . Hence the set $\overline{\mathcal{D}}_{(sing)}$ of singular points of $\overline{\mathcal{D}}$ has codimension ≥ 2 .

8.2. $\underline{g}$ *of type* B_ℓ *and* C_ℓ . For $m > 0$, let $R = R(B_m)$ be a root system of type B_m in an m-dimensional vector space E over k , let $W = W(B_m)$ by the Weyl group and let $\mathcal{B}_m = k[E^*]^W$. Then $\mathcal{B}_m$ is a graded polynomial algebra with algebraically independent homogeneous generators $P_1, \ldots, P_m$, where P_i is of degree $2i$, $i = 1, \ldots, m$.

Now let $\underline{g}$ be a simple Lie algebra of type B_ℓ and let J be a non-empty subset of the basis B of the root system R. Let J have n_i components of type A_{d_i-1}, $i = 1, \ldots, s$, and possibly one component of type B_j (set $j = 0$ if there is no such component). Type B_1 is distinguished from type A_1 by the root length. Set $r = \ell - j - \Sigma_{i=1}^{s} d_i n_i$. Then M_J acts on $k[\underline{z}_J]$ as a reflection group of type $B_{n_1} \times \ldots \times B_{n_s} \times B_r$. Hence $k[\underline{z}_J]^{M_J}$ is isomorphic to the tensor product $\mathcal{B}_{n_1} \otimes \ldots \otimes \mathcal{B}_{n_s} \otimes \mathcal{B}_r$. Let c be the number of non-trivial terms in this tensor product factorization; thus $c = s + 1$ if $r \neq 0$ and $c = s$ if $r = 0$. Then an easy argument shows that $\dim k[\underline{z}_J]_2^{M_J} = c$. In particular if $c > 1$, then ρ_J is not surjective. If $c = 1$, then one can show by a direct argument that ρ_J is surjective.

Exactly the same arguments work if $\underline{g}$ is simple of type C_ℓ. Thus we obtain:

Proposition 8.2.1. *Let* $\underline{g}$ *be simple of type* B_ℓ *(resp.* C_ℓ*) and let* J *be a non--empty subset of* B. *Then* ρ_J *is a surjective homomorphism if and only if one of the following three conditions holds:*

(a) J *has only one component, which is of type* B_j (resp. C_j), $j = 1, \ldots, \ell$;

(b) J *has* m *components, each of which is of type* A_{d-1}, *where* $md = \ell$; *or*

(c) J *has* $m + 1$ *components,* m *of type* A_{d-1} *and one of type* B_j *(resp.* C_j*),* *where* $md + j = \ell$.

As particular cases of Proposition 8.2.1, we record the following results, which we will need later:

8.2.2. *Let* $\underline{g}$ *be a simple Lie algebra of type* B_ℓ *or* C_ℓ, $\ell > 2$. *Let* $B = \{\alpha_1, \ldots, \alpha_\ell\}$, *where the roots are numbered as in* [4].

(a) *Let* $J = \{\alpha_1\}$. *Then* ρ_J *is not surjective.*

Let $J = \{\alpha_\ell\}$. *Then* ρ_J *is surjective and consequently* $\pi^{-1}(\pi(\underline{z}_J))$ *is a normal variety.*

8.2.3. *Let* $\underline{g}$ *be simple of type* B_2 *and let* J *denote either* $\{\alpha_1\}$ *or* $\{\alpha_2\}$. *Then* ρ_J *is surjective and consequently* $\pi^{-1}(\pi(z_J))$ *is a normal variety.*

8.3. $\underline{g}$ *of type* D_ℓ $(\ell \geq 4)$. In this case the situation is slightly more complicated since, in a number of cases, M_J does not act on $\underline{z}_J$ as a group generated by reflections. However an easy argument shows that if J has components of type A_i for more than one value of i, then $\dim k[\underline{z}_J]_2^{M_J} \geq 2$, so that ρ_J cannot be surjective. Since one has a simple description of the generators of $k[\underline{t}]^W$ (see [4, Chap. 6]) the remaining cases can be checked directly. We state the results without proof.

Proposition 8.3.1. *Let* $\underline{g}$ *be simple of type* D_ℓ, $\ell \geq 4$. *Let* J *be a non-empty subset of* B. *Then* ρ_J *is surjective if and only if one of the following two conditions is satisfied:*

(a) J *has* $m+1$ *components* $(m \geq 0)$, *one component of type* D_j $(j > 1)$ *and* m *of type* A_{d-1}, *where* $\ell = md + j$; *or*

(b) J *has* m *components, each of type* A_{d-1}, *where* $\ell = md$ *and* d *is even.*

In the above proposition if $\{\alpha_{\ell-1}, \alpha_\ell\}$ (resp. $\{\alpha_{\ell-2}, \alpha_{\ell-1}, \alpha_\ell\}$) is a component of J, it is considered to be of type D_2 (resp. D_3).

Remark 8.4. For each $m \geq 0$, let $\underline{g}_m = \{x \in \underline{g} \mid \operatorname{rank} \operatorname{ad}(x) \leq m\}$; $\underline{g}_m$ is closed G-stable subvariety of $\underline{g}$. In [10, p. 15], Procesi suggests that the properties of these varieties should be studied. If S is a sheet of $\underline{g}$ and if the G-orbits on S have dimension m, then it is easy to see that the closure of S is an irreducible component of $\underline{g}_m$. Moreover every irreducible component of $\underline{g}_m$ is the closure of some sheet of $\underline{g}$. It follows from Proposition 7.5 and the results of §8 that, in a large number of cases, the closure of a sheet is not normal. Thus we see that the geometric properties of the varieties $\underline{g}_m$ are not as nice as one might have hoped.

§9. Non-normality of the closure of the subregular sheet

Let $\underline{g}$ be simple of rank ℓ. An element $x \in \underline{g}$ is *subregular* if the dimension of the centralizer $\underline{g}^x$ is $\ell + 2$. The irreducible components of the set of subregular elements of $\underline{g}$ are the *subregular sheets* of $\underline{g}$. If all roots of R are of the same length there is only one subregular sheet and if there are two rooth lengths, then there are two subregular sheets. If $\alpha \in B$ and $J = \{\alpha\}$, then the closure of $G \cdot \underline{z}_J$ is the closure of a subregular sheet of $\underline{g}$ (see [19]). If condition $(N1)_J$ of Proposition 7.5 is not satisfied, then this closure is not a normal variety. Using the results of §8 and, for the exceptional groups, the tables of Howlett [11], it is an easy matter to check whether condition $(N1)_J$ is satisfied. The results are as follows:

9.1. (a) *Let* $\underline{g}$ *be simple of type* A_ℓ $(\ell \geq 3)$, B_ℓ $(\ell \geq 3)$, C_ℓ $(\ell \geq 3)$, D_ℓ $(\ell \geq 4)$,

E_6, E_7, E_8 or F_4 and let $J = \{\alpha_1\}$. Then condition $(N1)_J$ of Proposition 7.5 is not satisfied.

(b) *In the following cases, condition $(N1)_J$ is satisfied:*

(i) *rank* $\underline{g} = 2$ *and* $J = \{\alpha_1\}$ *or* $J = \{\alpha_2\}$ *;*

(ii) g *of type* B_ℓ *or* C_ℓ $(\ell \geq 2)$ *and* $J = \{\alpha_\ell\}$.

(c) *If* $\underline{g}$ *is of type* F_4 *, and* $J = \{\alpha_4\}$ *, then* $(N1)_J$ *is not satisfied.*

As a consequence of 9.1, we obtain:

Proposition 9.2.(a) *Let* $\underline{g}$ *be simple of type* A_ℓ $(\ell \geq 3)$ *,* D_ℓ $(\ell \geq 4)$ *,* E_6 *,* E_7 *, or* E_8 *. Then the clsoure of the subregular sheet of* $\underline{g}$ *is not a normal variety.*

(b) *Let* g *be simple of type* B_ℓ *or* C_ℓ *(*$\ell \geq 3$ *in both cases). Then the closure of the subregular sheet corresponding to* $J = \{\alpha_1\}$ *is not a normal variety.*

(c) *Let* $\underline{g}$ *be simple of type* F_4 *and let* S *be a subregular sheet of* $\underline{g}$ *. Then the closure of* S *is not a normal variety.*

Remark 9.3. If $\underline{g}$ is simple and if X is the closure of a subregular sheet in one of the cases not covered by Proposition 9.2 (i.e. $\underline{g}$ of rank two or $\underline{g}$ of type B_ℓ or C_ℓ and X corresponding to $J = \{\alpha_\ell\}$) , then we can conclude from Theorem A that X/G is normal, but not that X is normal. In certain of these cases, X is not a normal variety.

Now let $\underline{g}$ be simple and let X be the complement in $\underline{g}$ of the set of regular semisimple elements of $\underline{g}$. Then it is easy to see that the irreducible components of X are of the form $\pi^{-1}(\pi(\underline{z}_J))$, where J is a subset of B containing exactly one element. If $\underline{g}$ has one (resp. two) root lengths, then X has one (resp. two) irreducible components. As a consequence of Theorem B and the results of 9.1, we have:

Proposition 9.3. *Let* $\underline{g}$ *be simple and let* X *denote the complement in* $\underline{g}$ *of the set of regular semisimple elements.*

(a) *If* $\underline{g}$ *is of rank two, then each irreducible component of* X *is a normal variety.*

(b) *If* $\underline{g}$ *is of type* A_ℓ $(\ell \geq 3)$ *,* D_ℓ $(\ell \geq 4)$ *,* E_6 *,* E_7 *or* E_8 *, then* X *is irreducible and is not a normal variety.*

(c) *If* $\underline{g}$ *is of type* B_ℓ $(\ell \geq 3)$ *or* C_ℓ $(\ell \geq 3)$ *, then the irreducible component of* X *corresponding to* $J = \{\alpha_\ell\}$ *is a normal variety and the irreducible component corresponding to* $J = \{\alpha_1\}$ *is not a normal variety.*

(d) *If* $\underline{g}$ *is of type* F_4 *, then neither irreducible component of* *is a normal variety.*

§10. Normality of the G-orbit of a line.

As a last example of applications of Theorems A and B, we consider the following question:

Let L *be a line (through* 0*) in* $\underline{t}$ *and let* X *denote the closure of the orbit* $G \cdot L$ *. Is* X *a normal variety?*

In concrete examples it seems to be relatively easy to check whether condition (N1) of Theorem A holds in this situation. However we have not been able to formulate any reasonable sort of general theorem. We prove below a few easy results on the above question.

Proposition 10.1. *Let* $\underline{g}$ *be simple and assume that* -1 *does not belong to the Weyl group* W *. Let* $\underline{t}'$ *denote the set of non-zero elements* x *of* $\underline{t}$ *such that* $N_W(kx) = \{1\}$ *. Then* $\underline{t}'$ *is a non-empty open subset of* $\underline{t}$ *. If* $x \in \underline{t}'$ *and if* L *denotes the line* kx *, then the closure of the orbit* $G \cdot L$ *is not a normal variety.*

Proof. Let $\mathbb{P}(\underline{t})$ be the projective space corresponding to $\underline{t}$ and let $p : \underline{t} - \{0\} \to \mathbb{P}(\underline{t})$ be the canonical map. Since $-1 \notin W$, W acts faithfully on $\mathbb{P}(\underline{t})$. Let U be the set of all points a of $\mathbb{P}(\underline{t})$ such that stabilizer W_a is trivial. Then U is a non-empty open subset of $\mathbb{P}(\underline{t})$ and it is clear that $\underline{t}' = p^{-1}(U)$. Thus $\underline{t}'$ is non--empty and open. Let $x \in \underline{t}'$ and let $L = kx$. Then $W_0 = N_W(L)$ is equal to $\{1\}$ and hence $k[L]^{W_0} = k[L]$ is a graded polynomial algebra generated by an element of degree one. Thus the homomorhpism $k[\underline{t}]^W \to k[L]^{W_0}$ given by restriction is not surjective. It follows from Theorem A that the closure of the orbit $G \cdot L$ is not a normal variety.

Proposition 10.2. *Let* $\underline{g}$ *be simple and assume that* $-1 \in W$ *. Then there exists a non-empty open subset* $\underline{t}''$ *of* $\underline{t}$ *such that if* $x \in \underline{t}''$ *and if* $L = kx$ *, then the closure in* $\underline{g}$ *of the orbit* $G \cdot L$ *is a normal Cohen-Macaulay variety.*

Proof. Let V_1 denote the set of $x \in \underline{t}$ such that $N_W(kx) = \{\pm 1\}$; an argument similar to the one given in the previous proof shows that V_1 is a non-empty open subset of $\underline{t}$. Let β denote the Cartan Killing form of $\underline{g}$ and let $F_2 \in k[\underline{g}]_2^G$ be defined by $F_2(y) = \beta(y,y)$. Let $V_2 = \{x \in \underline{t} \mid F_2(x) \neq 0\}$ and let $\underline{t}'' = V_1 \cap V_2 \cap \underline{g}^{reg}$ Then $\underline{t}''$ is a non-empty open subset of $\underline{t}$. Let $x \in \underline{t}''$, let $L = kx$ and let $X = \pi^{-1}(\pi(L))$. Since x is a regular semisimple element of g, X is the closure of the orbit $G\cdot L$. Let $W_0 = N_W(L)/Z_W(L)$. Since $x \in V_1$, we see that $Z_W(L) = \{1\}$ and that $W_0 = N_W(L) = \{\pm 1\}$. It follows easily from this that $k[L]^{W_0}$ is a graded polynomial algebra generated by an element of degree two. Since $x \in V_2$, the restriction of F_2 to L is non-trivial. Thus the homomorphism $k[\underline{t}]^W \to k[L]^{W_0}$ is surjective. It follows from Theorem B that X is a normal Cohen-Macaulay variety.

Proposition 10.3. *Let* (x,h,y) *be an* $\underline{s\ell}_2(k)$ *triple in* g *and let* $L = kh$. *Then* $\pi(L)$ *is normal in* g/G *and* $\pi^{-1}(\pi(L))$ *is a normal Cohen-Macaulay variety.*

See Bourbaki [5, Chap. 8, §11) for the definition of an $\underline{s\ell}_2(k)$ triple.

Proof. We may assume that $h \in \underline{t}$. It follows from [5, Chap. 8, §11] that we may further assume that $\alpha(h)$ is equal to 0 , 1 , or 2 for every $\alpha \in B$. Let $\underline{t}_Q$ be the rational vector space spanned by the set of co-roots $\{H_\alpha \mid \alpha \in R\}$; then $h \in \underline{t}_Q$. It is known that the restriction of the Cartan Killing form β to $\underline{t}_Q$ is positive definite. Thus, if $F_2 \in k[g]^G$ is defined as in the previous proof, then $F_2(h) \neq 0$. Let $W_0 = N_W(L)/Z_W(L)$. Since $\underline{t}_Q$ is W stable, it follows easily that if $w \in N_W(L)$, then $w\cdot h = \pm h$. Since $F_2(h) \neq 0$, we see that the homomorphism $\varphi : k[\underline{t}]^W \to k[L]^{W_0}$ maps $k[\underline{t}]^W_2$ onto $k[L]^{W_0}_2$. Since L is a line, the algebra of invariants $k[L]^{W_0}$ is a graded polynomial algebra. In order to prove that φ is surjective it will suffice to show that there exists $w \in W$ such that $w\cdot h = -h$. Let $\underline{s}$ be the subalgebra spanned by $\{x,h,y\}$. Then L is a Cartan subalgebra of $\underline{s}$. Thus there exists an element a of the Weyl group of $(\underline{s},L)$ such that $a\cdot h = -h$. By a standard theorem, this implies that there exists $w \in W$ such that $w\cdot h = -h$. Consequently φ is surjective. The proof of Proposition 10.3 now follows from Theorem B.

§11. A generalization of Theorem A

The proof of Theorem A carries over to a number of similar situations. Roughly speaking, one has an analogue of Theorem A wherever an appropriate analogue of the Chevalley isomorphism $k[\underline{g}]^G \cong k[\underline{t}]^W$ holds. In order to make this precise, we make the following definition:

Definition 11.1. Let K be a linearly reductive algebraic group and let V be an irreducible affine K-variety. Let M be a closed irreducible subvariety of V and let $F = N_K(M)/Z_K(M)$. We say that M is a *Cartan subvariety* of V if the following four conditions hold:

(i) F is a finite group;

(ii) if $x \in M$, the orbit $K\cdot x$ is closed;

(iii) every closed K-orbit on V meets M; and

(iv) the homomorphism $k[V]^G \to k[M]^F$ given by restriction is an isomorphism.

We have the following generalization of Theorem A:

Theorem C. *Let K be a linearly reductive algebraic group and let V be an irreducible affine K-variety. Assume that there exists a Cartan subvariety M of the affine K-variety V and let $F = N_K(M)/Z_K(M)$. Let X be a closed irreducible K-stable subvariety of V and let D be an irreducible component of the intersection $X \cap M$. Let $F_0 = N_F(D)/Z_F(D)$. Consider the following three conditions:*

(N1) *The homomorphism $k[M]^F \to k[D]^{F_0}$ given by restriction is surjective.*

(N2) *The algebra of invariants $k[X]^G$ is integrally closed.*

(N3) *$\pi_{M,F}(D)$ is a normal subvariety of M/F.*

Then conditions (N2) *and* (N3) *are equivalent and* (N3) *implies* (N1). *If D is a normal variety, then* (N1) *implies* (N3) *so that the three conditions are equivalent.*

The proof of Theorem C follows from Lemmas 2.1 and 2.2 in exactly the same way as the proof of Theorem A. We omit further details.

We list below several examples of affine K-varieties which contain Cartan Subvarieties:

11.2. Let $\underline{k}$ be a reductive Lie algebra, let K be the adjoint group of $\underline{k}$ and let $\underline{t}$ be a Cartan subalgebra of $\underline{k}$. Then $\underline{t}$ is a Cartan subvariety of $\underline{k}$. If $\underline{k}$ is semisimple, then Theorem C applied to this case gives Theorem A.

11.3. Let K be a reductive algebraic group and let T be a maximal torus of K. Let K act on K by inner automorphisms. Then T is a Cartan subvariety of the affine K-variety K. This result is due to Steinberg [20]. In this case one has precise information on the algebra of invariants $k[T]^F$, where F denotes the Weyl group $N_K(T)/Z_K(T)$.

11.4. Let $\underline{g}$, Θ, $\underline{p}$ and $\underline{a}$ be as in §6. Let G be the adjoint group of $\underline{g}$ and let $K = \{g \in G \mid g \circ \Theta = \Theta \circ g\}$. Then it follows from the results of [13] that $\underline{a}$ is a Cartan subvariety for the affine K-variety $\underline{p}$. In this case the group $F=N_K(a)/Z_K(a)$ is the "little Weyl group" W_0. Let R_0 be the set of roots of $\underline{a}$ on $\underline{g}$. Then R_0 is a (not necessarily reduced) root system and W_0 is the corresponding Weyl group. In this situation we have analogues of most of the results of §7 and §8.

11.5. ($k = \mathbb{C}$) . The "polar representations" of Dadok and Kac [7] fit into our framework. Let $\rho : K \to GL(V)$ be a polar representation of the reductive algebraic group K and let E be a "Cartan subspace" of V in the sense of [7]. Then E is a Cartan subvariety of the affine K-variety V . These examples include the examples of 11.2 and 11.4 above. They also include the representations considered by Vinberg in [21]. Other examples of polar representations are given in [6] and [7].

11.6. (See [18]). Let $\theta : G \to G$ be an involutive automorphism and let $K = G$. Let $P = \{g\theta(g)^{-1} \mid g \in G\}$. Then P is a closed irreducible K-stable subvariety of G and is K-isomorphic to G/K . Let A be a maximal -anisotropic torus of G (see [18] for definition). Then A is a Cartan subvariety of the affine K-variety P .

§12. More on the De Concini-Procesi problem

After the manuscript for this paper had been typed, we received a letter from C. Procesi concerning the results of §6. In this letter, Procesi mentioned a suggestion of M. Kashiwara concerning the problem posed in [10, p. 8]. This suggestion led to a strengthening of the results of 6 which we indicate below.

Let $k = \mathbb{C}$. Let $\theta : G \to G$ be an involution of G . We also denote the corresponding involution of the Lie algebra $\underline{g}$ by θ . Let $\underline{p}$ denote the -1 eigenspace of θ on $\underline{g}$ and let X denote the closure of the orbit $G\cdot\underline{p}$ in $\underline{g}$. Let $K = G^{\theta}$. It is known that there exists a θ-stable real form $\underline{g}_0$ of $\underline{g}$ such that the restriction of θ to $\underline{g}_0$ is a Cartan involution of $\underline{g}_0$. Let G_0 be the connected real Lie subgroup of G with Lie algebra $\underline{g}_0$ and let $K_0 = G_0 \cap K$. Then the coset space G_0/K_0 is a Riemannian symmetric space. Let $D(G_0/K_0)$ denote the algebra of all C^{∞} linear partial differential operators on the symmetric space G_0/K_0 . Let $Z(\underline{g})$ denote the centre of the universal enveloping algebra $U(\underline{g})$. Then there exists a canonical homomorphism $\tau : Z(\underline{g}) \to D(G_0/K_0)$. It is known that there are pairs (G,θ) such that τ is not surjective. It was suggested by Kashiwara that the surjectivity of τ might be a test for the normality of X . In fact, it turns out that the surjectivity of τ is equivalent to our condition (N1). We have the following proposition:

Proposition 12.1. *Let the notation be as above. Let $\underline{a}$ be a Cartan subspace of $\underline{p}$ and let $\underline{t}$ be a Cartan subalgebra of $\underline{g}$ which contains $\underline{a}$. Let $\pi_1 : \underline{t} \to \underline{t}/W$ be the quotient morphism and let $W_0 = N_W(\underline{a})/Z_W(\underline{a})$. Then the following four conditions are equivalent:*

(N1) *The homomorphism $\mathbb{C}[\underline{t}]^W \to \mathbb{C}[\underline{a}]^{W_0}$ given by restriction is surjective.*

(N2) *$\pi_1(\underline{a})$ is a normal subvariety of $\underline{t}/W$.*

(N3) $\mathbb{C}[X]^G$ *is integrally closed.*

(N4) $\tau : Z(\underline{g}) \to D(G_0/K_0)$ *is surjective.*

Proof. The equivalence of (N1), (N2), and (N3) is given by Theorem A. The equivalence of (N1) and (N4) is an easy consequence of a theorem of Helgason [23, p. 590, Prop. 7.4].

REFERENCES

1. Borel, A.: Linear Algebraic Groups. New York: Benjamin 1969.
2. Borho, W.: Über Schichten halbeinfacher Lie-Algebren. Invent Math. 65, 283-317 (1981).
3. Borho, W., Kraft, H.: Über Bahnen und deren Deformationen bei linearen Aktionen reduktiver Gruppen. Comment. Math. Helv. 54, 61-104 (1979).
4. Bourbaki, N.: Groupes et algébres de Lie, Chapitres 4, 5, et 6. Paris: Hermann 1968.
5. Bourbaki, N.: Groupes et algébres de Lie, Chapitres 7 et 8. Paris: Hermann 1975.
6. Dadok, J.: Polar coordinates induced by actions of compact Lie groups. To appear.
7. Dadok, J., Kac, V.: Polar representations. To appear.
8. Dieudonné, J.: Cours de géométrie algébrique, 2. Presses universitaires de France 1974.
9. Helgason, S.: Differential Geomtry, Lie Groups and Symmetric Spaces. New York - San Francisco - London: Academic Press 1978.
10. Hotta, R., Kawanaka, N. (ed.): Open Problems in Algebraic Groups, Proceedings of the Twelfth International Symposium, Division of Mathematics, The Taniguchi Founction. Conference on "Algebraic Groups and their Representations," Kotata, Japan, Aug. 29 - Sept. 3, 1983. (Copies available from R. Hotta, Mathematical Institute, Tohuku University).
11. Howlett, R.: Normalizers of parabolic subgroups of reflection groups. J. London Math. Soc. (2) 21, 62-80 (1980).
12. Kostant, B.: Lie group representations on polynomial rings. Amer. J. Math. 85, 327-404 (1963).
13. Kostant, B., Rallis, S.: Orbits and representations associated with symmetric spaces. Amer. J. Math. 93, 753-809 (1971).
14. Kraft, H., Procesi, C.: Closures of conjugacy classes of matrices are normal. Invent. Math. 53, 227-247 (1979).
15. Kraft, H., Procesi, C.: On the geometry of conjugacy classes in classical groups. Comment. Math. Helv. 57, 539-602 (1982).
16. Peterson, D.: Geometry of the Adjoint Representation of a Complex Semisimple Lie Algebra. Ph.D. Thesis, Harvard University 1978.
17. Richardson, R.: An application of the Serre conjecture to semisimple algebraic groups. In: Algebra, Carbondale, 1980. Lecture Notes in Math. 848, 141-151 (1981).
18. Richardson, R.: Orbits, invariants and representations associated to involutions of reductive groups. Invent. Math. 66, 287-312 (1982).
19. Slodowy, P.: Simple singularities and simple algebraic groups. Lecture Notes in Math. 815 (1980).
20. Steinberg, R.: Regular elements of semisimple algebraic groups. Publ. Math. I.H.E.S. 25, 49-80 (1965).

21. Vinberg, E.: The Weyl group of a graded Lie algebra. Math. U.S.S.R. - Izv. 10, 463-495 (1976).

22. Vust, T.: Opération de groupes réductifs dans un type de cônes presque homogènes. Bull. Soc. Math. France 102, 317-334 (1974).

23. Helgason, S.: Fundamental solutions of invariant differential operators on symmetric spaces. Amer. J. Math. 86, 565-601 (1964).

UNIPOTENT ELEMENTS AND PARABOLIC SUBGROUPS OF REDUCTIVE GROUPS. II

Jacques TITS

1. Introduction

Let K be a field of characteristic p and G a reductive group over K. In [3], A. Borel and the author showed that if K is perfect, then

(U) every unipotent subgroup (i.e. subgroup consisting of unipotent elements) of G(K) is contained in the unipotent radical of a K-parabolic subgroup of G.

Furthermore, we conjectured that, if G is quasi-simple and simply connected, the same assertion holds when p is not a torsion prime for G (i.e. p is any prime if G has type A_n or C_n; $p \neq 2$ if G has type B_n, D_n or G_2; $p \neq 2, 3$ if G has type F_4, E_6 or E_7; $p \neq 2, 3, 5$ if G has type E_8). That conjecture will be proved in Section 2 (cf. Corollary 2.6).

Pairs (K,G) for which (U) is false are dealt with in the remaining sections, where we go a long way towards determining all of them in the case where G is split. (This restriction is less serious than it may seem; indeed, if (U) is false for G over K, it remains false over the separable closure of K, over which G splits: cf. [3], 3.6). Let us be more specific. We say that an element u of a reductive K-group H is anisotropic (in that group) if it is contained in no proper K-parabolic subgroup of H; by [2], (2.20), this is so if and only if the projections of Ad u in all K-simple factors of the adjoint group Ad H are anisotropic. It is not difficult to see that (U) is false if and only if there is a K-split torus in G whose centralizer possesses a K-rational anisotropic element of order p (cf. Corollary 3.3). Thus, the problem of determining all pairs (K,G) for which (U) does not hold is roughly equivalent to that of finding all (K,G) where G is K-simple and has a K-rational anisotropic element of order p. Until the end of this introduction, we shall assume that G is quasi-simple and K-split. Under these conditions, we conjecture that any anisotropic element of order p in G(K) normalizes a maximal K-split torus of G. Observe that if an element of the normalizer N of a maximal K-split torus T is anisotropic, it fixes no nontrivial rational character of T. Elements of N with that property have been studied by T.A. Springer [8]; we call them special. If the above conjecture is true, it reduces our problem (in the case of split groups) to that of determining, for all G and K, which special K-rational elements of order p are anisotropic. This question seems to be of the level of an exercise, which we solve here in the case where G is a classical group (cf. 3.5, 4.2, 4.3, 4.4). As for the conjecture itself, we are able to assert that it is indeed true, except possibly if G has type E_8 and $p = 3$ or 5. If one forgets about the easy case of type A_n (cf. 3.5) and the difficult (and unsolved) one of type E_8 in characteristic 5, the only characteristics to be considered are $p = 2$ and 3. For $p = 2$, the conjecture follows from a short and rather standard argument (cf. 4.1). The geometric proof I can propose for $p = 3$ (and, so far, $G \neq E_8$) is a case analysis and requires the knowledge of specific properties of the buildings of exceptional groups and special features of the triality in characteristic 3; it is too long to be given here and will be published later (unless a better proof is found meanwhile).

When dealing with anisotropic elements, it appears more natural to take a slightly more general viewpoint and to consider anisotropic automorphisms, that is, K-automorphisms which stabilize no proper K-parabolic subgroup. Our conjecture (and all we have said about it) extends to that situation. Similar generalizations could be contemplated for other results presented below, but they do not seem to bring much improvement. Furthermore, when our arguments do remain valid for arbitrary automorphisms (instead of inner ones), the prerequisite necessary to carry them out is not always readily available in the literature. So, we prefer to leave those generalizations aside.

The research which led to the results presented in this paper was motivated by a joint work with W. Kantor and R. Liebler [6]. I thank A. Borel who agreed to my using here the title of a common entreprise initiated in [3].

Throughout the paper, K, p have the same meaning as above, we suppose $p \neq 0$, G is a reductive group defined over K and $\overline{K}$ (resp. K_s) denotes an algebraic (resp. a separable) closure of K.

2. Good and very good unipotent elements

2.1 We say that a unipotent element of G is *very good* if its schematic centralizer is smooth; in simple terms, this means that its centralizer in Lie G is the Lie algebra of its reduced (= group-theoretical) centralizer in G. A K-rational unipotent element is called *good* if it is contained in the unipotent radical of a parabolic K-subgroup. By [3], 3.6, G possesses property (U) of the introduction if and only if all unipotent elements of G(K) are good.

2.2. The following easy proposition describes the behaviour of those notions under central isogenies. Let $\pi : \widetilde{G} \to G$ be a central isogeny, Let $\widetilde{T}$ be a maximal torus of $\widetilde{G}$, set $T = \pi(\widetilde{T})$ and let X, $\widetilde{X}$ be the (absolute) character groups of T, $\widetilde{T}$. The cokernel of the homomorphism $X \to \widetilde{X}$ induced by π is a finite group whose order c is called the *degree* of π. For any integer d and any algebraic group H, let $\mu(d,H)$ denote the d-th power morphism $x \mapsto x^d$ of H into itself.

PROPOSITION. (i) *If* n *is an integer divisible by* c, *there exists a unique* K-*morphism* $\varphi_n : G \to \widetilde{G}$ *such that* $\varphi_n \circ \pi = \mu(n,\widetilde{G})$; *one has* $\varphi_n(1_G) = 1_{\widetilde{G}}$, $\pi \circ \varphi_n = \mu(n,G)$ *and, for all* $d \in \mathbb{Z}$, $\varphi_n \circ \mu(d,G) = \mu(d,\widetilde{G}) \circ \varphi_n$.

(ii) *The map* π *injects the set* $\widetilde{G}(K)_u$ *of all unipotent elements of* $\widetilde{G}(K)$ *in the set* $G(K)_u$ *of all unipotent elements of* G(K) *and bijects the set of all good unipotent elements of* $\widetilde{G}(K)$ *onto the set of all good unipotent elements of* G(K).

(iii) *If* c *is prime to* p, $\pi(\widetilde{G}(K)_u) = G(K)_u$ *and* π *maps the set of all very good unipotent elements of* $\widetilde{G}$ *bijectively onto the set of all very good unipotent elements of* G.

(i) The first assertion follows from the fact that $\mu(n,\tilde{G})$ is constant on the (schematic) fibres of π, which is surjective. The relation $\varphi_n(1) = 1$ is obvious. Finally, applying the first and last terms of the following two sequences of equalities

$$\pi\circ\varphi_n\circ\pi = \pi\circ\mu(n,\tilde{G}) = \mu(n,G)\circ\pi$$

and

$$\varphi_n\circ\mu(d,G)\circ\pi = \varphi_n\circ\pi\circ\mu(d,\tilde{G}) = \mu(nd,\tilde{G}) = \mu(d,\tilde{G})\circ\varphi_n\circ\pi$$

to a generic point of $\tilde{G}$, whose image by π is a generic point of G, we get the two last assertions of (i).

(ii) The first assertion is clear since the kernel of π in $\tilde{G}(K)$ has order prime to p, and the second follows from [2], 2.15 and 2.20.

(iii) Suppose c prime to p, choose an integer n divisible by c and congruent 1 modulo the largest order of a unipotent element of G, and let φ_n be as in (i). By (i), the orders of the elements of $\varphi_n(G(K)_u)$ are powers of p; in other words, $\varphi_n(G(K)_u) \subset \tilde{G}(K)_u$. Since $\pi\circ\varphi_n = \mu(n,G)$ is the identity on $G(K)_u$, it follows, again by (i), that $G(K)_u \subset \pi(\tilde{G}(K)_u)$. The opposite inclusion being obvious, the first assertion of (iii) is proved. The second one readily ensues since, as a consequence of the assumption made on c, $(d\pi)_1$ is an isomorphism of Lie $\tilde{G}$ onto Lie G.

2.3. We recall that p is said to be *good for a quasi-simple K-group* H if it does not divide the coefficients of the basic roots in the dominant root of the root system of H; this means that $p \neq 2$ if H is not of type A_n, $p \neq 3$ if H is of exceptional type (G_2, F_4 or E_i) and $p \neq 5$ if H is of type E_8.

THEOREM (Richardson-Springer-Steinberg: cf. [9], §5). *Suppose p is good for all quasi-simple normal subgroups of G and no such subgroup has type A_{kp-1} for some $k \in \mathbb{N}$. Then, all unipotent elements of G are very good.*

2.4. PROPOSITION. *Let u be a unipotent element of G(K) and let P be a K-parabolic subgroup of G. Suppose P contains the reduced centralizer $Z_G(u)$ of u in*

G and its Lie algebra Lie P contains the centralizer $Z_{\mathrm{Lie}\ G}(u)$ of u in Lie G. Then, for $g \in G(\bar{K})$, if ${}^g u$ is K-rational, the subgroup ${}^g P$ is defined over K.

Let U be the unipotent radical of a K-parabolic subgroup opposite to both P and ${}^g P$, so that ${}^g P = {}^v P$ for some $v \in U(\bar{K})$, and let X be the conjugacy class of u in P. The set ${}^{UP}u$ is an open subvariety of the conjugacy class ${}^G u$ of u in G, and its contains ${}^g u$ since $g \in vP$. Let us show that the map φ: $(y,x) \mapsto {}^y x$ of $U \times X$ into ${}^{UP}u$ is an isomorphism of algebraic varieties. The group $U \times P$ operates transitively on $U \times X$ by $(u',p).(u'',x) = (u'u'', x^{p^{-1}})$ and on ${}^{UP}u$ by $(u',p).z = {}^{u'}z^{p^{-1}}$, and those actions are compatible with φ. Therefore, we only have to show that $\varphi^{-1}(u) = \{(1,u)\}$, which simply amounts to our hypthesis $Z_G(u) \subset P$, and that the differential of φ at the point (1,u) is injective. But if $\mathfrak{u} \in$ Lie U and $\mathfrak{p} \in$ Lie P are such that $u\mathfrak{p}$ is tangent to X at u and that

$$d\varphi_{(1,u)}(\mathfrak{u}, u\mathfrak{p}) = (\mathrm{Ad}\ u)\mathfrak{u} + \mathfrak{p} - \mathfrak{u} = 0,$$

we have $\mathfrak{p} = 0$ and $\mathfrak{u} \in Z_{\mathrm{Lie}\ U}(u) \subset \mathrm{Lie}\ U \cap \mathrm{Lie}\ P = \{0\}$. This establishes our assertion on φ. Now, it is clear that φ, hence also φ^{-1}, is a K-isomorphism and that v is the projection of $\varphi^{-1}({}^g u)$ in the factor U of $U \times X$. Therefore, if ${}^g u \in G(K)$, we have $v \in U(K)$ and the group ${}^g P = {}^v P$ is defined over K, q. e. d.

Remark. It seems plausible that the hypotheses made on u in the above proposition imply that the schematic centralizer $\mathfrak{Z}$ of u in G is contained in P. If it is so, the conclusion of the proposition immediately follows. Indeed, the inclusion of $\mathfrak{Z}$ in P induces a K-morphism of $G/\mathfrak{Z}$, hence of the conjugacy class of u, onto G/P, hence onto the conjugacy class of P, and that morphism is nothing else but ${}^g u \mapsto {}^q P$.

2.5. THEOREM. Let $u \in G(K)$ be a unipotent element. Suppose that one of the following conditions is satisfied:

(i) u is very good;

(ii) G is simply connected of type A;

(iii) G is simply connected of type C;

(iv) *$p \neq 2$ and G is of type G_2.*

Then, there exists a K-parabolic subgroup P of G, stable under all automorphisms of $G(\overline{K})$ fixing u (the group Aut $G(\overline{K})$ acts on the set of all $\overline{K}$-parabolic subgroups of G by the main theorem of [4]), and whose unipotent radical contains u. In particular, u is good. Any P with the above properties contains $Z_G(u)$.

The last assertion is obvious since, for any element z of $Z_G(u)$, the inner automorphism Inn z fixes u, therefore ${}^zP = P$, hence $z \in P$.

If there exists a K_s-parabolic subgroup P having the desired properties, it is defined over K; indeed, Aut (K_s/K), which operates on the whole situation, fixes u, hence P by hypothesis. Therefore, we may, and shall, assume that $K = K_s$, which implies that G is split. Without loss of generality, we also assume that G is defined and split over the prime field F of K, whose algebraic closure in K is denoted by $\overline{F}$; this is an algebraically closed field. Since G has only finitely many conjugacy classes of unipotent elements (cf. [7]), each one of them meets $G(\overline{F})$. In particular, u is conjugate (in $G(\overline{K})$) to an element u' of $G(\overline{F})$. Set $u = {}^gu'$ with $g \in G(\overline{K})$. By [3], 2.5, and the main theorem of [4], there exists a parabolic subgroup P' of G stable under all automorphisms of $G(\overline{K})$ which fix u' and whose unipotent radical contains u'. Being stable by Aut $(\overline{K}/\overline{F})$, P' is defined over $\overline{F}$. We now distinguish cases.

Case (i). By hypothesis, $Z_G(u')$ is contained in P'. Since u, hence also u', is very good, it follows that $Z_{\text{Lie }G}(u') = \text{Lie } Z_G(u') \subset \text{Lie } P'$. Proposition 2.4 now implies that $P = {}^gP'$ is defined over K and meets all our requirements.

Case (ii). The group G is F-isomorphic to SL_n for some n. By the Jordan normal form theorem, u and u' are conjugate in G(K). In other words, we can take g in G(K) and, again, $P = {}^gP'$ has all the desired properties.

Case (iii) (*resp.* (iv)). In this case, it is well-known that G can be embedded in a simply connected K-group H of type A (resp. D_4) as the fixed-point group of an outer automorphism σ of order 2 (resp. 3) such that the K-parabolic subgroups of G are precisely the intersections with G of the σ-stable K-parabolic subgroups of H and that every automorphism of $G(\overline{K})$ extends to an automorphism of $H(\overline{K})$ fixing σ. Now, the assertion follows from case (ii) (resp. case (i) and Theorem 2.3, which implies that u is very good in H) applied to H.

2.6. COROLLARY. Suppose G is semi-simple and simply connected. Then, if p is a torsion prime for no quasi-simple direct factor of G, all unipotent elements of G(K) are good.

This is an immediate consequence of Theorem 2.5, in view of Theorem 2.3.

2.7. Remark. If u is a good unipotent element of G(K), there exists a K-parabolic subgroup of G whose unipotent radical contains u and which is stable under all automorphisms of $G(K_s)$ fixing u (cf. [3], 2.5), but it may happen that no such K-parabolic subgroup is normalized by (i.e. contains) the centralizer of u in $G(\overline{K})$: an example will be seen in 4.3.5.

3. Bad and anisotropic unipotent elements

3.1. We say that a K-automorphism or a K-rational element of G is anisotropic if it normalizes no proper K-parabolic subgroup of G, and that a unipotent element of G(K) is bad if it is not good. We shall see (Corollary 3.3) that the existence of bad unipotent elements and the existence of anisotropic elements of order p are closely related phenomena. Clearly, any nontrivial anisotropic unipotent element is bad (and even especially bad !).

3.2. PROPOSITION. Let u be a unipotent element of G(K), let P be a K-parabolic subgroup containing u, let L be a Levi subgroup of P defined over K, so that $P = R_u(P) \rtimes L$, and let u' be the projection of u in L with respect to that product decomposition.

(A) The following properties are equivalent:

(i) u *is good in* G;

(ii) u' *is good in* G;

(iii) u' *is good in* L.

(B) *If* P *is minimal among all* K-*parabolic subgroups of* G *containing* u, *then* u' *is anisotropic and its order is the smallest power* q *of* p *such that* u^q *is good*.

(A) If Q is a K-parabolic subgroup of G whose unipotent radical contains u, u' is contained in the unipotent radical of $((Q \cap P).R_u(P)) \cap L$, which is a K-parabolic subgroup of L (cf. [1], 4.4, 4.7), hence the implication (i) ⇒ (iii) of which the implication (ii) ⇒ (iii) is a special case (taking u = u'). Conversely, (iii) implies (i) and (ii) because if u' is contained in the unipotent radical of a K-parabolic subgroup P_1 of L, both u and u' are contained in the unipotent radical of the K-parabolic subgroup $P_1.R_u(P)$ of G.

(B) Suppose the hypothesis of (B) satisfied. If P_1 is any K-parabolic subgroup of L containing u', the parabolic subgroup $P_1.R_u(P)$ of G contains u, and the minimality assumption implies that $P_1 = L$, hence the first assertion. Let q be any power of p. By (A), u^q is good if and only if u'^q is good in L, which happens only if $u'^q = 1$. Indeed, if u'^q was good and different from 1, its centralizer in L and, in particular u', would be contained in a proper K-parabolic subgroup of L (cf. [3], 3.1). This finishes the proof.

3.3. COROLLARY. *A necessary and sufficient condition for the group* G(K) *to contain a bad unipotent element is the existence of a split* K-*torus in* G *whose centralizer possesses an anisotropic element of order* p.

The condition is necessary by 3.2 (B), applied to any bad element of G(K) whose p-th power is good. The converse readily follows from 3.2 (A).

3.4. *Remark*. Since bad elements remain bad after separable extensions of the ground field and since G splits over such an extension, the investigation of bad unipotent elements in arbitrary reductive groups is, to a large extent, reduced by Corollary 3.3 to the investigation of anisotropic elements of order p in semi-simple groups.

3.5. *Example: split groups of type A.*

Suppose that the group G is split and quasi-simple of type A and that G(K) possesses an anisotropic element u of order p. The adjoint group of G is the group PGL(V) for some K-vector space V. Let $\tilde{u}$ be a representative in GL(V) of the canonical image of u in PGL(V). It is an anisotropic element of GL(V) whose p-th power $\tilde{u}^p$ is an element k of $K^\times$ (considered as a subgroup of GL(V)). Clearly, k does not belong to K^p, otherwise, dividing $\tilde{u}$ by $\sqrt[p]{k}$, we could assume that $\tilde{u}$ has order p and the stabilizer of the space of all fixed points of $\tilde{u}$ in V would be a k-parabolic subgroup of GL(V) containing $\tilde{u}$. The same argument shows that G cannot be simply connected (which also follows from 2.5 (ii)). Now, V has a structure of $K(\sqrt[p]{k})$-vector space defined by $\sqrt[p]{k}.v = \tilde{u}(v)$ for $v \in V$. This vector space must have dimension 1, otherwise the stabilizer in GL(V) of any nontrivial proper subspace of it would be a proper K-parabolic subgroup of GL(V) containing $\tilde{u}$. Therefore, dim V = p and G = PGL(V). Conversely, for any $k \in K - K^p$, the canonical image in $PGL_K(K(\sqrt[p]{k}))$ of the multiplication by $\sqrt[p]{k}$ is an anisotropic element of order p. We conclude that

a split quasi-simple group of type A_n *possesses anisotropic elements of order* p *if and only if* $n = p-1$, G *is adjoint and* K *is not perfect.*

3.6. The next lemma and the proposition which follows remain valid in characteristic zero.

LEMMA. *Let* H *be a reductive subgroup of* G *defined over* K.

(i) *For every* K-*parabolic subgroup* P *of* H, *there exists a* K-*parabolic subgroup* Q *of* G *whose unipotent radical contains that of* P *and such that every* K-*automorphism of* G *stabilizing* P *also stabilizes* Q.

(ii) <u>The automorphism of</u> H <u>induced by any anisotropic</u> K-<u>automorphism of</u> G <u>stabilizing</u> H <u>is also anisotropic</u>.

(i) Let S be a maximal split torus of the radical of P and let ψ be the set of all roots of G relative to S which are linear combinations with positive coefficients of the weights of S in the Lie algebra of $R_u(P)$. Then, the group $Q = G_\psi$, with the notation of [1], 3.8, clearly has the desired properties.

(ii) is an immediate consequence of (i).

4. <u>Anisotropic involutions</u>

4.1. PROPOSITION. <u>Let</u> α <u>be an anisotropic involutory</u> K-<u>automorphism of</u> G. <u>Then, for any</u> K-<u>parabolic subgroup</u> P <u>of</u> G, P <u>and</u> $\alpha(P)$ <u>are opposite. Every split torus stable by</u> α <u>is contained in a maximal split torus stable by</u> α; <u>in particular, there exist maximal split tori stable by</u> α. <u>If</u> G <u>is semi-simple, the automorphism induced by</u> α <u>on any such torus is</u> $t \mapsto t^{-1}$.

In the spherical building I of G over K, we introduce a distance invariant under G(K) and making each apartment into a Euclidean sphere of radius 1 (cf. [5], §8). Then, any two points of the building at distance strictly smaller than π are joined by a unique geodesic (<u>loc.cit</u>., 3.1, 8.1). To each facet of I, let us assign a "center of gravity" defined by some covariant process (a process which is usually not unique). If P and $\alpha(P)$ were not opposite (which implies that $P \neq G$), the middle point of the geodesic joining the centers of gravity of the corresponding facets would belong to the facet corresponding to a parabolic subgroup stable by α. Hence the first assertion.

If S is any split torus stable by α, and if P denotes a minimal K-parabolic subgroup containing S, the intersection $P \cap \alpha(P)$ is a Levi subgroup of P stable by α, whose center contains a unique maximal split torus which is stable by α and contains S.

Finally, if G is semi-simple and if S is a maximal split torus stable by α, the fact that α transforms each K-parabolic subgroup containing S into an opposite parabolic subgroup implies that α multiplies by -1 all relative roots of G with respect to S, hence all characters of S. This means that α transforms each element of S into its inverse. The proof is complete.

Throughout the remainder of this section, we suppose G quasi-simple and split, and $p = 2$, and we denote by T a maximal split torus of G. The above proposition suggests to study the involutory K-automorphisms of G stabilizing T and inducing the automorphism $t \mapsto t^{-1}$ on it, and to find out which one of them are anisotropic. We shall do that for all groups of classical type.

4.2. Groups of type A.

Suppose G of type A_{n-1} and set $I = \{1, \ldots, n\}$. We identify the adjoint group of G with the group PGL(V) of some vector space V in which we choose a coordinate system $(x_i)_{i\in I}$ such that the canonical image of T in PGL(V) is also the canonical image of the group of all invertible diagonal matrices in GL(V). Any involutory K-automorphism α of G stabilizing T and inverting its elements is "represented by" a nondegenerate symmetric bilinear form $a: V \times V \to K$ of the shape

$$a: ((x_i)_{i\in I},(y_i)_{i\in I}) \mapsto \Sigma\, a_i x_i y_i \qquad (\text{all } a_i \neq 0)$$

in the following sense: the form a defines an isomorphism of V onto its dual, hence an automorphism of PGL(V) which lifts uniquely to the automorphism α of G; proportional forms a define the same automorphism α. Now, consider a flag $V_1 \subsetneq V_2 \subsetneq \cdots \subsetneq V_r$ in V, with $V_1 \neq \{0\}$ and $V_r \neq V$. The K-parabolic subgroup of G defined by that flag is stable by α if and only if, for $j \in \{1, \ldots, r\}$, the space V_{r+1-j} is the orthogonal $V_j^{\perp}$ of V_j with respect to the form a, in which case the quadratic form $\Sigma\, a_i x_i^2$ vanishes on all V_j for $2j \leqslant r+1$. Conversely, if that form vanishes on some nontrivial proper subspace Y of V, the K-parabolic subgroup of G corresponding to the flag $\{Y, Y^{\perp}\}$ is stable by α. Consequently:

the involution α defined by the form a is anisotropic if and only if the coefficients a_i are linearly independent over K^2.

4.3. Groups of types B and D.

4.3.1. Suppose G of type B_m or D_m and let I be the set of all integers or all nonzero integers i with $|i| \leq m$ accordingly. We identify the adjoint group of G with a group $PGO°(V,q)$, where the exponent ° means "connected component of the identity", V is a (2m+1)- or 2m-dimensional vector space over K and q is a non-degenerate quadratic form of maximal Witt index in V. In V, we choose a coordinate system $(x_i)_{i\in I}$ with respect to which $q = \Sigma\, x_{-i}x_i$, where i runs from 0 or 1 to m, and such that the image of T in $PGO°(V,q)$ coincides with the image of the group of all invertible diagonal matrices preserving q. The K-automorphisms α of G stabilizing T and inverting its elements are induced, in an obvious sense, by the similitudes of the form

$$\tilde{\alpha}: (x_i)_{i\in I} \mapsto (y_i)_{i\in I} \quad \text{with } y_i = a_{-i}x_{-i},$$

where the a_i are nonvanishing constants such that the product $c = a_{-i}a_i$ (the ratio of similitude) does not depend on i. Let such α and $\tilde{\alpha}$ be given. If $0 \in I$ (case B_m), we assume, without loss of generality, that $a_0 = c = 1$.

4.3.2. The automorphism α is "algebraically inner", that is, it comes from an element u of the adjoint group, if and only if $\dim V \not\equiv 2 \pmod 4$. Assume that condition satisfied and suppose that G = Spin (V,q) (resp. $O°(V,q)$); then, u is the image of an element of G(K) - that is, α is the inner automorphism corresponding to such an element - if and only if c and the product of all a_i belong to K^2 (resp. if $c \in K^2$).

The first assertion simply recalls under which condition -1 belongs to the Weyl group.

Assume $\dim V \equiv 0 \pmod 4$ and G = Spin (V,q) (resp. $O°(V,q)$). The involution

$$\tilde{\alpha}_1 : (x_i)_{i\in I} \mapsto (y_i)_{i\in I} \quad \text{with } y_i = x_{-i}$$

is the image in $O°(V,q)$ of an element of Spin $(V,q)(K)$ since this is already true over the prime field. Therefore, u "comes from" G(K) if and only if the image t of

$$\tilde{\alpha}_1^{-1}\tilde{\alpha}: (x_i)_{i\in I} \mapsto (a_i x_i)_{i\in I}$$

in $PGO°(V,q)$ does. Observe that t belongs to the canonical image $\overline{T}$ of T in that group. Now, our assertion follows from the known fact that the character group of T is generated by the character group of $\overline{T}$ and two (resp. one) additional character(s) $\frac{1}{2}\chi$, $\frac{1}{2}\chi'$ (resp. $\frac{1}{2}\chi$) such that $\chi(t) = c$ and $\chi'(t) = \prod_{i=1}^{m} a_i$.

The case where $\dim V \equiv 1 \pmod 2$ is similar but simpler.

4.3.3. _The following properties are equivalent:_

(i) _the automorphism_ α _is isotropic (i.e. not anisotropic);_

(ii) _there exists a nonzero vector_ $\xi \in V$ _which is singular (i.e._ $q(\xi) = 0$) _and orthogonal to_ $\tilde{\alpha}(\xi)$ _with respect to the symmetric bilinear form associated to_ q;

(iii) _the equations_

$$(1) \quad \sum_{\substack{i\in I\\ i>0}} (\xi_i^2 + c\eta_i^2)a_i = \sum_{\substack{i\in I\\ i\geq 0}} \xi_i\eta_i a_i = 0$$

have a nontrivial solution with $\xi_0 = \eta_0$ (_if_ $0 \in I$).

They imply

(iii') _the_ a_i, _for_ $i \in I$ _and_ $i \geq 0$, _are linearly dependent over the field_ $K^2(c)$.

Suppose now that $c = 1$ _(i.e. that_ $\tilde{\alpha}$ _belongs to the orthogonal group). Then, the above conditions are also equivalent to:_

(ii_1) $\tilde{\alpha}$ <u>fixes a nonvanishing singular vector;</u>

(iii_1) <u>the</u> a_i, <u>for</u> $i \in I$ <u>and</u> $i \geq 0$, <u>are linearly dependent over the field</u> K^2.

If ξ is any element of $V - \{0\}$ contained in a totally singular subspace stable by $\tilde{\alpha}$, it satisfies (ii). Conversely, if ξ satisfies (ii), the linear span of ξ and $\tilde{\alpha}(\xi)$ is a totally singular subspace stable by $\tilde{\alpha}$. This proves the equivalence (i)$\Leftrightarrow$(ii). Assertion (iii) is just a reformulation of (ii), setting $\xi = (\xi_i)_{i\in I}$ and $\eta_i = a_i^{-1}\xi_{-i}$, and the implication (iii) $\Rightarrow$ (iii') is clear: just observe that if all $\xi_i^2 + c\eta_i^2$ vanish, then c is a square, say $c = c'^2$, and the second equality (i) becomes $c'.\Sigma\xi_i^2 a_i = 0$.

Now, suppose c = 1, that is, $\tilde{\alpha}^2 = 1$. Then, if ξ satisfies (ii), either it is fixed by $\tilde{\alpha}$, or $\xi + \tilde{\alpha}(\xi)$ is a nonvanishing singular vector fixed by $\tilde{\alpha}$; therefore, (ii) implies (ii_1), and the converse is obvious. Clearly, (iii') and (iii_1) are now equivalent; in particular, (iii) implies (iii_1). To prove the converse, suppose that $\Sigma\xi_i^2 a_i = 0$ for some $\xi_i \in K$, not all zero, with $i \in I$, $i \geq 0$. Then (iii) is satisfied by setting $\eta_i = \xi_i$ for all i. The proof is complete.

4.3.4. LEMMA. <u>Suppose</u> $\dim V \equiv 0 \pmod 4$ <u>and</u> $c = 1$. <u>In</u> $V \otimes \bar{K}$, <u>let</u> Y <u>be the space of all singular vectors fixed by</u> $\tilde{\alpha}$ <u>and let</u> Y_0 <u>be the space of all</u> $y + \tilde{\alpha}(y)$ <u>for</u> $y \in Y^\perp$. <u>Then</u> $\dim Y = m-1$, $\dim Y_0 = 1$, $Y_0 \subset Y$ <u>and the only totally singular subspaces stable by the centralizer of</u> $\tilde{\alpha}$ <u>in</u> G <u>are</u> $\{0\}$, Y_0, Y, <u>and the two maximal totally singular subspaces containing</u> Y.

We assume, without loss of generality, that $K = \bar{K}$, that $G = O°(V,q)$ and, by an appropriate change of coordinates, that all a_i are equal to 1. Then, Y is defined by the equations $x_{-i} = x_i$ ($i \in \{1, \ldots, m\}$) and $\sum_{i=1}^{m} x_i = 0$, its orthogonal $Y^\perp$ is defined by $x_{-i} + x_i = x_{-j} + x_j$ for all i, j, and Y_0 is the space of all vectors all of whose coordinates are equal. Hence the three first assertions of the lemma. For the remainder of the proof, we content ourselves with two simple observations from which the last assertion follows by routine arguments, the detail of which is left to the reader. Let Z denote the centralizer of α (or $\tilde{\alpha}$) in G.

1) The group H of all linear transformations of V expressed by an arbitrary invertible self-contragredient substitution of the x_i ($i > 0$) and "the same" substitution of the x_{-i} is contained in Z; the only nontrivial proper linear subspace of Y stable by H is Y_0.

2) Let V' be a maximal totally singular subspace such that $V' \cap Y = \{0\}$ and let $b: V \times V \to K$ be the symmetric bilinear form defined by $b(v,v') = q(v + \tilde{\alpha}(v'))$ for $v,v' \in V'$. Then, the system (V',b) entirely determines the system $(V,q,\tilde{\alpha},V')$ up to unique isomorphism. Since all pairs consisting of a vector space of dimension m and a nondegenerate, nonalternating symmetric bilinear form on that space are isomorphic, it follows that Z permutes transitively each one of the two classes of maximal totally singular subspaces intersecting Y only at $\{0\}$.

4.3.5. <u>An example</u>. Suppose $\dim V = 4m'$ and, for $j \in \{1, \ldots, m'\}$, let V_j be the 4-dimensional subspace of V on which all coordinates vanish except x_{-2j}, x_{-2j+1}, x_{2j-1} and x_{2j}. Thus, V is the direct sum of the V_j's. Suppose further that $c = 1$ and that, for all $j \in \{1, \ldots, m'\}$, $a_{2j-1} = a_{2j}$. By 4.3.2, this implies that α is the inner automorphism of G corresponding to an element u of G(K) which belongs to the canonical image of Spin(V,q)(K) in G(K) (if $G = O°(V,q)$, $u = \tilde{\alpha}$). In V_j, the space of all singular vectors fixed by $\tilde{\alpha}$ is the one-dimensional subspace Z_j defined by the equations $x_{-2j} = x_{-2j+1} = a_{2j}x_{2j-1} = a_{2j}x_{2j}$. Let Z be the sum of all Z_j's. It is easily checked that if $v \in V_j$ is orthogonal to Z_j, then $\tilde{\alpha}(v) \in Z_j + v$. Therefore, if $v \in V$ is orthogonal to Z, then $\tilde{\alpha}(v) \in Z + v$. This shows that all subspaces of V orthogonal to Z and containing Z are stable by $\tilde{\alpha}$. Consequently, there exist maximal flags of totally singular subspaces stable by $\tilde{\alpha}$ (just take the union of a maximal flag of Z and a maximal flag of totally singular subspaces orthogonal to Z and containing Z). In other words, u <u>is good</u>, which implies that the centralizer of u in G(K) is contained in a proper K-parabolic subgroup of G. On the other hand, if not all a_i belong to $a_1.K^2$, the spaces Y_0 and Y of Lemma 4.3.4 and the two maximal totally singular subspaces containing Y are not "defined over K", therefore the Lemma implies that no proper K-parabolic subgroup contains the centralizer of u in G. In particular, u cannot be very good, by Theorem 2.5; this could of course also be checked by direct computation, or deduced from Proposition 5.3 below.

4.4. Groups of type C.

Now, suppose G of type C_m. We can repeat the preliminaries of 4.3.1, replacing $PGO°(V,q)$ by $PSp(V,a)$, where a is the alternating form $\sum_{i=1}^{m} x_{-i} \wedge x_i$. Let $\tilde{\alpha}$ be defined by the same equations as in 4.3 and let α be the automorphism of G that it induces. The proofs of the following assertions are similar to the proofs in 4.3.2 and 4.3.3 but simpler.

The involution α is an inner automorphism corresponding to an element of $Sp(V,a)(K)$ if and only if $c \in K^2$. In order that α be anisotropic, it is necessary and sufficient that $c \notin K^2$ and that the a_i be linearly independent over $K^2(c)$.

4.5. The case $[K:K^2] = 2$.

PROPOSITION. Suppose $[K:K^2] = 2$. Then G possesses an anisotropic K-automorphism of order 2 if and only if it is adjoint of type A_1. If the group G is simply connected, all unipotent elements of G(K) are good.

Suppose G possesses an anisotropic K-automorphism α of order 2. We may and shall assume that α stabilizes the torus T and inverts its elements. For any closed symmetric subset Ψ of the root system Φ of G relative to T, let G_Ψ denote the corresponding semisimple subgroup of G normalized by T (this notation is not that of [1] used in 3.6 above). By 3.6 (ii), α induces an anisotropic automorphism of G_Ψ; therefore, Ψ cannot be of type A_2 or B_2 (here, we use 4.2, 4.3.3 (iii_1) and the hypothesis made on K). Since the root system Φ has no subsystem of type A_2 or B_2, it must be of type A_1 and, by 3.5, G must be adjoint. Conversely, 3.5 also implies that "the" split adjoint group of type A_1 over K does have anisotropic automorphisms of order 2.

Now, suppose that G is simply connected and let L be the centralizer of a split torus in G. It is well-known that the derived group of L is simply connected, hence a direct product of simply connected quasi-simple groups. From the first part of the proposition, already proved, it follows that L contains no K-rational anisotropic element of order 2, and Corollary 3.3 implies our second assertion.

5. Special elements

Until the end of the paper, we assume G quasi-simple as well as defined and split over the prime field $\mathbb{F}_p$ of K; as before, T denotes a maximal split torus of G and X is the character group of T.

5.1. Let $\hat{W}$ be the group of all automorphisms of X preserving the root system of G; it contains the Weyl group W. We say that an element of $\hat{W}$ is <u>special</u> if it has no eigenvalue 1 (that is, if it fixes no nontrivial element of X) and that an automorphism of G normalizing T or an element of the normalizer of T in G is <u>special</u> (with respect to T) if its canonical image in $\hat{W}$ is special.

For any $w \in W$, let N_w denote the corresponding coset of T in its normalizer. Suppose w is special. Then, the endomorphism $t \mapsto w(t).t^{-1}$ of T is surjective, therefore all elements of $N_w(\overline{K})$ are conjugate under $T(\overline{K})$. It follows that any prime ℓ dividing the order of an element n of N_w also divides the order of w, otherwise there would exist a multiple k of ℓ congruent 1 modulo the order of w, we would have $w^k = w$, hence $n^k \in N_w$, contradicting the fact that the orders of n and n^k are different. If z is any central element of G, we have $z.N_w = N_w$, therefore the order of z divides that of the elements of N_w. In view of the preceding remark, this implies that any prime number dividing the order of the center of G - or of any group having the same Weyl group - divides the order of any special element of W. In particular,

<u>if</u> $w \in W$ <u>is a special element of order a power of</u> p, <u>the center of</u> G <u>(and that of the universal covering of</u> G) <u>is purely infinitesimal and the elements of</u> N_w <u>have the same order as</u> w.

We observe that all examples of anisotropic unipotent elements we have met thus far (cf. 3.5 and § 4) turn out to be special elements. Two more facts expressed by the following two propositions also point towards special elements as a source of bad unipotent elements.

5.2. PROPOSITION. Let $\varphi: \widetilde{G} \to G$ be a central isogeny, let w be a special element of W of order a power of p and let u be an element of $N_w(\mathbb{F}_p)$, set $\widetilde{T} = \varphi^{-1}(T)$ and $T_1 = \varphi(\widetilde{T}(K)) - T(K)$. Then, the elements of uT_1 (which are unipotent) are all bad. The set T_1 is empty if and only if K is perfect or φ is an isomorphism.

Since $\mathbb{F}_p$ is perfect, u is good (cf. [3], 3.1), hence contained in $\varphi(\widetilde{G}(K))$ (cf. Proposition 2.2. (ii)). Therefore $uT_1 \cap \varphi(\widetilde{G}(K)) = \emptyset$ and loc.cit. implies that no element of uT_1 is good.

Let $\widetilde{X}$ be the character group of $\widetilde{T}$. From the discussion in 5.1, it follows that the cokernel of the canonical map $X \to \widetilde{X}$ is a p-group, and this group is trivial only if φ is an isomorphism; the second assertion of the proposition readily follows, since $T(K) = \mathrm{Hom}\,(X,K^{\times})$ and $\widetilde{T}(K) = \mathrm{Hom}(\widetilde{X},K^{\times})$.

5.3. PROPOSITION. The centralizer in G of a special unipotent automorphism is not reduced.

We omit the proof, except for making the following, trivial but suggestive observation, which can be considered as its first step: if a unipotent automorphism of G is special with respect to T, its group-theoretical centralizer in T is finite whereas its centralizer in Lie T is at least one-dimensional, therefore its schematic centralizer in the torus T is not reduced.

5.4. CONJECTURE. All anisotropic K-automorphisms of G are special.

By Corollary 2.6, Theorem 4.1, the discussion in 3.5 and the results announced in the Introduction concerning the case $p = 3$, the above conjecture is proved in all cases except when $G = E_8$ and $p = 3$ or 5.

Remembering Proposition 3.2, we may conclude in heuristic terms, and assuming the truth of the above conjecture, that special elements of order p are "essentially the only source" of bad unipotent elements.

Collège de France, 11 Place Marcelin-Berthelot, 75231 Paris Cedex 05.

REFERENCES

[1] A. Borel et J. Tits, Groupes réductifs, Publ. Math. I. H. E. S. 27 (1965), 55-150.

[2] —— ——, Compléments à l'article "Groupes réductifs", Publ. Math. I. H. E. S. 41 (1972), 253-276.

[3] —— ——, Eléments unipotents et sous-groupes paraboliques de groupes réductifs. I, Inventiones Math. 12 (1971), 95-104.

[4] —— ——, Homomorphismes "abstraits" de groupes algébriques simples, Annals of Math. 97 (1973), 499-571.

[5] C.W. Curtis, G.I. Lehrer and J. Tits, Spherical buildings and the character of the Steinberg representation, Inventiones Math. 58 (1980), 201-210.

[6] W.M. Kantor, R.A. Liebler and J. Tits, On discrete chamber-transitive automorphism groups of affine buildings, Bull. Amer. Math. Soc. 16 (1987), 129-133.

[7] G. Lusztig, On the finiteness of the number of unipotent classes, Inventiones Math. 34 (1976), 201-213.

[8] T.A. Springer, Regular elements of finite reflection groups, Inventiones Math. 25 (1974), 159-198.

[9] T.A. Springer and R. Steinberg, Conjugacy classes, <u>in</u> Seminar in Algebraic groups and related fields, ed. A. Borel, Springer Lecture Notes in Math. n° 131 (1970), 167-266.

MIX
Papier aus verantwortungsvollen Quellen
Paper from responsible sources
FSC® C105338

If you have any concerns about our products, you can contact us on
ProductSafety@springernature.com

In case Publisher is established outside the EU, the EU authorized representative is:
Springer Nature Customer Service Center GmbH
Europaplatz 3, 69115 Heidelberg, Germany

Printed by Libri Plureos GmbH
in Hamburg, Germany